Evaluation, Maintenance and Upgrading of Wood Structures

A Guide and Commentary

Prepared by the Subcommittee on Evaluation, Maintenance and Upgrading of Timber Structures of the Committee on Wood of the Structural Division of the American Society of Civil Engineers

Alan Freas, Chairman

Published by the
American Society of Civil Engineers
345 East 47th Street
New York, New York 10017

The material presented in this publication has been prepared in accordance with generally recognized engineering principles and practices, and is for general information only. This information should not be used without first securing competent advice with respect to its suitability for any general or specific application.

The contents of this publication are not intended to be and should not be construed to be a standard of the American Society of Civil Engineers (ASCE) and are not intended for use as a reference in purchase specifications, contracts, regulations, statutes, or any other legal document.

No reference made in this publication to any specific method, product, process, or service constitutes or implies an endorsement, recommendation, or warranty thereof by ASCE.

ASCE makes no representation or warranty of any kind, whether express or implied, concerning the accuracy, completeness, suitability or utility of any information, apparatus, product, or process discussed in this publication, and assumes no liability therefor.

Anyone utilizing this information assumes all liability arising from such use, including but not limited to infringement of any patent or patents.

Copyright © 1982 by the American Society of Civil Engineers,
All Rights Reserved.
Library of Congress Catalog Card No. 82-72779
ISBN 0-87262-317-3
Manufactured in the United States of America.

PREFACE

The Technical Committee on Wood of the Structural Division of ASCE is charged, as a part of its purpose, "to disseminate knowledge concerning wood as an engineering material." In part, this can be done by means of papers presented at technical meetings, at seminars, and at workshops. Such papers, valuable as they may be in presenting information of value to the civil engineering profession, tend to become lost to the practicing engineer because they are scattered in a variety of sources.

The collection and interpretation of the results of research and practice into a single volume as a guide to a specific area of practice has an obvious advantage to the practicing professional by providing, in a single source, the basic information needed to solve a specific problem. In addition, however, it provides a set of references to more in-depth information on specific aspects if they are not treated as fully as needed in the guide.

It was this concept which led the Committee on Wood to prepare the volume "Wood Structures: A Design Guide and Commentary," which, after several years of work, was published by the Society in 1975. This volume has been a best seller.

More recently, the Committee on Wood recognized the increasing potential for rehabilitation of existing structures as opposed to replacement with totally new structures as a conservation measure. This recognition led the Committee to visualize the desirability of a document which would bring together, for the professional faced with a rehabilitation, renovation, or repair problem, the basic information needed to solve the problem. This publication, the result of work by many people over a period of several years, is again presented as a guide and commentary rather than as a textbook.

The undersigned wish to express their appreciation for his efforts to Thomas Williamson, former chairman of the Committee on Wood in whose term the book was conceived. In addition, however, their thanks and appreciation go, as well, to the authors, coordinators, and committee members who gave so freely of their time and expertise.

ALAN FREAS, Chairman
Subcommittee on Evaluation,
  Maintenance and Upgrading of
  Timber Structures

SHERMAN NELSON, Chairman
Committee on Wood

ASCE COMMITTEE ON WOOD

Nelson, Sherman A.	Chairman	
Lyons, Bruce	Secretary Chairman	Technical Publications Subcommittee
Faherty, Keith F.	Chairman	Technical Sessions Subcommittee
Freas, Alan D.	Chairman	Evaluation, Maintenance and Upgrading of Timber Structures Subcommittee
Tuomi, Roger L.	Vice-Chairman	Evaluation, Maintenance and Upgrading of Timber Structures Subcommittee
Goodman, James R.	Chairman	Reliability and Limit States Design for Wood Structures Subcommittee
Williamson, Thomas G.	Chairman	Liaison to the ALSC, ASTM & Other ASCE Committees Subcommittee
Domer, Ronald G.		Executive Committee Contact Member

Membership

Adams, Noel
Arnold, Orville E.
Avent, R. Richard
Bohannan, Billy
Bower, Warren H.
Crandall, Lee W.
Criswell, Marvin E.
Dietz, Albert G.
Duke, Theodore J.
Ebeling, Richard W.
Gabrielsen, B. L.
Gangarao, Hota
Gutkowski, Richard
Hoyle, Robert J.
Hurlbut, Ben
Itani, Rafik Y.
Krueger, Gordon P.
Lanius, Ross, Jr.
Madsen, Borg
Malhotra, S. K.
Mangus, Alfred R.

Montrey, Henry
Moody, Russell C.
Nicholas, Paul T.
O'Halloran, Michael
Onysko, Donald M.
Oviatt, Alfred E., Jr.
Pashina, Brian J.
Peterson, John
Peyrot, Alain H.
Pincus, George
Polensek, Anton
Sack, Ronald L.
Schaffer, Erwin L.
Sharp, Donald J.
Stadelmann, Douglas A.
Stern, E. George
Suddarth, Stanley K.
Taylor, Raymond J.
Van Dalen, Karl
Zahn, John J.

TABLE OF CONTENTS

Chapter	Page

1 INTRODUCTION

    1.1 Purpose of the Publication 1
    1.2 Scope and Organization 2
    1.3 Wood as a Structural Material 3
    1.4 World War II Structures 20
    1.5 Old Wood Structures 24
    1.6 Overview of Building Regulations That Relate to Rehabilitation 39
    1.7 Special Considerations for Historic Structures 48

2 FACTORS WHICH INFLUENCE SERVICEABILITY OF WOOD STRUCTURES

    2.1 Loadings 56
    2.2 Duration of Load 64
    2.3 Temperature 69
    2.4 Moisture 92
    2.5 Chemicals 97
    2.6 Weathering 106
    2.7 Fire 120
    2.8 Insects, Fungi, and Other Organisms That Attack Wood 126

3 INSPECTION

    3.1 Introduction 136
    3.2 Where to Look and What to Look For 138
    3.3 Techniques and Equipment 146
    3.4 Inspection of Specific Structural Types 151
    3.5 References for Chapter 3 157

4 EVALUATION

    4.1 General 159
    4.2 Inspection 159
    4.3 Determination of Loads 160
    4.4 Structural Analysis 163
    4.5 Estimating Load-Carrying Capacity of Structural Elements 165
    4.6 Estimating Load-Carrying Capacity of Connections 179
    4.7 Estimating Load-Carrying Potential of Overall Structure 189
    4.8 Reporting on a Structural Evaluation 190
    4.9 References for Chapter 4 191

5 METHODS OF REPAIR

    5.1 Considerations in Selecting the Right Repair Procedure 194
    5.2 Description of Alternative Repair Methods 196
    5.3 Quality Control to Insure Adequate Repair 221

| Chapter | Page |

    5.4 Examples of Successful Repairs 223

6   MAINTENANCE

    6.1 Introduction 254
    6.2 Condensation 261
    6.3 Ventilation 264
    6.4 Maintenance and Repair of Roofs on Wood Structures 268
    6.5 Water Repellents 271
    6.6 Finishes 277
    6.7 Heavy Timber (Solid-Sawn Timber) 297
    6.8 Other Wood Structures 303

7   DESIGNING TO AVOID PROBLEMS

    7.1 Loadings 305
    7.2 Weather Exposure 307
    7.3 High-Humidity Environments 312
    7.4 Fire 319
    7.5 Decay Fungi, Insects, Marine Organisms 338
    7.6 Mechanical Damage 362

8   RECYCLING OLD STRUCTURES (CASE HISTORIES)

    8.1 Introduction 366
    8.2 Reconstruction of Wooden Cathedral 366
    8.3 Hi-Rise in Wood (Butler Square) 377
    8.4 Dome Strengthening at Quincy Market (Boston, Massachusetts) 382
    8.5 Historic Restoration and Adaptive Use of Old Burke County Courthouse,
        Morganton, N.C. 386
    8.6 Restoring Covered Bridges 395
    8.7 Repairs to Wooden Trusses in Elastomers Finishing Building—Exxon
        Chemical Company 407
    8.8 Rehabilitation of Fire-Damaged Timber—The Filene Center 417
    8.9 Reconstruction of Two Wood Bridges 422

CHAPTER 1   INTRODUCTION*

1.1   Purpose of the Publication**

Any structure, regardless of the material from which it is made may, from time to time, be subject to review of its ability to perform a specific function or functions. Then a decision must be made, after the review, as to whether the structure is suitable as it is or whether it requires modification, reinforcement, rebuilding, or renovation. In some instances, replacement may be required. Some of the circumstances under which a structure may require such a review include:

--a change in the use of the structure;
--damage to a substantial portion of the structural elements as a result of fire, earthquake, windstorm, or other catastrophe;
--partial or complete collapse of the structure or development of a state of unserviceability as a result of foundation settlement, excessive deflection, vibration, or other cause;
--deterioration of structural elements due, in the case of wood elements, to such causes as decay, insect attack, or the like, with resultant weakening;
--changes in the applicable building code.

Regardless of the reason, owners are, from time to time, faced with the problem of evaluating an existing timber structure and making a decision as to its future. Such an evaluation probably will require the services of a consultant and may well be more difficult than would the designing of a completely new structure. With new structures, the engineer selects the components needed based on the results of an analysis and on available options. With existing structures, however, the elements are already in place, have had a loading history probably not fully known, and may have sustained some damage or deterioration. The engineer must, therefore, determine whether the elements, as they exist, are adequate for safety and serviceability or whether modifications are needed to insure that the structure will perform its intended function. The engineer faced with this problem may find it difficult to locate the information needed to make the evaluation.

It is the purpose of this publication to bring together, from experts in the field, information relating to the evaluation, maintenance, and

---

* Chapter coordinator: Alan D. Freas, former Assistant Director, Forest Products Laboratory, Madison, Wisconsin.

**Author: Of this and of other sections of this chapter, except as indicated at appropriate section headings, Alan D. Freas.

upgrading of wood structures. Such information will assist the practicing professional who is faced with the problem of evaluating an existing structure and making a decision as to its future.

It is well worth noting that rehabilitation of wood structures is an effective conservation measure. If a structure has outlived its usefulness because of obsolescence, change in occupancy, damage, or any other cause, it may be replaced by razing the old and building a new structure or by saving as much as possible of the original structure and rebuilding to the minimum required to permit it to serve its new purpose. Every cubic foot of wood--in whatever form--which can be reused either in place or in a new location saves the equivalent amount of wood in the timber resource. In addition, it saves the energy, labor, shipping, storage, and money required to convert the logs to lumber, or plywood, or particleboard, or flooring, as well as the energy, time, and expense to put them in place. Sound evaluation of the potential of an old building for a new use is, therefore, an extremely important base for conservation, and rehabilitation should always be looked at carefully as a possible alternative to building a new structure.

1.2 Scope and Organization

This guide has been prepared to present, to the professional, information on technical aspects of inspection, evaluation, reinforcement, repair, and rehabilitation of timber structures. Aesthetic considerations are not covered. The restoration of historic structures requires special care, and Section 1.7 offers guidance in this area.

The guide is organized to cover, in logical order, the information needed by a professional faced with the necessity for evaluating and making a decision on the renovation of an existing structure. Factors which influence serviceability of wood structures are reviewed in Chapter 2, with consideration of such factors as temperature, moisture, weather, and microorganisms as background for that which follows. Chapter 3 covers the inspection of wood structures, including the tools and techniques and a discussion of vulnerable points in wood structures to aid the inspector in where to look for. Evaluation of structural adequacy is covered in Chapter 4. Repair procedures, including both descriptions of techniques and examples of successful repairs, are covered in Chapter 5. Maintenance can be an important tool in preventing or reducing deterioration of a structure and is covered in Chapter 6. The guide, to this point, has been concerned with the techniques to be employed in evaluating and rehabilitating wood structures. Chapter 7 is devoted to design considerations which, had they been employed in the original design, might have reduced the incidence of problems necessitating rehabilitation, and is offered as an assistance to the designer in avoiding future problems. The last chapter offers a number of case histories in recycling old buildings with the thought that examples of past work may offer important clues to the solution of a present problem.

INTRODUCTION

1.3 Wood as a Structural Material

1.3.1 Background

Wood is man's oldest structural material and it is only in the relatively recent past that steel, aluminum, concrete, and structural plastics have assumed their present importance. As a matter of fact, it was not much more than a century ago that there was great controversy over the introduction of "iron-clads" into our Navy.

Data do not seem to be available to provide anything approaching an accurate picture of the amounts of wood used in structural elements in comparison to metals (primarily steel), concrete, and structural plastics. The importance of wood in the overall economic picture of the United States is, however, indicated by the fact that, in the early 1970's, the tonnage of wood produced for industrial uses (including construction) was greater than that of steel and cement combined (17). In 1972, the per capita use of metals was about 2/3 ton as compared with about 1-1/3 tons of forest products (22).

The increasing importance of metals and concrete has led to an increasing emphasis on them in engineering curricula and to a corresponding decrease, in the curricula, of emphasis on the structural use of wood. As a consequence, many engineers have little training in the structural use of wood and little feeling for its potential in such applications. Many may not be familiar with the forms of wood-based products available for use in structures. It is beyond the scope of this publication to present a thorough discussion on wood as an engineering material. However, brief information is provided on some of wood's unique characteristics, on some of the products available, and on some of the important aspects of the structural use of wood.

Broad-scale information on wood as an engineering material is presented in "Wood Structures: A Design Guide and Commentary" (18) published by the American Society of Civil Engineers. Similarly, the "Wood Handbook: Wood as an Engineering Material" (26), provides both background on the nature of wood and its products and detailed technical information on their properties. A number of excellent design handbooks relating to wood structures have been published (3,19,20,21,23, 24).

In addition, information on wood products is available from the following groups:

Structural Glued Laminated Timber

American Institute of Timber
  Construction
333 West Hampden Avenue
Englewood, CO 80110

Lumber

Northeastern Lumber Manufacturers
  Association, Inc.
4 Fundy Road
Falmouth, ME 04105

Northern Hardwood and Pine
  Manufacturers Association, Inc.
Northern Building
Green Bay, WI 54301

National Lumber Grades Authority
  (Canada)
P.O. Box 97
Ganges, British Columbia
CANADA
VOS 1E0

Redwood Inspection Service
One Lombard Street
San Francisco, CA 94111

Southern Pine Inspection Bureau
4709 Scenic Highway
Pensacola, FL 32504

West Coast Lumber Inspection
  Bureau
6980 SW. Varnes Road
(P.O. Box 23145)
Portland, OR 97223

Western Wood Products
  Association
1500 Yeon Building
Portland, OR 97204

National Hardwood Lumber
  Association
59 East Van Buren Street
Chicago, IL 60605

Forest Products Generally

National Forest Products Association
1619 Massachusetts Avenue, NW.
Washington, D.C. 20036

Preservative and Fire Retardant Treatments

American Wood-Preservers' Association
Suite 444, 7735 Old Georgetown Road
Bethesda, MD 20014

American Wood Preservers Institute
1651 Old Meadow Road
McLean, VA 22102

American Wood Preservers Bureau
2772 South Randolph Street
(P.O. Box 6085)
Arlington, VA 22206

Panel Products

American Plywood Association
7011 South 19th Street
(P.O. Box 11700)
Tacoma, WA 98411

American Hardboard Association
887-B Wilmette Road
Palatine, IL 60067

Hardwood Plywood Manufacturers
  Association
P.O. Box 2789
Reston, VA 22090

National Particleboard
  Association
2306 Perkins Place
Silver Spring, MD 20910

INTRODUCTION

1.3.2 Wood--An Overview

The following brief presentation of a few salient points about wood may be of assistance as background for those who need to search for more detailed information on wood and its characteristics.

1.3.2.1 Species

Hundreds of species of trees grow in the United States, but only a relatively few of them are of commercial importance. The species may be broadly classified as softwood or hardwood, two somewhat less-than-descriptive terms, since some of the softwoods are harder than some of the hardwoods. The softwoods come from the coniferous species such as the pines, spruces, hemlocks, and Douglas-fir. The hardwoods come from the broadleaved species such as the oaks, ashes, and hickories, as well as some of the softer hardwoods such as the poplars. Of the many species grown in the United States, more than a hundred are available to the prospective user, but, commercially, only about sixty are important--about one-half hardwoods and one-half softwoods.

By far the majority of structural elements--joists, studs, beams, posts, columns--come from the softwoods, with Douglas-fir, southern yellow pine, redwood, eastern and western hemlocks, and some of the firs being especially important. Hardwoods currently find little use as structural elements--only about three having allowable stresses assigned to them--but it is quite possible that hardwood structural members will be found in some older buildings. This is not to say that hardwoods cannot be used structurally, but the stronger hardwood species such as the oaks and hickories are heavy, hard to handle, and hard to nail. Hardwoods, of course, find significant usage in structures in such applications as flooring, cabinetry, and furniture.

1.3.2.2 Structure

Wood is composed mostly of hollow, elongated cells most of which are arranged parallel with each other and with the long dimension parallel to the axis of the trunk of the tree. Such an arrangement, as might be expected, results in differences in physical and mechanical properties associated with the different directions relative to the axis of the tree. Thus, properties parallel to the length of the fibers (the longitudinal or "parallel-to-grain" direction) are quite different from those at right angles to the axis of the tree. Thus wood is said to be anisotropic, rather than isotropic as is characteristic of metals. Actually, wood is orthotropic; that is, it has three mutually perpendicular axes of symmetry--longitudinal, radial, and tangential. This orthotropy calls for special care in design to take account of the differences in properties among the three directions.

1.3.2.3 Density

The specific gravity of wood substance is on the order of 1.5 regardless of species, but the apparent specific gravity of a piece of wood varies greatly between species and somewhat less within species depending upon growing conditions. The specific gravity of balsa, for

6 WOOD STRUCTURES

example, is about 0.17 based on ovendry weight and volume at 12 percent moisture content while other imported species may have average values of specific gravity higher than unity. Lignum vitae, for example, has an average specific gravity of 1.09 on the same basis as that quoted above for balsa. Converted to "density", which includes the weight of water in the wood, balsa at 12 percent moisture content will have an average density of about 11 or 12 pounds-mass per cubic foot, while lignum vitae will have an average density on the order of 76 pounds-mass per cubic foot. Most species, however, range in density from about 20 to 45 pounds-mass per cubic foot. Obviously, density (or specific gravity) is a measure of the amount of wood substance in a given volume of wood, the rest of the volume being taken up by the voids present, as in the interior of the hollow cells. Since the strength of a piece of wood is furnished by wood substance and not by voids, the more dense woods would be expected to be stronger and stiffer.

1.3.2.4 Moisture

Wood in the living tree contains variable amounts of moisture, depending upon the species, and can contain up to twice as much water as wood substance. When a piece of wood is cut from the log and exposed to the air, the moisture in the wood evaporates until the moisture content comes to equilibrium with the surrounding atmosphere. The "equilibrium moisture content" depends upon the temperature and relative humidity of the surrounding air but may be expected to range from about 4 percent in the arid Southwest to as much as 14 percent in some coastal areas and perhaps 8 or 10 percent in a heated building in the northern United States during the winter.

As the wood dries, two things happen which affect its serviceability. First, as wood dries below the "fiber saturation point" (about 30 percent on the average) it becomes stronger and stiffer. This is taken into account by allowing higher design stresses for wood which will be used under relatively dry conditions, such as a girder in a building, than for wood which will be used under relatively moist conditions such as in a waterfront structure or in a bridge. Second, as wood dries below the fiber saturation point, its dimensions are reduced--it shrinks. And it shrinks differently in the three principal directions--hardly at all in the longitudinal direction, more in the radial direction (say 3 to 5 percent for most softwoods), and still more in the tangential direction (say 6 to 9 percent). It is, obviously, desirable to have the dimensions of a piece of wood fairly constant before putting it into service. In addition, warping may result from shrinkage, which could cause problems if it occurred after the wood was in place. It is, therefore, common practice to dry a piece of wood, whether it be in the form of lumber or veneer, to approximately the moisture content to be expected in service prior to use.

1.3.2.5 Grain Direction

The fiber direction in a piece of lumber or veneer may or may not be exactly parallel to the principal axis of the piece. In the usual case, the principal stress in, say, a joist, acts parallel to the axis

INTRODUCTION

of the piece and, thus, if the fiber direction is not parallel to the axis, the stress has a component parallel to the fiber direction (parallel to grain) and one perpendicular to grain. Since the strength in the latter direction is considerably less than in the fiber direction, the net strength parallel to the axis of the piece is lower than if the piece had been "straight-grained." For this reason, the deviation of grain direction with respect to the axis of a piece of lumber is limited if the piece is to be assigned an allowable stress.

1.3.2.6 Knots

Knots are the remnants of branches left in a piece of lumber or piece of veneer when it is cut from the log. Since the branches are oriented in the tree approximately at right angles to the length of the trunk, fibers in the knot in a piece of lumber are also oriented at approximately right angles to the axis of the piece of lumber. Thus, a force applied in the longitudinal direction acts on an area of low strength (the knot) and the overall strength is reduced. In addition, the fibers around a knot are frequently distorted, causing local cross grain and thus weakness. For these reasons, knots are always limited in structural grades, with the higher grades having smaller knots.

1.3.2.7 Checks and Splits

As wood dries, the moisture evaporates first from the surface of the piece and moisture from the interior moves outward. Thus the surface may be at a relatively low moisture content while the interior is at a substantially higher moisture content. Depending upon the conditions in the surrounding atmosphere, the surface moisture content might be on the order of 10 or 12 percent while the interior is at or above fiber saturation (i.e., at 30 percent or more). The outer shell, therefore, is trying to shrink, while the inner core stays at its original dimension, and a tensile stress perpendicular to grain is set up in the shell. If this stress becomes large enough, it may exceed the tensile strength perpendicular to grain and a fracture extending along the grain occurs; these fractures are called "checks." If the fracture extends through the full cross sectional dimension, it is called a split. Both reduce resistance to shear stress and are thus limited in structural lumber grades. They may also reduce the ability of fastenings to resist load.

1.3.2.8 Decay Resistance

Decay resistance is a function of species, the heartwood of some species being much more resistant to decay than that of others; the sapwood of all species deteriorates rapidly when exposed to conditions conducive to decay. Wood kept continuously dry--below about 20 percent moisture content--will not decay. Wood continuously submerged in water is not subject to attack by ordinary wood-decaying fungi because there is insufficient air for them to be active. Soft-rot fungi, however, can attack submerged wood, as can bacteria, but the resulting deterioration is quite slow. If wood is to be used in applications involving high moisture contents, either the heartwood of decay-resistant species should be used or the wood should be treated with preservative chemicals which are toxic to fungi. Conditions which cause decay--moisture content above 20 percent, moderate temperatures, access to air--will also result in attack by other fungi, such as the staining fungi and molds. Since these do not attack the wood substance, they generally are not harmful to strength, but their presence indicates that conditions suitable for decay have occurred and thus decay should be suspected.

1.3.2.9 Resistance to Chemicals

Wood is resistant to many chemicals. In the chemical processing industry, for example, it is preferred for many applications, including various types of tanks and other containers and for structures adjacent to or housing chemical equipment. Wood is widely used in cooling towers and in the fabrication of buildings for bulk chemical storage where the wood may be in direct contact with chemicals.

Wood may, however, be degraded by the action of chemicals, the degree of degradation depending upon a number of factors. Among these factors are species, whether the wood is heartwood or sapwood, the specific chemical or chemical type involved, and moisture content. Where necessary, degradation can be prevented or reduced by impregnation of the wood with a variety of chemicals including coke-oven coal tar, phenolic resin solutions cured in the wood, and furfuryl alcohol.

1.3.2.10 Electrical Resistance

Electrical conductivity of wood varies greatly with moisture content, being from 10^{10} to 10^{13} times as great at fiber saturation as at a moisture content near zero percent. This effect can be a matter of concern in certain uses such as utility poles. The effect is put to use in electric moisture meters, using the change in electrical resistance to indicate the moisture content of a piece of wood.

1.3.2.11 Mechanical Properties/Design Allowables

Information on the mechanical properties of wood may be found in the literature in several forms. In one form, average properties of a species, together with information on variability, are derived from tests of "small clear" specimens, free of strength-reducing characteristics and, thus, presumably reflecting the intrinsic characteristics of the species. A second form is derived from tests of lumber, structural timbers, plywood, laminated wood, poles, and so on, and gives information on the properties of wood products of a specific grade or grades, and specific constructions. A third form represents the properties which may be assumed as a basis for design; these may be derived from data in either of the other two forms.

Intrinsic (small clear) properties vary greatly between species, much of the difference being attributable to differences in density. The Wood Handbook (26, Table 4-4), for example, gives an average value of modulus of rupture at 12 percent moisture content for balsa (specific gravity, 0.17) as 2,800 psi while the same source gives the modulus of

INTRODUCTION 9

rupture of greenheart (specific gravity, 0.93) as 25,500 psi. Douglas-fir, a species in the more normal range of density, is shown by Table 4-2 of the Wood Handbook to have an average modulus of rupture of 12,400 psi and an average value of specific gravity of 0.48.

The intrinsic (small clear) properties vary also with moisture content. For example, the Wood Handbook shows modulus of rupture for Douglas-fir at 12 percent moisture content as 12,400 psi, while that at high moisture content (green) is only 7,700 psi. The dry/green ratio varies considerably with species and the range differs among properties. For example, the ratio for modulus of rupture varies from about 1-1/4 to 2 while the range for modulus of elasticity is from just over one to about 2. ASTM D 2555 (12) gives data on dry/green ratios for a number of species.

Design allowables are, in general, only a fraction of the average strength values reported from tests of small clear specimens. The large difference results from the number of adjustments necessary to convert from a laboratory basis to a service basis. References will be given in a later section to sources of information on this conversion for various products.

1.3.2.12 Heat

Mechanical properties of wood decrease when the wood is heated and increase when it is cooled. This is an "immediate" effect which disappears on return to the original temperature. Prolonged exposure to high temperature, however, causes an irreversible decrease in properties. Thus the effect of heat is a function not only of temperature but also of the length of time the wood is exposed to heat.

Wood is a poor conductor of heat, its thermal conductivity being only a fraction of that of metals and only about two to four times that of common insulating materials. Thermal diffusivity is low and thus wood does not feel extremely warm or cold to the touch. The coefficient of thermal expansion along the grain is quite low and usually temperature changes do not produce significant changes in length. Coefficients across the grain range on the order of 5 to 10 times that parallel to grain.

1.3.2.13 Fire Resistance

Because of its low thermal conductivity, the high temperatures of a fire cause a temperature rise for only a short distance into the wood from the fire-exposed surface. Thus the effect of heat is confined to a very small depth from the surface and, after a fire, large members may often be reused after char is scraped from the surfaces.

1.3.2.14 Time-dependent Effects

Wood is subject to a number of time-dependent effects. For example, wood loaded over a long period will incur deformation beyond the elastic deformation incurred as load is applied. At typical design levels, the additional deformation due to creep may equal the initial, instantaneous deformations. Where deformation is critical as, say, in a header over a large door, the effect of creep must be considered in design.

It has long been known that wood can withstand higher stresses for short than for long periods and the effect of long durations of load is considered in establishing design stresses. A wood beam loaded for ten years will carry only about 60 percent of the load required to cause failure in a standard test having a duration of only a few minutes. There is some evidence that the duration effect is related to the grade of lumber, with the effect being smaller for the lower grades.

There is a fatigue effect resulting from repetitive loading on wood. Unfortunately the data available are extremely sparse and are insufficient to permit establishing general rules for considering the effect of fatigue in design.

Time alone, where wood is kept dry, at normal temperatures, and protected from the deteriorating effects of such agencies as decay, has little or no effect on mechanical properties. Test results on very old timbers suggest that significant losses in strength occur only after several centuries of normal aging conditions.

1.3.3 Forms of Wood Used in Structures

1.3.3.1 Lumber

The form most familiar is lumber, a material which has been sawed from the log generally to rectangular cross section, planed, and cross-cut to length. Lumber for use in engineered structures must be stress-graded, either visually or by mechanical means. Visually graded lumber is that in which knots, cross grain, checks and other strength-reducing characteristics are limited by visual examination of each piece in relation to the requirements of a grading rule. Mechanically graded lumber is tested in some fashion, commonly by measuring the deflection resulting from application of a known bending load and relating this to strength. This is done most commonly at high speed in a continuous operation. In many cases, the pieces are, in addition, examined visually to assure limitation of strength-reducing characteristics near the ends of the piece.

Lumber for structural applications may be in one of several forms. Dimension lumber is nominally 2 to 4 inches thick and may or may not be graded to limit strength-reducing characteristics. Beams and stringers are nominally 5 by 8 inches in cross section and larger and are graded for strength in bending when loaded on the narrow face. Posts and timbers are square or nearly square in cross section, are 5 by 5 inches or larger in size and are graded primarily for use as posts. Decking is used in floors or roofs in conjunction with supporting members such as joists, purlins, or beams. Scaffold plank is graded specifically for use flatwise (i.e., with the larger dimension horizontal). Stadium plank is also graded for flatwise use and generally has special appearance requirements. The thinner material such as dimension, decking,

INTRODUCTION

and scaffold plank is commonly dried before use to a moisture content approximating that to be expected in service. Material thicker than about 4 inches, such as beams and stringers, is not dried before being put into service. Boards are not stress graded.

1.3.3.2 Glued Laminated Timber

Glued laminated timber refers to an assembly of three or more layers of sawn lumber bonded together with the grain direction of all layers approximately parallel. The timbers may be either straight or curved in form and may serve as a beam (either straight or curved), as an arch, as a column, as a truss chord, or in any of a variety of applications. Laminations are generally not over 2 inches thick and may be thinner if the member is sharply curved. Lamination lengths are not limited by available lengths of lumber, since available lengths may be end-jointed to form laminations of any desired length. Similarly, standard lumber widths may be placed side by side (bonded or unbonded) to produce members of required widths. Adhesives used in the manufacture of laminated members used in exterior applications such as bridge stringers should be waterproof and commonly are of the phenol-resorcinol type which, when properly cured, are not affected by moisture, high temperature, fungi, or other such deteriorating influences. For interior application, water-resistant adhesives such as the caseins may be used.

The control of the fabrication process needed to assure good quality requires that structural glulam members be manufactured in a shop. Size limitations on such members, therefore, are generally the result of limitations in the fabrication process, limits set by transportation problems such as overhead clearance along the delivery route and, if preservative treatment is required, by the size of the available treating equipment. However, arches spanning on the order of 300 feet and simple beams with spans over 150 feet--and depths of as much as 7 feet--are in service. Laminated members have been used in domes with spans exceeding 500 feet.

1.3.3.3 Plywood

Plywood is a glued wood panel made up of relatively thin layers bonded together with an adhesive and with the grain direction of adjacent layers at an angle, usually 90°. Usual constructions have an odd number of layers. In some instances, a layer may be made of two or three plies, the grain directions of the plies being parallel. Thus a 3-layer plywood with a thick center layer made of two or three plies would actually contain four or five plies. The layers may vary as to number, thickness, species, and grade of veneer. Some panels may be surfaced with overlays for special uses such as concrete forms.

Because of its special cross-laminated construction, plywood tends to have properties somewhat the same in the face-grain direction and in the across-the-face-grain direction. Similarly, dimensional change perpendicular to face grain is much reduced as compared with lumber. Plywood has greater resistance to splitting, and its availability in large sheets makes it suitable for applications where lumber may have limited application, as in coverings for frameworks of walls, floors, and roofs.

Plywood is made from both softwoods and hardwoods. The former generally finds application in construction uses and the latter is more frequently used where appearance is important. The quality (grade) of the various layers may vary considerably. The adhesives used to bind the layers together may also differ considerably in their resistance to the effects of moisture and thus in their ability to withstand exterior exposure. Plywoods can no longer be accurately classified as "softwood" or "hardwood", since both the construction plywoods and those used for decorative purposes may have both softwoods and hardwoods used in their manufacture.

Plywood is classified as to grade, depending upon the quality of the veneers, and as to type, depending upon whether or not the adhesive is suitable only for interior applications or whether it is suitable for either interior or exterior use.

Strength and stiffness properties of construction plywood depend upon the species used, and the construction of the panel; that is, the relative thicknesses in the parallel and perpendicular directions, and the number of layers.

1.3.3.4 Fiber and Particle Panels

This class of panel materials includes such familiar products as insulation boards, hardboards, particleboards, waferboards, and laminated paperboards. These panels are made from reconstituted wood (or some other lignocellulosic material like bagasse), the wood being reduced to small fractions and then put back together by special forms of manufacture into panels of relatively large size and moderate thickness. In some cases, the wood is reduced essentially to fibers and in others is cut by mechanical means into discrete particles.

The fibrous materials are reconstituted by interfelting the fibers, the bond between the fibers commonly being that resulting from the interfelting. Some binder or other materials may be added to increase strength, resistance to fire, moisture, or decay or to improve some other property. The particleboards are manufactured from the discrete particles mentioned earlier, which may range from specially cut flakes to fine particles. The particles must be bound together with binders such as urea-formaldehyde, phenol-formaldehyde and, sometimes, melamine-urea-formaldehyde resins. The phenol resins are used where maximum resistance to severe exposure is required. Strength and stiffness properties vary greatly depending upon the nature of the fibers or particles and the way they are combined, as well as on the density of the finished panel.

1.3.3.5 Poles and Piles

Poles and piles are sufficiently familiar that, perhaps, little or no explanation is needed. The most common use of poles, of course, is as support for conductors in utility lines, either as individual poles or

INTRODUCTION 13

as pole frames. Some poles are used in building construction, including residences, especially where the building site is very steep and it is necessary to avoid disturbing the soil, or in areas where wind or water can damage the underpinning of a structure; poles are especially valuable in such instances, since they furnish both foundation and vertical framing. Piles are used in bridge, wharf, and building construction, primarily as foundations.

Poles and piles are commonly of softwood species. Hardwood species can be used, but this is uncommon. Because of the conditions of exposure found in both applications--below-ground burial of all or part of the pole or pile, or exposure to high moisture and thus hazards of decay or exposure to other organisms like marine organisms--poles and piles are normally treated with preservatives, usually by pressure methods.

1.3.4   Processes Related to Wood Products

1.3.4.1 Seasoning

For most uses, it is necessary to reduce the moisture content of the wood below that commonly found in the tree. Lumber, for example, is commonly required to be dried to moisture contents not above 19 percent, although for some applications and for use in some areas it is desirable to dry it to a still lower level. Flooring and trim, for example, are subject to long periods of low humidity during winter in northern areas of the United States and relatively low moisture contents are needed. Lumber may be air-dried by exposure to open air, but the degree of drying and the rate of drying are subject to prevailing conditions in the area of production; long periods are normally needed. More commonly, lumber is dried in a kiln by the application of controlled conditions of temperature and humidity; attainment of considerably lower moisture contents than by air drying are possible and in a shorter time. Veneer is commonly dried in veneer driers to very low levels of moisture content, only a few percent.

Not all lumber is seasoned before use. In some areas of the United States, unseasoned lumber is successfully used in house construction.

1.3.4.2 Wood Preservation

For uses where the hazard of decay is present, lumber, poles, piles, ties, and even millwork may be treated with preservatives to extend the service life. For the more severe exposures, and those applications where cost of replacement is high in terms of both money and periods out of service, treatment is commonly by a procedure in which the preservative chemicals are injected into the wood under pressure. Such procedures may be expected to give the deepest penetration and greatest retention of the preservative. At the other extreme are the more superficial methods which may involve simply dipping the material in the preservative solution or brushing the preservative onto the surface; penetration and retention are limited but may be adequate for mild hazards. Window sash, for example, are commonly dipped for a few seconds into solutions of such preservative chemicals as pentachlorophenol.

The types of preservatives vary greatly. Creosote, for example, is a chemical commonly used for severe exposures such as piles or utility poles. It has the disadvantage of being dirty and nonpaintable, and having a disagreeable odor, which makes it unsuitable for some uses such as residences. A number of water-soluble chemicals in various combinations are frequently used. Some are not suitable for use where a high degree of protection is required. The waterborne preservatives are clean and paintable. In some uses, where the hazard of damage from marine borers is high, a dual treatment involving copper-containing preservatives followed by treatment with coal-tar creosote may be used. Pentachlorophenol in solvents of the mineral spirits type is common in treatments of millwork. Pentachlorophenol in heavier oils is used to treat poles. More recently, treatment with pentachlorophenol dissolved in liquid petroleum gas or other gases has come into use. Water-repellent preservatives containing pentachlorophenol in a volatile solvent with water-repellent components find considerable use in treatment of millwork.

All preservative chemicals have some degree of toxicity. It is for this reason that only chemicals registered by the U.S. Environmental Protection Agency should be used, and then only in applications prescribed in the registration and in the manner and at the concentrations prescribed. The list of registered chemicals varies from time to time and prospective users should be sure to obtain current information.

1.3.4.3 Fire-Retardant Treatments

Building codes or other requirements may dictate that wood products are to be treated with fire retardants to improve their resistance to flame spread. The fire-retardant materials are generally water soluble and are commonly applied under pressure. Obviously, large amounts of water are inserted into the wood and, for the majority of uses, the procedures result in moisture contents well above those desired; therefore, redrying is necessary. If such treatments are applied to timbers of large cross section, such as a large laminated beam, it is essentially impossible to dry the member to desired levels of moisture content and shrinking, warping, checking, and other things may occur. Thus, treatment of large glued laminated timbers with water-soluble preservatives after fabrication is not recommended. If such treatment is required, the individual laminations should be treated before they are glued together. Special care must be taken in the gluing process to assure high-quality bonds.

1.3.5 Standards for Wood Products

Many, if not most, wood-based products are manufactured to a standard which contains provisions relating to the manufacture of the product or to its performance. The following discussions contain guides to the various standards and to their sources. It should be understood that, in general, standards have no legal force unless they are incorporated into building codes or other legislation by reference. They

INTRODUCTION

are, however, valuable guides and, at the very least, indicate industry practice with respect to the product.

1.3.5.1 Lumber

Manufacturers of lumber are generally grouped into associations, many of which develop and maintain "grading rules". Such rules describe each of the individual grades (levels of quality) which they produce. The rules describe not only those characteristics permitted in a grade--i.e., knot sizes, slope of grain, sizes of checks, and amount of wane--and those prohibited in a grade but, in addition, describe standard sizes, levels of seasoning, and the like. If they include stress grades, they will give the allowable design stresses and values of modulus of elasticity applicable to each of the grades. Any grade purporting to be in accordance with American lumber standards is required to be based on Voluntary Product Standard PS20-70, "American Softwood Lumber Standard," developed by the National Bureau of Standards through the American Lumber Standards (ALS) Committee (31). All proposed rules and the related design allowables are subject to review by the ALS committee. The stress grades available from some six United States lumber manufacturers' associations and from a Canadian lumber grades agency are shown, together with appropriate design values, in a supplement to the National Design Specification for Wood Construction (24). Most of the grades shown are for softwoods, with structural grades included for only three hardwoods.

Most associations issuing grading rules also provide supervision over the grading in the lumber mill. That is, lumber graders in each mill are visited periodically by supervisory graders from the association and the accuracy of their grading checked. A number of groups provide grading service but do not issue grading rules.

Many lumber associations as well as associations of producers of other wood-based products are members of the National Forest Products Association. This association does not issue grading rules, but does provide technical, economic, and other information.

Standards of the American Society for Testing and Materials which relate to the establishment of structural grades of lumber and design allowables for lumber are ASTM D 245, D 2555, and D 2915 (10,12,14).

1.3.5.2 Glued Laminated Timber

Many of the producers of glued laminated timbers are members of the American Institute of Timber Construction, which develops standards for its products. The principal standard in this field is Voluntary Product Standard PS 56, "Structural Glued Laminated Timber" (32). The Institute also publishes a number of internal standards which constitute recommendations on design, including design properties, on typical construction details, on treating, and on other factors important to quality and serviceability. The Institute has developed inspection procedures for use in laminating plants and provides supervisory inspection of quality of in-plant inspection and quality control through its Inspection Bureau. Several independent inspection agencies provide the same service. Guidance on inspection procedures is given in an inspection manual published by the American Institute of Timber Construction (2).

Procedures for establishing design stresses for glued laminated members are contained in ASTM D 3737 (16).

1.3.5.3 Plywood

Construction-type plywood, made largely from softwoods, is the subject of a voluntary product standard, PS 1, "Construction and Industrial Plywood" (30). Most manufacturers of this type of plywood are members of the American Plywood Association (APA), which has been active in developing the product standard. Hardwood and decorative plywood is also covered by a product standard, PS 51, "Hardwood and Decorative Plywood" (33). Many producers of this type of plywood are members of the Hardwood Plywood Manufacturers Association (HPMA). Both associations offer supervisory inspection of quality control and laboratory service and issue technical literature related to the use of their products. Where applicable, they have developed design stresses.

1.3.5.4 Fiber and Particle Panel Materials

Many producers of these materials have joined into trade associations--the American Hardboard Association, and the National Particleboard Association. Each has been active in developing standards and technical data relating to its products. Among these are ASTM C208, Standard Specification for Structural Insulating Board Made from Vegetable Fibers (7), ASTM D2277, Standard Specification for Fiberboard Nail-Base Sheathing (11), ASTM C532, Standard Specification for Structural Insulating Formboard Made from Vegetable Fibers (8), and the American Hardboard Association's Hardboard Industry Standard, AHA IS-71 (1). Several product standards cover materials of these types (27,34-36).

1.3.5.5 Poles and Piles

A widely used standard for wood poles is issued by the American National Standards Institute--ANSI Standard 05.1, Standard Specifications and Dimensions for Wood Poles (4). Although this standard includes "design stresses", it should be noted that the stresses therein are not on the same basis as those generally given for design of wood structures. Requirements for wood poles for utility use are sometimes covered by internal standards of individual utilities. In addition, some groups such as the Edison Electric Institute have developed their own standards. The American Society of Agricultural Engineers has issued a standard, ASAE Engineering Practice ASAE EP388, for round, sawn, and laminated construction poles which includes design stresses (6).

The American Society for Testing and Materials has a standard for wood piles, ASTM D25, Standard Specification for Round Timber Piles (9). Pressure-treated marine piles are covered by two product standards (28,29).

INTRODUCTION

1.3.5.6 Preservative and Fire-Retardant Treatment

The American Wood-Preservers' Association has a series of standards dealing with preservatives and their application, as does the American Wood Preservers Bureau. The Bureau licenses testing agencies which inspect and test preservatively treated products to be sure that they conform to standards. The Bureau maintains a laboratory to which agency inspectors may send samples of treated products for analysis. The American Society for Testing and Materials has standards for both preservative and fire-retardant materials and their application.

1.3.5.7 Other Sources of Standards

Standards for many of the products used in construction and, in some cases, methods of testing the products, are available from sources other than those mentioned above. For example, Federal and Military specifications cover a variety of wood products and processes. Trade associations frequently have their own standards and specifications, as have some individual manufacturers.

1.3.6 References for Section 1.3

1. American Hardboard Association.
   n.d. Hardboard Industry Standard, AHA-IS-71, Palatine, Ill.

2. American Institute of Timber Construction.
   1973. Inspection manual, AITC 200-73, Englewood, Colo.

3. American Institute of Timber Construction.
   1974. Timber Construction Manual. John Wiley and Sons, Inc., New York. Second edition.

4. American National Standards Institute.
   n.d. Specifications and Dimensions for Wood Poles. Am. Natl. Stand. ANSI O5.1 (see current edition), New York.

5. American Plywood Association.
   1978. Plywood design specification. Am. Plywood Assoc. Tacoma, Wash.

6. American Society of Agricultural Engineers.
   1977. Design properties of round, sawn and laminated preservatively treated construction poles. ASAE Eng. Practice: ASAE EP 388, St. Joseph, Mich.

7. American Society for Testing and Materials.
   n.d. Standard specification for structural insulating board made from vegetable fibers. ASTM C 208, Philadelphia, Pa.

8. American Society for Testing and Materials.
   n.d. Standard specification for structural insulating formboard made from vegetable fibers. ASTM C 532.

9. American Society for Testing and Materials.
   n.d. Standard specification for round timber piles. ASTM D 25.

10. American Society for Testing and Materials.
    n.d. Standard methods for establishing structural grades and related allowable properties for visually graded lumber. ASTM D245.

11. American Society for Testing and Materials.
    n.d. Standard specification for fiberboard nail-base sheathing. ASTM D 2277.

12. American Society for Testing and Materials.
    n.d. Standard method for establishing clear wood strength values. ASTM D 2555.

13. American Society for Testing and Materials.
    n.d. Standard methods for establishing design stresses for round timber piles. ASTM D 2899.

14. American Society for Testing and Materials.
    n.d. Standard method for evaluating allowable properties for grades of structural lumber. ASTM D 2915.

15. American Society for Testing and Materials.
    n.d. Standard specifications and methods for establishing recommended design stresses for round timber construction poles. ASTM D 3200.

16. American Society for Testing and materials.
    n.d. Standard method for establishing stresses for structural glued laminated timber (glulam) manufactured from visually graded lumber. ASTM D 3737.

17. Brown, Harrison.
    1970. Production of materials in U.S. Sci. Am., vol. 203. No. 3. Sept.

18. Committee on Wood.
    1975. Wood structures: A design guide and commentary. Am. Soc. Civ. Eng.

19. Gurfinkel, German.
    1973. Wood engineering. So. For. Prod. Assoc., New Orleans, La.

20. Hansen, Howard J.
    1948. Timber engineers' handbook. John Wiley and Sons, Inc., New York.

21. Hoyle, R. J., Jr.
    1973. Wood technology in the design of structures. Mountain Press Publishing Co., Missoula, Mont.

22. National Committee of Materials Policy.
    1973. Material needs and the environment. U.S. Gov. Print. Off. June.
23. National Forest Products Association.
    1982. National design specification for wood construction. Natl. For. Prod. Assoc., Washington, D.C.
24. National Forest Products Association.
    1982. Design values for wood construction: A supplement to the 1977 edition of National Design Specification for Wood Construction. Natl. For. Prod. Assoc. April.
25. Timber Engineering Company.
    1956. Timber design and construction handbook. F. W. Dodge Corp., New York.
26. U.S. Department of Agriculture.
    1974. Wood handbook: Wood as an engineering material. Agric. Handb. No. 72, rev., U.S. Gov. Print. Off.
27. U.S. Department of Commerce.
    n.d. Mat-formed wood particleboard. Commercial Standard CS 236 (see current edition), U.S. Dep. Commer., Natl. Bur. Stand.
28. U.S. Department of Commerce.
    n.d. Pressure-treated Douglas-fir marine piles. Commercial Standard CS 249 (see current edition), U.S. Dep. Commer., Natl. Bur. of Stand.
29. U.S. Department of Commerce.
    n.d. Pressure-treated Southern Pine marine piles. Commercial Standard CS 250 (see current edition), U.S. Dep. Commer., Natl. Bur. of Stand.
30. U.S. Department of Commerce.
    n.d. Construction and industrial plywood. Product Standard PS 1 (see current edition), U.S. Dep. Commer., Natl. Bur. Stand.
31. U.S. Department of Commerce.
    n.d. American softwood lumber standard. Product standard PS 20 (see current edition), U.S. Dep. Commer., Natl. Bur. Stand.
32. U.S. Department of Commerce.
    n.d. Structural glued laminated timber. Product Standard PS 56 (see current edition), U.S. Dep. Commer., Natl. Bur. Stand.
33. U.S. Department of Commerce.
    n.d. Hardwood and decorative plywood. Product Standard PS 51 (see current edition), U.S. Dep. Commer., Natl. Bur. Stand.

34. U.S. Department of Commerce.
    n.d. Cellulosic fiber insulating board. Product Standard PS 57
    (see current edition), U.S. Dep. Commer., Natl. Bur. Stand.

35. U.S. Department of Commerce.
    n.d. Basic hardboard. Product Standard PS 58 (see current
    edition), U.S. Dep. Commer., Natl. Bur. Stand.

36. U.S. Department of Commerce.
    n.d. Prefinished hardboard panelling. Product Standard PS 59
    (see current edition), U.S. Dep. Commer., Natl. Bur. Stand.

1.4 World War II Structures

1.4.1 Background

When it became evident that the United States was going to be involved in World War II, it became necessary to construct facilities for a rapid expansion of the military forces in terms of barracks, warehouses, and the like. There was an enormous expansion of production of war material requiring a corresponding expansion in the buildings needed to house the production lines. As the war material was produced, untold quantities of material were needed to provide shipping containers. Not all of the World War II structures, however, were of the relatively simple barracks, warehouse types. For example, the hangars built to house the Navy's blimps were on the order of 1,000 feet long, with a span of about 265 feet and a center height of about 120 feet. The roof supports were framed arches using ring connectors and bolts. These hangars certainly must be among the largest timber structures ever built.

And, as all these needs developed, the use of metals was restricted to the production of machinery, tanks, and ships. Wood, therefore, was called on to provide the bulk of the raw material for building construction. Eventually the supply of wood became a problem and steps were taken to stretch the supply. Some of these steps were the following:

> Designs were based on a roof load of 15 pounds per square foot although in more normal times the design load would have been 20, 30, or even more pounds per square foot.

> Designs were based on the assumption that 1,200f (1,200 psi allowable flexural stress) lumber would be used, but this grade was assigned a stress of 1,800 psi.

> The supply of 1,200f lumber was inadequate to supply the needs. So, in the auctions at which lumber was purchased, large quantities of non-stress-graded material were purchased and used--but with no revision of designs.

> Because of the need for large quantities of lumber in a hurry, considerable quantities of unseasoned lumber were purchased and used.

INTRODUCTION

Despite the abuses and the short design life, many of these 40-year-old buildings still exist, either in their original locations or after dismantling and moving to a new location. Although we have no statistics on the number of these buildings still in use, it may be that some of the problem structures we hear about are the result of the limitations of the World War II period, of changes in use, and of long years of neglect.

1.4.2 Some World War II Experiences

Abuses such as those outlined above, together with design for a 5-year life and the lack of experienced construction workers and supervisors could be expected to lead to problems of the type with which this publication is concerned. This and the following section provide a brief review of typical problems encountered with WWII structures, their causes, and their results. This information should be helpful to the professional faced with the necessity for evaluating a problem structure. Not only will it provide clues to causes of problems but it will, in addition, indicate the degree of abuse which a timber structure can sustain without a catastrophic failure.

In a large industrial building, parallel-chord Warren trusses of considerable span were used in the roof system. After the roof sheathing and the roof covering were in place, inspection revealed that the trusses had been installed upside down, so that stresses in all members were reversed from those for which they had been designed. To further complicate matters, monorails were designed to be supported at a panel point. With the truss in inverted position, and with the position of the monorails fixed with respect to the wall, the rails ended up being supported between, rather than at, panel points.

Excessive deflection of roof trusses in buildings at a large installation was finally traced to the fact that, although the trusses had been designed for use of split rings at the joints, none had been installed. The bolts were in place, but obviously they could not carry the applied loads without deformations much greater than would have occurred with the rings in place.

Trouble was reported with an important joint in a 100-foot truss. Inspection of the joint showed that only two of the three load-carrying members entering the joint were attached to the gusset plates; this was in accordance with the working drawings! And, apparently, someone had slipped a decimal point, for the joint was designed with enough rings to carry a 10,000-pound load although the load to be transferred was 100,000 pounds.

Excessive sagging of flat roofs was reported on several sites in a certain area. The design of the flat wood roof trusses was reasonable, camber had been specified, and the contractor swore that he had put in the specified camber. It turned out, however, that during fabrication, the member lengths had not been adjusted to provide camber. So, the contractor had assembled the trusses loosely, jacked in the camber, and tightened the bolts. A little shrinkage took out the friction imposed by the bolts and a series of sagging trusses resulted.

At another site, the roof truss design required the use of 4-inch nominal members. However, there was not enough 4-inch material on hand at the fabrication site, so that the fabricator decided to use two 2-inch-thick members, reasoning that twice 1-5/8 inches came close enough to the actual size of the 4-inch material. With this construction, however, the connectors were in only the outer one of the two thicknesses and so there was nothing to tie the two 2-inch layers together at the joints. Further, the length/thickness ratio was about doubled and the load-carrying capacity of the members loaded in compression was seriously reduced.

The more undesirable features resulting from this accelerated building program may be summarized about as follows:

With respect to design--

poor joint design,

failure to provide sufficient camber, particularly when unseasoned lumber was to be used,

failure to design doors (such as those on hangars) so that their operation would not be affected by truss deflection,

use of trusses with too small a height/span ratio,

improper design of compression members such as assuming that nailed members acted as solid members, and

failure to recognize that the nonhomogenous nature of wood requires special attention, so that direct transfer from a design meant for metal is not possible.

With respect to lumber--

Use of ungraded and nonstructural lumber where stress-graded lumber was needed, and

use of unseasoned lumber.

With respect to fabrication and erection--

Misfabrication and sloppy workmanship, as in cases where ring grooves and bolt holes were placed in the wrong location and then rebored nearby to correct the error,

using wider or deeper ring grooves than specified,

mauling members into place when misfabrication had occurred,

omission of fastenings,

improper handling, such as picking up trusses at a single point or dropping them during erection,

INTRODUCTION

improper storage of lumber awaiting fabrication,

resawing lumber without regrading, and

omission of needed truss bracing.

With respect to maintenance--

bolts not retightened as lumber seasoned in place, frequently necessitating repair or replacement,

inadequate reinforcement or repair, and

ponding from poor drainage which, in turn, resulted from lack of camber, neglect in tightening bolts, poor fabrication, etc.

1.4.3 Performance of Wartime Structures

The recital in the preceding section may well sound like a prescription for ulcers on the part of owners and engineers. In spite of the problems described and the thousands of buildings erected during the wartime period, very few instances of actual collapse were reported. In part, of course, this means that someone found a deficiency at an earlier point and corrected it--by repair, replacement of members, reinforcement, or some other means.

F. J. Hanrahan, in a discussion of performance of wartime buildings ($\underline{2}$) reported that, as of about July 1944, the Army had something on the order of 1-1/4 billion square feet of buildings country-wide. Four of nine service commands reported that they had experienced serious problems in seven buildings. In five of the seven cases, the trusses were all of a single design--a deficient design--and although the trusses had to be shored or reinforced, none actually collapsed. Further investigation indicated that there had actually been four collapses. In part, this record may be laid to the fact that timber structures give evidence of distress, in many cases, before actual failure, so that it is possible to take corrective measures. Hanrahan comments also that a contractor retained to inspect and service timber structures reported that, based on inspection, stress analysis, and supervision of repair on about 500 buildings with timber trusses ranging from 20 to 200 feet in span, about 95 percent were in very good structural condition, needing only minor servicing such as bolt tightening and a limited amount of stitch-bolting or clamping of splits. The contractor felt that, with the exception of a few failures which were obviously the result of the use of low-grade lumber, most difficulties were the result of faulty design.

1.4.4 Summary

From the preceding discussion, the engineer faced with a problem relating to a timber structure may get clues as to possible causes of some of the problems he faces. He may get some reassurance, as well, that in spite of deficiencies he does not necessarily face the immediate collapse of the structure. That is, he may have time to make corrections. This is not to say that he should become euphoric, but that he can approach the problem, in many cases, with some deliberation.

Much of the foregoing was adapted from the article by F. J. Hanrahan (2). Other authors, however, have reviewed the wartime structural problems and the reader may be interested in reviewing them (1,3).

1.4.5   References for Section 1.4

1. Carter, A. N.
    1944. Wood in war construction. Sci. Am. May.

2. Hanrahan, F. J.
    1945. Performance of wartime wood buildings--and its postwar application. So. Lbrmn., July 1, pp. 57, 58, 60, 62.

3. Salgo, M. N.
    1954. Examples of timber structure failures. Proc. Struc. Div., ASCE, November.

1.5   Old Wood Structures

1.5.1   A Historical Overview*

1.5.1.1 Introduction

The history of civilization has been symbolized by man's achievements as a builder of cities, transportation systems, and the utilization of water resources. Archeologists are well aware of the importance of man's activities as a builder, since the record of ancient and prehistoric civilizations has been interpreted from the remains of buildings and other structures.

From the ruins of prehistoric sites to the modern skyscraper the artistic, technical, and historical interpretation of buildings has been firmly in the hands of art and architectural historians. A concern for the history of technology is of very recent origin and hence brings a fresh new viewpoint to the evaluation of "man the builder."

Traditional architectural educational programs have featured courses in the history of architecture. Generations of students have used Banister Fletcher's famous text on the history of architecture by the comparative method (2). From its first appearance in 1896 to 1954 this book has been published in 16 editions. Its influence has been immense. Despite the "battle of the styles" in the 19th century, architects, in general, were committed to the idea that not all building was architecture. Architecture was understood in monumental terms. Thus, a book like Fletcher's is almost entirely devoted to

---

*Author: Emory L. Kemp, Professor of the History of Science and Technology and of Civil Engineering, West Virginia University.

INTRODUCTION

building in stone, with a brief nod to other building materials, particularly brick. As a result, the Crystal Palace, arguably the most important building of the 19th century, is not mentioned because it was built of glass and iron and "really" engineering and not architecture!

In a similar way wood was not considered a suitable material for monumental architecture and its use begrudgingly admitted by architectural historians. A closer look, however, reveals the extensive use and influence of wood throughout the history of building. It is the purpose of this section to give a brief summary of how wood has played a continuous and highly significant role in the evolution of architecture and engineering, particularly in 19th century America.

1.5.1.2 An Ancient Lineage

The roots of Western architecture, a term which until modern times must include engineering works, are firmly planted in three distinct geographical areas. Each of them made distinctive contributions to the art of building. Perhaps the oldest is the tradition of mud brick building in Mesopotamia, an area devoid of both timber and stone for construction purposes. It was here, by all accounts, that the true arch was invented. This tradition of arcuated architecture reached towering heights architecturally and technically in the hands of first the Romans and later the master builders of the Middle Ages. And yet, as we shall see, timber played an important role in what was essentially a triumph in stone.

The Egyptians were the first to use stone for large-scale engineering works at the temple associated with the great step pyramid at Sakkara, ca. B.C. 3900. Closer examination of the temple reveals that its origins lie in a domestic architecture based upon wood and reeds. The stone columns are imitations of early bundled reed posts, and the roof stone slabs unmistakably represent palm trunks. Thus, we see a very early influence of wood in the establishment of an architectural tradition.

The post-and-beam framing tradition did not continue from the early Egyptian precedent but owes its development to the Greeks. These northern people brought into the Mediterranean basin a vernacular architectural tradition using wood and featuring a sloping or pitched roof. To many the Parthenon represents a perfection in architecture that has never been equaled. It is the epitome of trabeated (the Latin for beam is trabs) architecture and is based upon the earlier timber-framed megaron house, illustrated in figure 1.5.1. Thus, the floor plan of the humble house gave rise to the Greek temple. The temple was conceptually a megaron on a grand scale with unsurpassed refinement of geometrical detail. The timber posts were replaced with stone columns and the beams with stone lintels. From outward appearances the transformation from wood to stone was complete. The sloping roof of the megaron, however, was not only retained but was supported by timber joists and rafters.

The Greeks remained true to the trabeated tradition and did not experiment with or use the truss. As a result, the ceiling joists and roof

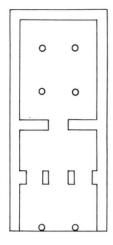

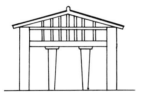

Figure 1.5.1--Greek Megaron house.

rafters worked independently of each other and thus necessitated an interior colonnade to support the joists and on top of this a second story colonnade to support the roof rafters. One of the earliest extant examples of the classic Greek style is the Temple of Poseidon at Paestum, ca. 450 B.C. It serves as a symbol of Greek architecture (fig. 1.5.2).

1.5.1.3 Roman and Medieval Building

Just as trabeated architecture is associated with the Greeks, the arcuated tradition is firmly linked to Roman architecture and engineering works on an unprecedented scale. The use of the arch and its three-dimensional manifestation, the dome, implies the use of materials possessing compressive strength but not necessarily tensile strength in either the material itself or the joints. This, of course, means the use of stone, tile, brick, concrete, or a combination of these materials. Not all Roman buildings were enclosed with vaults or domes--a surprising number used wood roof trusses. Vitruvius (5) describes a truss he designed for the roof of a basilica at Fano in the first

INTRODUCTION

Figure 1.5.2--Temple of Poseidon at Paestum, ca. 450 B.C.

century B.C. None of this early work survives, but later fourth and fifth century A.D. truss work does and is illustrated in figure 1.5.3. This use of trusses influenced later Italian Renaissance architects and this in turn influenced 18th century American builders.

In the hands of Medieval builders the Romanesque building tradition was transformed into high Gothic architecture with soaring columns and ribs and supporting buttresses. This tour-de-force in stone represents one of the most revolutionary advances in the history of building. Timber played a most significant supporting role in providing the roof structure above the vaulting. From humble parish churches to towering cathedrals a great deal of timber truss work survives from the Middle Ages (fig. 1.5.4). One of the most spectacular uses of timber in Medieval building is the stone-sheathed timber spire on Salisbury Cathedral which rises to the height of 404 feet (fig. 1.5.5). Until the 1960's this was the tallest building in Britain.

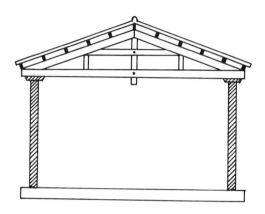

Figure 1.5.3--Typical Roman Timber trusses.

In domestic building, many examples of the timber post-and-beam system remain as reminders of the skill of Medieval builders and the durability of wood. This framing system resulted in the familiar half-timbered construction with nonstructural infilling of daub and wattle or bricks or the application of wood siding. This system was also used for barns and mills throughout Britain and on the Continent.

1.5.1.4 American Wood

It was the post and beam vernacular architecture that the Pilgrim Fathers transplanted to America. This later evolved into the simple and yet elegant Colonial version of Georgian architecture which often found expression in all wood structures. Exquisite examples of houses, churches, mills, and other commercial structures dating from the 18th century can be found all along the East Coast of the United States.

When the tide of the Industrial Revolution reached the shores of the New World at the beginning of the 19th century, timber was the primary material used to build an amazing array of structures for transportation, industry and commerce, and for housing. As an example, White (6) points out that the American railway system was largely built of wood. Buildings, trestles, rolling stock, fuel, and crossties were of wood and in the early days even the rails were of wood protected with a strap-iron wearing surface. It may be correct to speak of the iron horse, but nearly everything else related to the 19th century American railway was built of wood.

Figure 1.5.4--Monks' dormitory, Durham Cathedral (built 1298-1404). From "A History of Architecture" by Banister Fletcher. Raphael Tuck and Sons Ltd., England. Batsford, London, 1954, 16th Edition, p. 327.

Numerous other examples of the use of wood could be cited, but it seems more appropriate to discuss just two developments which represent important American contributions to the art of building in wood. In Britain as early as the Tudor era wood was reserved for ship building and truss work, where its strength in bending, tension, and compression, coupled with light weight could not be matched by any other common structural material. Thus, the building of the world's first urban industrial society in Britain during the 18th and 19th centuries was largely accomplished using brick and stone. Just the opposite was true in America where wood was abundant and cheap. The two examples of American innovation in the use of wood are the covered bridge and the balloon frame for houses and buildings.

1.5.1.5 The Covered Bridge

No 19th century structure has captured the attention and affection of Americans like the covered wood bridge which was built by the thousands throughout the earlier half of the 19th century in North America. Although it came to be regarded as a uniquely American genre its antecedents were in Europe. As mentioned above, knowledge and use of the truss passed from the Romans to Renaissance Italy. The architect Andrea Palladio published a book on architecture in which he showed wood truss bridges (fig. 1.5.6). During the first half of the 18th century, his work was translated into English and became very

Figure 1.5.5--Spire of Salisbury Cathedral.

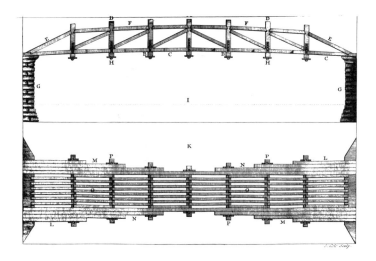

Figure 1.5.6--Palladio's truss bridge design. Science Museum Collection, London.

popular. It was also at this time (1767) that Grubenmann built a noteworthy bridge in Oberglatt, Switzerland--a timber bridge (fig. 1.5.7) which showed remarkable similarities to the Burr truss which was patented by Theodore Burr at the beginning of the 19th century.

The use of wood for bridges eventually took the form of an adaptation of the traditional King and Queen Post roof trusses. The multiple Kingpost truss was found to provide an efficient structural system for spans up to about 80 feet. Not content with mere adaptation, American bridge builders developed and patented a large number of truss types, the most significant being the Burr (1817), Town Lattice (1820), the Long (1830), and Pratt (1844), as shown in figure 1.5.8. Except for the Long and Pratt trusses these patented systems were developed by bridge builders without engineering training and in a spirit of competition rather than cooperation. Thus, the field was advanced by empirical designs which were proven by experience.

It is instructive to look at the evolution of the Burr truss as a case study of technological development on an empirical basis. In an effort to find a structural system capable of providing both strength and stiffness for long-span timber bridges, Burr experimented with a combination suspension-multiple Kingpost system and also with a bowstring arch with iron suspenders to support the deck. The final design, which was later patented and became known as the Burr truss combines a timber arch and a multiple Kingpost truss. These bridge types are shown in figure 1.5.8. A handsome example of a Burr truss is the Barrackville

Figure 1.5.7--Grubenmann's bridge, 1767.

bridge completed by Lemuel Chenoweth in 1853 (fig. 1.5.9). The bridge survived Jones' raid during the Civil War and serves local traffic to this day in substantially original condition. Its longevity is largely the result of the siding and roof which have protected the deck and truss work for more than a century. Without such covering, unprotected wood bridges had a life expectancy of 15 or 20 years at the most.

In 1829 Col. Stephen Long introduced his truss on the Baltimore and Ohio railroad. Long, a West Point graduate, is perhaps the first covered bridge builder to apply engineering principles to a patented design. His patent of 1830 featured a preloaded counter system which was quite sophisticated in concept. Pratt's patented truss of 1844, which later became the most popular truss type in the country until the First World War, was also based upon the principles of structural mechanics. Earlier, in 1840, Howe patented a combination truss which utilized wrought iron tension rods and cast iron joint blocks, all the rest of the members being timber. This marked the beginning of the transition from the timber truss to the all-metal truss, which was largely accomplished just before the Civil War. It was also a transition from an empirical craft-orientated technology to bridge building based upon mechanics and in the hands of professional engineers. Thus, timber became associated with an earlier mode of building and was henceforth not in the forefront of developments in structural engineering, its place being taken first by iron and later by steel and reinforced concrete.

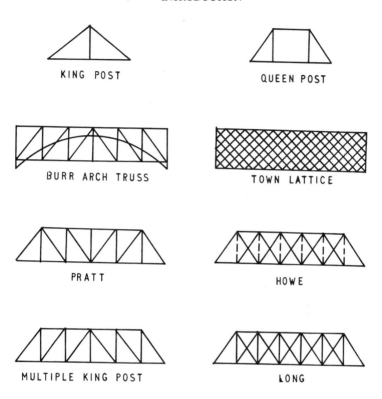

Figure 1.5.8--Timber bridge truss systems.

1.5.1.6 The Balloon Frame

Post and beam framing was imported from Europe, as indicated above, and provided the skeleton for many notable structures. In figure 1.5.10 the truss work in Decatur House, early 19th century, shows a traditional use of heavy timber framing and indicates the durability of timber if protected from moisture. Timber was also widely employed in commercial and industrial structures. A notable example is the extant truss arch in the President Street Station in Baltimore (fig. 1.5.11). When applied to large industrial building with the concentration of the structural members in heavy timbers, the post and beam system gave considerable fire protection, and this "heavy mill construction" utilizing masonry load-bearing walls and interior timber floors and framing was very popular throughout 19th century America. From midcentury a rival emerged in terms of iron framing and brick arch floors. This led

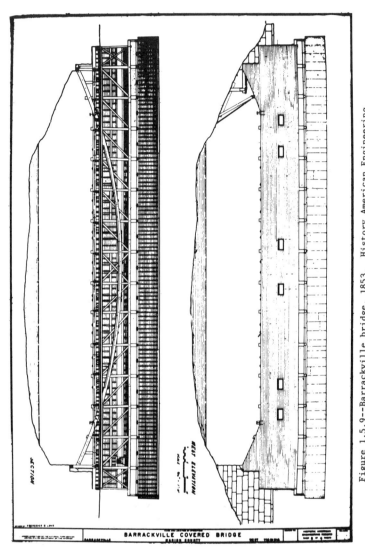

Figure 1.5.9--Barrackville bridge, 1853. History American Engineering Record, U.S. Department of the Interior.

INTRODUCTION

Figure 1.5.10--Decatur house, Washington. Photograph by W. Edmund Barrett.

Figure 1.5.11--President Street Station, Baltimore. Photograph by W. Edmund Barrett.

36   WOOD STRUCTURES

directly to the development of the skyscraper and the early 20th century concept of a fireproof building. The virtual eclipse of wood as a structural material for monumental architecture and major civil engineering works did not mark its end as a building material, for wood was to find application on an immense scale in vernacular architecture as the result of the "Balloon Frame" which was introduced in Chicago by A. D. Taylor in 1833.

The "Balloon Frame," derisively named because of its ethereal appearance, constituted a revolution in building which was ideal for American conditions, and its use persists to the present day. The traditional post and beam method required heavy timbers and considerable millwright skills in hewing the timber and preparing the joints. With balloon framing (fig. 1.5.12), only elementary skills were required to nail

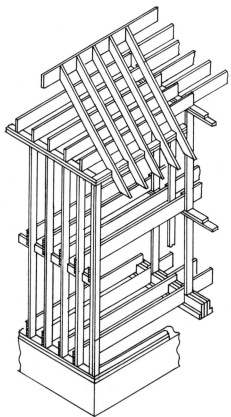

Figure 1.5.12--Balloon framing.

INTRODUCTION 37

together relatively small sticks of standard dimension lumber. This simple and versatile framing system has become the accepted way of building houses in America, even those with brick veneer instead of wood siding. Many houses dating from the 19th century continue to serve their original function and silently witness to the durability of wood.

1.5.1.7 Conclusion

Wood has played a continuous and significant role in the history of building, sometimes as the primary material and often in a secondary role. In addition, it has been through the utilization of the inherent advantages of wood that major architectural traditions have been established.

In examining historic structures, one is struck by the durability of wood if adequately protected from the weather. Many examples of Medieval timber work 800 years old exist in Europe. Many extant 18th and 19th century structures in America give further verification of the versatility and durability of wood as a building material.

1.5.2   Age and Strength

There appears to be a common misconception that old wood is automatically deficient in strength and stiffness. In part this may stem from the fact that wood may be subjected to deteriorating influences such as decay and, if such exposure continues over a long period, the wood will be weakened. In part, too, it may stem from the fact that wood may have become dirty and unattractive over the years and the assumption that this indicates deterioration. These are legitimate concerns, and several investigators have studied old structural members to determine whether, in fact, there has been any loss in strength aside from that which might have resulted from exposure to deteriorating factors.

Dismantling of a number of railroad bridges gave the Forest Products Laboratory an opportunity to evaluate strength of the bridge timbers after 23 years of service (4). The timbers, rough-sawn Douglas-fir, were about 8 by 18 inches in cross section and about 30 feet long. The stringers themselves were tested, and samples were taken from them for evaluation of material not influenced by checks, shakes, or decay. The latter, small-clear tests indicated that the wood itself was not affected by the 23 years of service. Decay, deep checks, crushing over pile caps, and the like influenced the strength of the timbers, but even the weakest (marginal in quality) gave strengths about 2-1/2 times the stress recommended for design. The authors concluded that some of these timbers could have been reused if certain precautions were observed.

The Association of American Railroads and the Forest Products Laboratory collaborated on tests of 50-year-old southern pine stringers taken from a bridge which was dismantled in order that it might be replaced with a culvert (1). The number of tests was too small to permit firm conclusions. In addition, only one of eight stringers was tested statically, the rest being tested under repeated loading. Tests of small clear specimens taken from the stringers indicated that there was no serious reduction in the strength, stiffness, or shock resistance of the wood in the stringers. The stringers themselves appeared to have lost strength due largely to the development of seasoning checks. Test values, however, were generally adequate in relation to the stresses imposed in service on the bridge.

Unpublished data at the Forest Products Laboratory describes tests on two 101-year-old girders which had supported the floor of the sanctuary of St. Raphael's cathedral in Madison, Wisconsin. Remodeling of the cathedral made the eastern white pine beams available for study. One of the beams developed, in a flexural test, about six times the stress estimated to be imposed on it by service conditions. The second failed in horizontal shear through deep checks which had developed from drying in service; it developed a flexural stress on the order of three times the stress imposed by service conditions. Shear stress at failure calculated from the estimated net area at the deeply checked section was well below that expected from tests of small clear specimens, but in line with stresses developed at failure in comparable tests of large members. Tests of small clear specimens taken from the beams indicated that one was somewhat denser and thus somewhat stronger than the species average. The other was somewhat lower in density than the species average and this was reflected in low values for some properties; the important properties of modulus of rupture, maximum crushing strength, toughness, shear, and modulus of elasticity were above the average. All things considered, there was no evidence of deterioration in strength from aging alone. Thus the data provide additional assurance of the lasting qualities of structural wood.

The University of Oregon has reported on tests of old bridge timbers, some of which had been in service more than 50 years and some others which had been in service more than 40 years (3). The report compares the data with the results of tests made on "new" timbers, both green and dry. Without going into detail, the author states that the results indicated that the "old" timbers were "undoubtedly stronger than they were when new." The author points out that the factors that make for strength or weakness are the same in old as in new timbers--a high strength depends mainly on a high percentage of summerwood (wood grown in the latter part of the growing season), reasonable freedom from defects, and thorough seasoning. Again, the results confirm a lack of effect of age alone.

All in all, the data seem clear that, without some form of deterioration, strength properties do not decrease with age (7). It is known, however, as explained in the discussion of time-dependent effects (Section 1.3.2), that, depending on the stress level, a load which can be carried by a wood member over a short period of time may cause failure if the load is carried for a long period. Therefore, it seems only prudent that, when the loading history of an old structure is not known, the stresses to be used in design of a rehabilitated structure should be taken at a level suitable for long-time loading rather than for normal loading. In addition, it seems desirable, in such instances, to forego the increases recommended for short-time loading such as for wind and snow.

1.5.3 References for Section 1.5

1. Association of American Railroads.
    1961. Investigation of fatigue and static strength of 50-year-old timber stringers from A.T. and S.F. Ry. Bridge near La Junta, Colorado. Eng. Res. Div., AAR Res. Center, Rep. No. ER-2. January. Chicago, Ill.

2. Fletcher, Banister.
    n.d. A history of architecture, on the comparative method. 16 ed.

3. McAlister, E. H.
    1930. Strength tests of old Douglas fir timbers. Univ. of Oregon Publ. Mathematics Ser., vol. 1, No. 2.

4. Newlin, J. A. and George E. Heck.
    1934. Tests show strength of Douglas fir stringers after 23 years service. Railway Engineering and Maintenance, August.

5. Vitruvius, Pollo.
    1914. De architectura. Translated by M. H. Morgan, pub. Harvard University Press, Cambridge, Mass.

6. White, John H., Jr.
    1976. Tracks and timber, IA. J. of the Soc. for Ind. Archeol. 2(1), Morgantown, W. Va.

7. Wood, Lyman W.
    1954. Structural values in old lumber. So. Lbrmn. Dec. 15.

1.6 Overview of Building Regulations That Relate to Rehabilitation*

1.6.1 Introduction

Throughout the United States, increasing concern is being expressed for the need to more fully utilize the existing building stock. The reuse of older buildings has grown far beyond the preservation movement which spawned it with all existing buildings now viewed as assets that can be reclaimed. In early 1980, the U.S. Department of Commerce projected that the seasonally adjusted annual expenditures for residential alterations and repair will reach a new high of $49.9 billion. This will exceed or equal new-construction spending. The expenditures for non-residential rehabilitation are expected to exceed $15.0 billion in the same year.

There are many forces at work which potentially impede the full utilization of existing buildings--market forces, financial considerations, government policy, industry structure, and regulatory activities are

---

*Author: James H. Pielert, M. ASCE, Group Leader - Building Rehabilitation Technology, Building Economics and Regulatory Technology Division, Center for Building Technology, National Bureau of Standards, Washington, D.C.

but a few. The purpose of this section is to present an overview of
the building regulatory system in the United States with emphasis on
its impact on the rehabilitation of existing buildings. A report of
the National Commission on Neighborhoods (10) states that current code
enforcement practices can add as much as 10 percent to the cost of
building rehabilitation.

1.6.2   Overview of the Current Building Regulatory Process

1.6.2.1 Development of Building Codes in the United States

A building code is a legal document which sets forth requirements to
protect the public health, safety, and general welfare as they relate
to the construction and occupancy of buildings and structures. The
building code development process in the United States is quite complex. The police power of the state generally includes the development
and enforcement of building codes. In some cases, local governments
have developed building codes when states have not chosen to exercise
this power. However, within the past several years, many states have
assumed a more active role in writing, promulgating and enforcing
building codes. A growing number of states have withdrawn virtually
all authority to enact building codes from their local municipalities.
There are now 25 states, containing over one-half of the Nation's population (4), with some type of statewide building code authority. There
are about 14,000 jurisdictions issuing building permits in this
country. One consequence of this widely decentralized regulatory
system is a considerable diversity of substantive provisions among the
thousands of locally enacted codes. This is true even though three-quarters of locally enacted codes are based on one of the four model
building codes discussed below. Possible reasons for this diversity
are: (1) provisions of the model codes are frequently altered due to
local interest; (2) local codes are infrequently updated; (3) model
codes are not uniform; and (4) some municipalities write their own
codes.

There are four model building codes which are available for use by
states and local jurisdictions. The first model building code was
published in 1905 by the National Board of Fire Underwriters to guide
municipalities concerned with reducing the fire hazard in and about
buildings. This is now known as the National Building Code which was
drafted by the American Insurance Association. In 1980, the National
Conference of States on Building Codes and Standards, Inc. (NCSBCS)
assumed sponsorship of the National Building Code. The other three
model code organizations have membership controlled by local government
code enforcement officials. These are: Building Officials and Code
Administrators International, Inc. (BOCA), which maintains the Basic
Building Code (3); International Conference of Building Officials
(ICBO), which maintains the Uniform Building Code (8); and Southern
Building Code Congress International, Inc. (SBCCI), which maintains
the Standard Building Code (14).

INTRODUCTION

1.6.2.2 Building Code-Related Organizations

In addition to the model building code groups, there are other organizations concerned with building code development. These include:

CABO--Council of American Building Officials
CABO was formed in 1971 by BOCA, ICBO, and SBCCI to work toward uniformity in building codes. CABO provides coordination between these three model code organizations.

BCMC--Board for the Coordination of the Model Codes
This organization functions under CABO and addresses specific model code problem areas such as energy conservation and emergency egress requirements.

NACA--National Academy of Code Administration
NACA was established in 1970 for the purpose of developing building regulatory training and education programs to promote certification of enforcement in order that code administration can become a recognized profession in the United States.

AMCBO--Association of Major City Building Officials
AMCBO is an organization established as a forum for 30 major U.S. cities involved with the enforcement of building codes. The member cities contain approximately one-third of the U.S. population.

NCSBCS--National Conference of States on Building Codes and Standards, Inc.
NCSBCS was established in 1967 with the goal of "achieving reasonable national uniformity in the regulation of buildings and the acceptance of industrialized buildings across State lines." NCSBCS currently undertakes studies in many regulatory and code development areas such as mobile homes, energy conservation, and building rehabilitation.

NIBS--National Institute of Building Sciences
The establishment of the National Institute of Building Sciences was authorized by the Housing and Community Development Act of 1974 (Public Law 93-383). Section 809(b)(2) of this Act for "the development of an organization framework to encourage and provide for the maximum feasible participation of public and private, scientific, technical and financial organizations, institutions and agencies now engaged in activities pertinent to the development, promulgation and maintenance of performance criteria, standards and other technical provisions for building codes and other regulations."

1.6.3 Building Regulations and Existing Buildings

1.6.3.1 Building Codes Applied to Existing Buildings

Studies have indicated that the current building regulatory process discriminates against the rehabilitation of existing buildings (1,5,6, 7,12,13,16). Some of the reasons for this include:

A. Inflexible Provisions Requiring That Existing Buildings Comply With Codes for New Construction

Most codes contain provisions, variously worded, requiring that a building's conformance with the requirements of a currently enforced building code for new construction should increase in relation to the dollar amount of the rehabilitation planned. An example is the requirement generally known as the 25-50 percent rule which is illustrated in figure 1.6.1. The application of these administrative provisions is based on the following economic hierarchy which is itself awkward and difficult to enforce in actual practice.

--The alteration must be restored to at least its original condition for renovation which costs less than 25 percent of the value of the building, or

--the alteration may, at the discretion of the building official, be required to comply with the building code requirements for new construction when the cost is between 25 and 50 percent of the building's value, or

--the entire building must comply with the building code requirements for new construction when alterations exceed 50 percent of the building's value.

The provisions are difficult to enforce at the local level, and may have negative effects on both public safety and the quantity and quality of the building stock. This is because there is a considerable period of time during which a building is not worth the expense of rehabilitation under current codes for new buildings but is not sufficiently hazardous to justify condemnation and demolition. Moreover, the effects of building deterioration are typically not confined to an

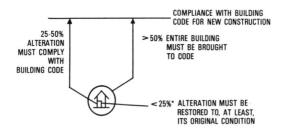

* Percentages refer to cost of rehabilitation related to value of the structure before rehabilitation

Figure 1.6.1--Application of 25-50 percent rule.

INTRODUCTION

individual structure and have an areawide effect with accompanying social, political, and economic problems.

A specific example of the problem discussed above is the situation which occurs when an existing timber building which has performed well over time is evaluated against modern seismic design criteria. Even though the use of the building will not be changed and a record of good structural performance exists, strict application of the 25-50 percent rule could require costly seismic upgrading of the structure to comply with code requirements for new construction. For example, partitions and load bearing walls may have to be extensively strengthened to withstand horizontal seismic forces.

B. Codes and Standards Fail to Address Types of Construction Represented in Many Older Buildings

Previously accepted construction assemblies may be dropped from a building code or standard, not because they are undesirable, but because they are no longer a part of the current catalog of construction materials and techniques. The situation occurs frequently in the area of fire resistance. Fire testing of construction assemblies has been performed in essentially the same way for many years. However, as materials and techniques become obsolete, they cease to be listed in new code editions. Unless assemblies involving such materials and techniques have deteriorated, however, there is no reason to believe they will not perform as previously rated and approved for use.

C. Methods for Assessing the Condition of Existing Building Systems Are Lacking

The process of rehabilitating an existing building can present unique construction problems since many of the systems may be affected. Thus, it becomes necessary to assess the condition of the systems in order to determine their ability to support the modifications. Condition assessment techniques vary from rather simple visual inspection to more complicated laboratory, or in-situ, testing which may require sophisticated equipment and highly skilled operating personnel.

The National Bureau of Standards has recently published a report (9) on the condition assessment of structural, HVAC, plumbing and electrical systems in existing buildings. A chapter on structural timber includes evaluation techniques for assessing timber properties such as strength and grade, extent of decay, moisture content, etc.

D. Code Enforcement Process for Existing Buildings Differs From That for New Buildings

Some regulatory agencies tend to use the same administrative and enforcement processes for additions and alterations to existing buildings as are used for new construction. Forcing rehabilitation projects into a model established for new construction presents unique difficulties such as: (1) the technical capabilities of plan reviewers and inspectors may be inadequate; (2) the appeals process may be time-consuming and costly; and (3) the project may proceed too far before it is determined that extensive technical and economic barriers exist. These frequently imposed regulatory barriers make building rehabilitation projects expensive, difficult, and in some cases impossible.

1.6.3.2 Guidelines for Building Rehabilitation

To date, there have been limited attempts to formulate code provisions specifically related to the unique needs of building rehabilitation. The following is a brief discussion of five such documents which address the subject.

A. Guidelines for HUD Rehabilitation Projects--Guidelines setting forth basic objectives and provisions specifically related to building rehabilitation have been issued by the Department of Housing and Urban Development-Federal Housing Administration (HUD-FHA) for use under various mortgage insurance programs dealing with property maintenance and rehabilitation. These HUD guidelines, contained in Handbook 4940.4 entitled "Minimum Design Standards for Rehabilitation for Residential Properties," were originally issued in September 1973 and there have been several revisions since (17). The guidelines apply to the physical rehabilitation of existing residential properties from single family dwellings to large multistory, elevator-type apartment buildings in all HUD field jurisdictions. The structure and format of the guidelines are such that the "minimum" requirements are stated mostly in performance terms, with supplementary guides to good practice contained in accompanying, but separate, sections.

B. Rehabilitation Guidelines Developed by Model Code Organizations-- In another endeavor, BOCA developed a set of rehabilitation provisions and published them under the title "Code Enforcement Guidelines for Residential Rehabilitation" (2). The studies that formed the basis for these guidelines were conducted under a contract with HUD. These code enforcement guidelines have not been widely adopted by State and local jurisdictions. The guidelines are intended only for residential buildings and specifically exclude hotels, motels, and other housing for transients.

C. Department of the Interior Standards for Rehabilitation--The Department of the Interior has issued a publication entitled "The Secretary of Interior Standards for Rehabilitation and Guidelines for Rehabilitating Historic Buildings" (19) as a means of determining if a rehabilitation project qualifies as a certified rehabilitation pursuant to the Tax Reform Act of 1976. The publication contains guidelines which are designed to help individual property owners formulate plans for the rehabilitation, preservation, and continued use of old buildings. The guidelines pertain to buildings of all occupancy and construction types, sizes, and materials.

D. HUD Rehabilitation Guidelines--The Department of Housing and Urban Development has developed rehabilitation guidelines (18) in conjunction with NIBS in response to Section 903 of the Housing and Community Development Amendments of 1978. The purpose of the guidelines is to

INTRODUCTION 45

help communities resolve building code problems encountered when implementing rehabilitation programs. The eight specific guidelines are as follows:

    Guideline 1 - Setting and Adopting Standards for Building Rehabilitation
    Guideline 2 - Approval of Building Rehabilitation
    Guideline 3 - Statutory Guideline for Building Rehabilitation
    Guideline 4 - Managing Official Liability Associated with Building Rehabilitation
    Guideline 5 - Egress Guidelines for Residential Rehabilitation
    Guideline 6 - Electrical Guidelines for Residential Rehabilitation
    Guideline 7 - Plumbing DWV Guideline for Residential Rehabilitation
    Guideline 8 - Fire Ratings of Archaic Materials and Assemblies

E. Massachusetts Article 22--The State of Massachusetts recognized that the State Building Code, which contained the "25-50 percent rule," was a constraint to building rehabilitation (1). In response, Massachusetts has developed a new Article 22 of the State Building Code entitled "Repair, Alterations, Additions and Change in Use of Existing Buildings" (15). Article 22 incorporates a code concept to replace the "25-50 percent rule" which allows building rehabilitation without necessarily meeting all new construction code requirements; the conditions are shown in figure 1.6.2. Compliance alternatives are allowed in lieu of strict compliance with the prescriptive provisions of the code for new construction. The 1979 edition of the code, including Article 22, was promulgated in June 1979, was revised slightly in 1980, and is now in extensive use. As an indication of the impact of Article 22, appeals for variances for buildings being rehabilitated were reduced almost 80 percent during the first 6 months of its use (11).

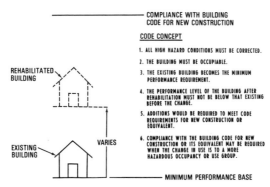

Figure 1.6.2--New code concept for repairs, alterations, additions to, and changes in use of existing buildings.

1.6.4 Technical Concerns Relative to Timber Design in the Regulation of Building Rehabilitation

As has been discussed, building rehabilitation presents technical concerns which may differ from those associated with new construction. Some specific technical concerns relative to timber are:

   a. The adequacy of available condition assessment techniques to accurately predict the structural performance of existing timber;

   b. techniques for reconstructing or strengthening timber systems which must be made to comply with stricter requirements of modern codes;

   c. methods for improving the fire performance of timber systems when a building is brought up to more stringent fire safety criteria.

1.6.5 References for Section 1.6

1. Berry, S. A., Ed.
    1979. "Proceedings of the national conference on regulatory aspects of building rehabilitation. Special Publ. 549, Natl. Bur. Stand., August.

2. Building Officials and Code Administrators International, Inc.
    1975. Code enforcement guidelines for residential rehabilitation. Homewood, Ill.

3. Building Officials and Code Administrators International, Inc.
    1978. Basic building code. Homewood, Ill.

4. Cooke, P. W., and R. M. Eisenhard.
    1977. A preliminary examination of building regulations adopted by the states and major cities. NBSIR 77-1390, Natl. Bur. Stand. November.

5. Cooke, P. W.
    1979. Selected papers dealing with regulatory concerns of building rehabilitation. Special Publ. 536, Natl. Bur. Stand. February.

6. Gross, J., J. Pielert, and P. Cooke.
    1979. Impact of building regulations on rehabilitation - status and technical needs. Tech. Note 998, Nat. Bur. Stand. January.

7. Habraken, N. J., et al.
    1979. An investigation of regulatory barriers to the re-use of existing buildings. GCR 78-139, Nat. Bur. Stand. January.

8. International Conference of Building Officials.
    1979. Uniform building code. Whittier, Calif.

9. Lerchen, F., J. Pielert, and T. Faison.
    1980. Selected methods for condition assessment of structural, HVAC, plumbing and electrical systems in existing buildings. NBSIR 80-2171. Natl. Bur. Stand. November.

10. National Commission on Neighborhoods.
    1979. People, buildings, neighborhoods. March.

11. National Conference of States on Building Codes and Standards, Inc.
    1980. Building rehabilitation research and technology for the 1980's. Kendall/Hunt Publ. Co. (B402253 01).

12. National Trust for Historic Preservation.
    1978. Assessment of current building regulatory methods as applied to the needs of historic preservation projects. Special Publ. 524. Nat. Bur. Stand. October.

13. Rogg, N. H.
    1977. Urban housing rehabilitation in the United States. United States League of Savings Assoc. Chicago, Ill.

14. Southern Building Code Congress International, Inc.
    Standard building code. Birmingham, Ala.

15. State of Massachusetts.
    1980. Article 22 - Repair, alteration, additions and change in use of existing buildings. Massachusetts state building code. September.

16. United States Senate Committee on Banking, Housing, and Urban Affairs.
    1978. Impact of building codes on rehabilitation. Hearings of March 24, 1978.

17. U.S. Department of Housing and Urban Development.
    1978. Minimum design standards for rehabilitation for residential properties. HUD Circ. 4940.4, February (revised Aug.)

18. U.S. Department of Housing and Urban Development.
    1980. Rehabilitation guidelines - 1980 (vols. 1-8). HUD-PDR 613-1 through 613-8, October.

19. U.S. Department of Interior.
    1978. The Secretary of Interior standards for rehabilitation and guidelines for rehabilitating historic buildings. February.

1.7 Special Considerations for Historic Structures*

1.7.1 Background Information

The identification, protection, maintenance, and upgrading of historic structures, including, but not limited to, timber structures, involves the efforts of both the private and the public sector. Referred to as historic preservation, these activities include several which should be of particular interest to civil engineers and their clients:

1. Inventories of surviving historic structures and delineation of historic districts have taken, or are taking place at the local, state, and federal levels of government.

2. Financial incentives to encourage sensitive rehabilitation exist for many identified historic structures or buildings within historic districts.

3. General design controls, often in the form of architectural review commissions, are also in existence for many locally designated landmarks.

4. Considerable research into archaic structural systems in North America and the technical problems inherent in the conservation of historic properties is currently being undertaken.

1.7.2 Identification and Inventories of Historic Properties

The National Register of Historic Places, the official list of the Nation's cultural resources worthy of preservation, is maintained by the U.S. Department of the Interior. Composed of buildings, sites, objects, structures, and districts, the Register includes many timber structures and is constantly being expanded. Listing in the National Register:

1. Makes private property owners eligible to be considered for federal grants-in-aid for historic preservation through state programs.

2. Provides protection by requiring comment from the Advisory Council on Historic Preservation when federally assisted projects affect these resources.

3. Makes many owners who rehabilitate certified income-producing historic properties eligible for federal tax benefits.

Information concerning the National Register of Historic Places is available from the National Register Division, National Park Service,

---

*Author: Charles Fisher, Technical Preservation Services Division, National Park Service, U.S. Department of the Interior, Washington, D.C. Section 1.7.4 was prepared with the assistance of N. L. Gianopulos, P.E., Keast and Hood Company, Structural Engineers, Philadelphia, Pennsylvania.

U.S. Department of the Interior, 440 G Street N.W.,
Washington, D.C. 20243.

Other inventories of historic structures have been undertaken by many municipal and state governments. In addition to local designation of historic structures, many municipal governments have passed ordinances requiring that rehabilitation work on locally designated historic structures comply with established preservation standards. Architectural plans often have to be approved by locally created boards of architectural review prior to issuance of necessary building permits.

Various state and local government agencies make available funds for engineering and architectural fees and for preservation and rehabilitation work, on a selective basis, in order to promote the preservation of historic structures. In addition, some jurisdictions have passed tax abatement programs in order to encourage sensitive rehabilitation work.

Information concerning local or state inventories, applicable review board roles, and available financial assistance may be obtained from either the specific municipal government or the State Historic Preservation Office which has been established in each state.

1.7.3 The Secretary of the Interior's Standards for Historic Preservation Projects

For the past 75 years, the U.S. Department of the Interior has taken the leadership role on the federal level in promoting the preservation of our country's rich cultural heritage. As a result of this involvement, the Secretary of the Interior has made available to the general public recommended Standards for Historic Preservation Projects. The standards consist of eight broadly-worded statements that stress the need for sensitivity toward, and retention of, significant historic and architectural characteristics of a building during the course of preservation and rehabilitation work.[1]

    1. Every reasonable effort shall be made to provide a compatible use for a property which requires minimal alteration of the building, structure, or site and its environment, or to use a property for its originally intended purpose.

    2. The distinguishing original qualities or character of a building, structure, or site and its environment shall not be destroyed. The removal or alteration of any historic material or distinctive architectural features should be avoided when possible.

    3. All buildings, structures, and sites shall be recognized as products of their own time. Alterations that have no historical basis and which seek to create an earlier appearance shall be discouraged.

---

[1] These standards also address five other work treatments: acquisition, protection, stabilization, restoration, and reconstruction of historic properties.

4. Changes which may have taken place in the course of time are evidence of the history and development of a building, structure, or site and its environment. These changes may have acquired significance in their own right, and this significance shall be recognized and respected.

5. Distinctive stylistic features or examples of skilled craftsmanship which characterize a building, structure, or site shall be treated with sensitivity.

6. Deteriorated architectural features shall be repaired rather than replaced, wherever possible. In the event replacement is necessary, the new material should match the material being replaced in composition, design, color, texture, and other visual qualities. Repair or replacement of missing architectural features should be based on accurate duplications of features, substantiated by historic, physical, or pictorial evidence rather than on conjectural designs or the availability of different architectural elements from other buildings or structures.

7. The surface cleaning of structures shall be undertaken with the gentlest means possible. Sandblasting and other cleaning methods that will damage the historic building material shall not be undertaken.

8. Every reasonable effort shall be made to protect and preserve archeological resources affected by, or adjacent to, any project.

Initially published in 1977 in the "Federal Register" (36 CFR 1207 and 1208) as requirements for the Department of the Interior's Federal Tax Incentives Program for Historic Structures and in 1978 expanded to include the Historic Preservation Fund Grant-In-Aid Program and the Federal Surplus Historic Property Transfer Program, the Secretary of the Interior's Standards for Historic Preservation Projects are now used by other federal agencies in cases where their activities may affect historic resources. The standards have also been adopted by numerous state and local government agencies.[2]

1.7.4 Guidelines for Applying the Secretary of the Interior's Standards

In order to assist engineers, architects, and property owners in formulating plans for rehabilitation and restoration of historic properties in a manner consistent with the standards, a set of Guidelines for Applying the Secretary of the Interior's Standards was additionally

---

[2]Architectural review board preservation criteria vary from community to community; some address only street facades while others encompass the entire rehabilitation project and its effect on the historic resource. For specific local architectural review board criteria, contact the specific municipal government.

INTRODUCTION

established in 1977. In the following guidelines for evaluating, maintaining, and upgrading historic timber structures,[3] approaches, materials, and methods consistent with the standards for historic preservation are listed in the Recommended column on the left; approaches, materials, and methods which may adversely affect a property's significant historical and architectural characteristics are listed in the Not Recommended column on the right.

Recommended	Not Recommended
EVALUATION OF HISTORIC TIMBER STRUCTURES	
Recognizing that design decisions and repair methods for restorations of important historic structures may require more considerations than are normally associated with rehabilitation work in order to maximize the preservation of historic fabric, features, and spaces while meeting local building codes and functional needs.	
Recognizing and accepting, rather than correcting, routine deflection in historic structures that is attributed to such factors as normal drying and shrinkage and loss of tightness of joints.	
Identifying the species of wood and the nature of the structural system and establishing limitations of the various components of the system (such as joints) in evaluating the condition of a historic structure and its adaptability for repair or remedial work relative to functional need.	Identifying only a single solution to a structural problem where other alternatives exist, especially where such alternatives are more sensitive to the preservation needs of the structure.

---

[3]These guidelines were compiled especially for inclusion in this ASCE publication. For a complete copy of the "Secretary of the Interior's Standards for Historic Preservation Projects with Guidelines for Applying the Standards" write: Technical Preservation Services Division, National Park Service, 440 G Street NW., Washington, D.C. 20243.

Recommended	Not Recommended

EVALUATION OF HISTORIC TIMBER STRUCTURES--con.

Examining and evaluating the condition of a structure in a manner that minimizes loss of historic fabric through use of nondestructive techniques such as X-ray photography.	Utilizing destructive probing techniques which damage or destroy significant historic fabric such as decorative stencil work on plaster walls.
Assessing the structural system in order to identify weaker portions that are causing overstress in other areas.	
Undertaking a material testing program to establish actual physical condition especially where the preservation of significant architectural features is jeopardized by apparent structural deficiencies.	
Assessing the rigidity of light frame structures where interior wet plaster finish is to be replaced with drywall finish.	
Examining the local building code for special provisions dealing with historic structures.	

MAINTENANCE AND UPGRADING OF HISTORIC TIMBER STRUCTURES

Replacing historic structural members only when necessary; supplementing existing structural systems when damaged or inadequate.	Attempting to relevel complex structures unless joint relief and reinforcement can be performed with a full understanding of the consequences.
Correcting structural deficiencies without losing decorative features or substantially altering historic spaces.	Leaving structural problems untreated that will cause continuing deterioration and will shorten the life of the structure.

Recommended	Not Recommended

MAINTENANCE AND UPGRADING OF HISTORIC TIMBER STRUCTURES--con.

Where relief of load is necessary, reducing the imposed load through removal of nonsignificant and non-functioning material or features and replacing where necessary with lighter-weight material. Examples include removing an obsolete cooling tower or multiple layers of old roofing material which are historically insignificant and do not contribute to the architectural character of the building.	Removing rather than repairing historically important features which are structurally deficient.
	Redistributing live loads within the building if the redistribution may cause overstressing in other areas.
Upgrading structures to meet seismic code in a manner which minimizes loss of historic fabric (such as historic floor framing systems) and changes in historic appearance (such as strapping channels to the exterior of a building or removing a decorative cornice).	
	Disturbing existing foundations with new excavation, including archeological investigation, that may undermine the structural stability of the building.
	Changing the grade or drainage pattern around foundations when it adversely affects below-grade structural elements.
Maintaining the structure in essentially a dry condition by providing sound roof covering, adequate ventilation and flashing, proper drainage, and other such practices.	
Maintaining the soundness of the structural system through providing proper ventilation and vapor barriers, as necessary, when installing insulation.	Discontinuing ventilation in such areas as basements, crawl spaces, and attics without providing for dehumidification or evacuation fans.

Recommended	Not Recommended
MAINTENANCE AND UPGRADING OF HISTORIC TIMBER STRUCTURES--con.	
Respecting the structural integrity of the historic building by minimizing cutaways or holes in structural members.	Installing mechanical and electrical systems which result in numerous cuts, splices, or alterations to the building's historic structural system.
	Installing mechanical systems within the historic resource which require additional structural work due to increased loading where alternate systems exist which allow placement of component units inconspicuously on the outside.
Installing fire suppression systems in lieu of applying fire-resistant coverings such as gypsum board when dealing with significant wooden architectural features or structural members.	
Applying fire-retardant coatings in lieu of fire-resistant coverings if the coatings do not obscure significant architectural features and where local building codes permit such application.	
Applying preservatives to wooden structural elements that are exposed to decay hazards; selecting preservatives which do not change the appearance of significant wooden architectural features.	Using preservatives such as creosote which may change the appearance of significant wooden architectural features.

1.7.5 Historic Preservation Technology

The Technical Preservation Services Division of the National Park Service and the Association for Preservation Technology are two organizations involved with the important problems encountered in the evaluation, maintenance, and upgrading of historic structures. Both organizations have regular publication series dealing with preservation technology for historic structures which cover such topics as "Epoxies for Wood Repairs in Historic Buildings," published by the Technical Preservation Services Division, and "The W.E.R.--System Manual for Structural Rehabilitation of Deteriorated Timber," published by the Association for Preservation Technology.

INTRODUCTION

In addition, there are various other national and regional preservation centers, such as the Society for the Preservation of New England Antiquities, and within each state, the State Historic Preservation Office, which are routinely concerned with preservation technology issues.

Further information about the historic preservation programs of the U.S. Department of the Interior, including available publications, or the addresses of State Historic Preservation Offices or regional preservation centers, may be obtained from the National Park Service, 440 G Street NW., Washington, D.C. 20243. Information regarding the Association for Preservation Technology is available from the Association at Box 2487, Station D, Ottawa, Canada KIP 5W6.

CHAPTER 2   FACTORS WHICH INFLUENCE SERVICEABILITY OF WOOD STRUCTURES*

2.1   Loads**

2.1.1   Introduction

A timber structure must be designed in such a way as to provide adequate resistance to all loads which may reasonably affect safety and serviceability during its intended life. These include both man-made loads and those resulting from natural hazards. Local building authorities specify the design loads to be used, either in their own building codes or by reference to national codes. In the absence of such a code, the American National Standard, Building Code Requirements for Minimum Design Loads in Buildings and Other Structures, ANSI A58.1 (5) is widely accepted; the current edition should be consulted. The comments included herein are intended to supplement rather than to supplant the information in a building code or an applicable standard.

It is the unanticipated or not-so-apparent loads and conditions which are the main concern of this section, since these are often among the causes requiring evaluation of the condition of an existing structure. For example, two areas requiring careful analysis are load combinations and construction loads. The former requires superposition of loads having a reasonable likelihood of simultaneous occurrence during the lifetime of a structure. The latter includes the effect of concentrated dead loads and natural forces on the stability and anchorage of structural elements during the erection process. In addition, consideration must be given to what might be termed "internal functions" which cause stress and deformation. Among these functions are moisture content changes which may cause shrinkage and warping, settlement of supports, eccentricity, or partial internal failure (as in connectors.)

The following considerations merit special attention with timber construction:

  1--Members tend to creep under constant or repeated service loads and thus to accumulate residual deformations, possibly necessitating a nonlinear structural analysis;

---

*Chapter coordinator: T. G. Williamson, Vice President, Engineering, Laminated Fabricators, Inc., Indianapolis, Indiana.

**Authors: Ross B. Corotis, Prof. of Civil Engineering, Northwestern University, Evanston, Illinois and Ben F. Hurlbut, Consulting Engineer, HKM Associates, Billings, Montana.

2--Environmental factors can lead to a long-term reduction in load-carrying capacity as a result of shrinkage and splitting of members and possible loosening of connections;

3--Efficient lightweight structural systems exhibit a high ratio of live to dead load and hence are sensitive to variations in live load as well as to uplift and overturning;

4--Design for serviceability may be controlled by stiffness requirements in order to limit transverse deflections, lateral drift, and transmission of vibrations.

2.1.2 Man-Made Loads

Loads generally included under this category are the dead and live gravity loads, constraint forces, and accidental loadings such as from fire and blast. Forces due to accidental loads are usually accounted for indirectly through structural integrity requirements.

Dead load consists of the weight of the structure and permanently attached items, such as plumbing, ductwork, and ventilating and air conditioning systems. Approximations are generally made for plumbing and ductwork, as well as for doors, openings in walls, etc. Some fixed loads, such as roofing, partitions, hung ceilings, and service equipment may actually be added, removed, or relocated during the life of the structure. These should be treated as dead loads when computing accumulated deformations, settlement, etc., but combined with live loads when computing stresses. Subsequent changes in dead load may necessitate structural modification.

Live loads are related to the intended use of the structure as well as to occasional extraordinary loading situations. It is often appropriate to design for possible changes in occupancy that might introduce heavier loads than does the initial use. Other sources of variation include remodeling, additional wearing and protective surfaces, and temporary changes in occupancy such as for meetings.

Additional considerations in the assessment of live loads include concentrated loads, impact, and live load reduction. It is generally advisable for any panel point of the lower chord of an accessible roof truss or other primary structural member spanning manufacturing or storage type occupancies to be designed to carry a 2,000-pound suspended concentrated load; for residential or office space a concentrated load of 200 pounds appears appropriate. For the effect of impact, design strengths of floors supporting machinery, and hangers for floors and balconies are increased about one-third. A reduction in uniform live load is generally taken with large floor areas to reflect the reduced likelihood of large-scale overload. Exceptions include areas of public assembly and storage, and one-way floor systems.

Provision should be made for constraint forces arising from differential settlement and from restrained deformations due to temperature, creep, and similar effects.

The addition of more roofing than allowed for in the original design may significantly reduce the strength available for support of live loads. The 1979 Uniform Building Code (6) permits a design roof live load of 12 psf in regions where snow is not considered. If additional layers of roofing are applied later, the available live-load strength is reduced by the weight of added roofing. A similar reduction in live-load strength might be caused by addition of a fire-protective ceiling or a sprinkler system. These conditions of adding dead loads become most critical when the allowable stresses have been increased by 25 percent for the 7-day load-duration factor.

2.1.3   Natural Hazards

Depending on the site, design recognition for timber structures is generally made for forces due to wind, snow, flood, earthquake, ponding, and expansive soil.

2.1.3.1 Wind

Critical areas for design against wind forces are attachment of cladding and windows and the anchorage of the roof to the walls, and the walls and floor system to the foundation. Tie-downs must resist uplift, overturning, and sliding. Since wind is not a static load, its dynamic forces may induce unexpected forces in a structure such as harmonic oscillations acting in a tension strut. A dynamic analysis or wind tunnel test may be advisable for a structure potentially sensitive to wind forces.

2.1.3.2 Snow

Snow loading is a function not only of geographical location but also of local terrain effects and structure geometry. Non-uniform loading due to drifting and to melting and ponding in deflected areas of flat roofs can be of significant importance. While it generally takes several decades to accumulate a 2-month duration of full design-level snow, some areas may experience 2 months of moderate snow cover in a single year.

The unit density of snow can vary considerably and in some locations may approach half the density of water. This can be brought about by weather conditions or by heat escaping through the roof structure causing the snow to either melt or pack to a higher than usual density. This may or may not result in an increased roof loading depending on the overall depth of snow accumulation and on the resultant density.

Wind-blown snow can cause large accumulations or drifts to develop which, in turn, produce unit loads far in excess of those normally anticipated from annual precipitation records for a region. A taller structure close to the roof being considered, combined with wind, will sometimes result in heavy snow loads in limited areas. The National Building Code of Canada (8) has recognized this phenomenon for many years and has developed guidelines for the designer to consider. The American National Standards Institute, in its publication

SERVICEABILITY FACTORS

ANSI A58.1-1972 (5), suggests snow load distributions similar to those presented in the Canadian code.

A relatively wet snow combined with wind can produce an extensive buildup in the form of a snow cornice which can extend beyond the edge of a roof for 2 or 3 feet.

In some instances snow accumulations on a parking structure may be improperly removed. Parking bays are often cleared with the excess snow piled high and left in one place. When removing snow from large roof areas the snow is moved progressively to one side of the roof before it is disposed of, resulting in actual loads three or four times those used in the original design. Unbalanced snow loads may, in some cases, produce unit stresses considerably greater than those resulting from a uniform load. This condition is particularly critical in structures involving trusses, various arch configurations, or beams continuous over two or more spans.

Bridges are usually designed for vehicular or pedestrian traffic but, in high elevations where snow has accumulated for many months, snow loads may be considerably greater than the normal bridge design load and these must be considered, particularly for long spans. As an example, the American Association of State Highway and Transportation Officials (AASHTO) list in their standards an equivalent lane load for H15 loads as 480 pounds per linear foot of span plus a concentrated load of 13,500 pounds for maximum moment. If the span is 100 feet and the lane is 14 feet wide, the equivalent uniform load for bending is 54 psf. This load is frequently exceeded, in the western states at higher elevations and in some northern and eastern states, by snow accumulation and this should be considered in a structural evaluation.

2.1.3.3 Combined Wind and Snow

The probability of unusual circumstances occurring which might produce unexpected loads and stresses on a structure should be carefully considered. However, consideration of the simultaneous occurrence of high wind force coupled with a maximum snow load may result in an overly conservative design. On the other hand, if the snow load is, in reality, an ice load, it may be proper to investigate such a combination. The occurrence of rain on snow accumulations may also result in unusually heavy loads.

2.1.3.4 Flood

Structural flood damage occurs predominantly at the anchorage system of a building. Calculations of buoyancy during flooding should be based on framing, sheathing, roofing, partitions and finishings, plus 6 to 8 inches of air entrapment below each interior level. Flooding is often associated with high wind loads such as may occur in hurricanes and this combination of wind and water action can result in extreme loadings of wood structures, particularly at the foundation anchorages and at the wall-roof interface.

WOOD STRUCTURES

2.1.3.5 Ponding

Ponding can be a serious hazard with flat roofs when there is inadequate strength or stiffness and insufficient roof drainage. Long-term creep under sustained load and potential environmental deterioration of timber structures are of special concern here, as water trapped in low spots can develop into roof leaks and lead to eventual decay of the wood roof deck. Roofs should be designed either to ensure adequate drainage or to support the additional loads due to ponding, including the effect of snow load deflections when considering rain or snow or snow meltwater. Relatively flat roofs and roofs designed for small live loads are especially sensitive to stresses caused by ponding. This is a complex situation to analyze and is beyond the scope of this section. The reader is referred to the "Timber Construction Manual" (1) and Technical Note No. 5 "Roof Slope and Drainage for Flat or Nearly Flat Roofs" (3) both developed by the American Institute of Timber Construction, for a more comprehensive discussion of this subject. The effects of ponding may be magnified in the long term action of a roof structure as it is related to settlement of supports, long term deformation of structural components such as parallel-chord trusses, and its effects in conjunction with other loads.

2.1.3.6 Earthquakes

Wood framing provides a rigid yet ductile unit to resist earthquake forces, with exterior walls and sheathing limiting racking. Wall stiffness may be significantly reduced, however, by corner windows. The high rigidity of floor diaphragms tends to distribute lateral load to the stiffest supporting elements. Points to be considered in aseismic design include ductility in members and connections, limited strength loss with load reversal, compatible ductility between elements, and alignment of the mass and rigidity centers to avoid torsion.

2.1.3.7 Expansive Soil

The effects of expansive soil are especially important for timber structures because of their relatively light structural weight. Appropriate soil preparation and foundation design should be considered where the potential for this hazard exists.

2.1.4 Other Considerations

2.1.4.1 Construction Overloads

Localized overloads may occur during construction. The most obvious one of these results from storing roof sheathing or other construction materials at concentrated locations, usually near a convenient location for lifting. Construction equipment which is used on a structural floor or roof may also cause excessive stress. The use of heavy machinery for placing aggregate on a built-up gravel roof or for placing earth on the roof of an earth structure are examples of this type of unusual loading.

Those persons responsible for handling, shipping, and erection of timber members should be cognizant of possible overstress that may result from these operations. For example, when a glulam member is supported or lifted in a flat position, two things must be considered. First, the member is being stressed in a direction for which it probably was not designed. Second, glulam beams attain their bending strength by utilizing higher-strength wood in the outer tension and compression zones. The inner portion of the cross section utilizes lower-strength laminations. Thus, when the member is stressed in its flat direction, the advantage of selective lamination assembly is negated. Reference is made to "Standard Specifications for Structural Glued Laminated Timber of Softwood Species" AITC 117-79 ($\underline{4}$).

2.1.4.2 Failure of an Individual Element

Overloading, if it causes overstress in any component of the structure, may reduce the structural integrity of that component with, as a result, excessive stress in another component. This may lead to eventual failure under what would normally be a safe load. For example, consider a timber stringer bridge with a plank deck. If one stringer is loaded to a point of excessive deflection, perhaps caused by a lower modulus of elasticity in one member than originally assumed, the adjacent stringers must carry additional load beyond their original design level. Similarly, for a roof structure consisting of beams supported by girders, if the beams deflect below level from an overload such as ponding, more water will collect over them resulting in higher loads to the girders than originally anticipated. Trusses are another example of this phenomenon. The failure or overstress of one component or connection generally will cause excessive stress in another component.

2.1.4.3 Anisotropic Nature of Wood

Timber has special characteristics which may induce stresses that are not usually anticipated. Timber does not exhibit ductility in the sense that mild steel does. It is, in addition, orthotropic in nature, with different strengths in the three different directions--parallel to the fiber direction and perpendicular to it in directions both tangential and radial to the annual rings. Timber has its greatest strength in tension and compression parallel to the grain of the wood. Conversely, wood is relatively weak in compression across the grain and is particularly so in tension across the grain. Thus an understanding of the directionally oriented strength characteristics of timber is important in evaluating its resistance to loading.

2.1.4.4 Yielding of Supports

Yielding of supports may be of concern as a cause of excessive stress when beams are continuous over two or more spans. When the relative elevation of beam supports changes, the reactions change. This in turn changes the moments and the shear forces. For instance, with a beam continuous over two 30-foot spans, a settlement of 1 inch in each of the end supports can increase the negative moment by as much as 25 percent.

2.1.4.5 Member Continuity

Continuity of framing members frequently occurs in timber trusses. For many years bowstring trusses have been used very effectively as low-cost structural framing members, particularly when relatively long clear spans are required. Usually the bottom chord is fabricated with a single member spanning over several panels or, in the case of glulam, spanning the entire length of the truss. In time, the wood framing members of the truss may change dimensions due to both across-the-grain and longitudinal shrinkage of the wood although longitudinal shrinkage is usually unimportant. Shrinkage is most pronounced in large solid timber members and is virtually negligible in glulam members. The top chord can tend to shorten because it is in compression and the connectors can also slip a small amount; similarly, the bottom chord can lengthen. These changes in dimension may result in the bottom chord being subjected to vertical forces from the web members which cause bending stresses combined with the normally anticipated tensile stresses. Depending on how the chord members are assembled and how the connections are made, a potential serious overstress in the bottom chord may result. This is similar to the previous discussion of continuity in beams where yielding of intermediate supports can result in increased moments.

2.1.4.6 Eccentric Loads on Compression Members

In the design or investigation of axially loaded members such as columns or struts, the method of applying the load and of resisting that load are of utmost importance. The National Design Specification (NDS) ($\underline{7}$), published by the National Forest Products Association, 1977 Edition, specifies the allowable concentric load that may be taken by an axially loaded member. Other formulas take into account the axial load combined with a bending stress. Bending can be induced in a column through an eccentrically applied load or an eccentrically placed resisting element or connection. For example, for a nominal 6 x 6 post with an effective length of 10 feet, an eccentricity of 1 inch will reduce the allowable axial load to about one-half of the load which would have been allowed had it been applied concentrically.

2.1.4.7 Shrinkage Effects

Wood shrinks with a decrease or swells with an increase in moisture content. This phenomenon must be accounted for in the design of a structure and in the planning of structural connections. Connections which will not permit free shrinking or swelling of wood may cause unanticipated stresses in the members being joined. The designer is referred to Technical Note No. 3, "Connections in Glued Laminated Timber" ($\underline{2}$), published by the American Institute of Timber Construction for specific detailing recommendations.

2.1.4.8 Eccentricity at Connections

Previous discussions point out tht the method of assembly of a timber structure and its fastenings can result in stresses and deformations that may not have been considered in the original design. In timber

SERVICEABILITY FACTORS

truss design, connections at joints are frequently considered to be pinned, while, in reality these joints may be almost fully fixed. Because of this fixity, secondary bending stresses which were not considered in the truss design may occur in individual truss members. Similarly, three-hinged arches may tend to function as two-hinged frames in service because of excessive rigidity in the peak connection and this joint fixity could induce stresses which the arch was not designed to resist.

2.1.5 Summary

It is obvious from the preceding discussion that in the design or evaluation of a structure, one of the first tasks for the designer is to determine the loads or forces that have induced stress or deformation in the structure. Very seldom is this a problem with the usual dead, live, wind, or earthquake loadings that are called for in the various building codes, governmental standards or other applicable design specifications.

It is the not-so-apparent loads or unanticipated loads that may have caused undue stressing of a structural member that may be a problem. For example, live loads do not always occur as uniformly distributed forces as may be assumed in the original design. Often stresses may be induced in a structure because of the way it is assembled. These stresses can be the result of improper handling during erection, temporary bracing not properly installed, or improper alignment of members or their connections during installation. Other possible stress-inducing conditions include moisture content change, settlement of supports, partial internal failure of connections, abuse, unintended eccentricities, or unexpected movement.

It is obvious that judgment and experience are important tools for the investigator or designer. The preceding discussion has been intended to highlight some of the potential stress-inducing load factors that the designer may need to evaluate.

2.1.6 References for Section 2.1

1. American Institute of Timber Construction.
   1974. Timber Construction Manual. Second edition. John Wiley and Sons, Inc.

2. American Institute of Timber Construction.
   1977. Connections in glued laminated timbers. Tech. Note No. 3. Englewood, Colo.

3. American Institute of Timber Construction.
   1978. Roof slope and drainage for flat or nearly flat roofs. Tech. Note No. 5. Englewood, Colo.

4. American Institute of Timber Construction.
   1979. Standard specifications for structural glued laminated timber of softwood species. AITC Std. No. 117-79. In two parts: manufacturing; design. Englewood, Colo.

5. American National Standards Institute.
    1972. Building code requirements for minimum design loads in buildings and other constructions. Am. Natl. Stand. ANSI A58.1. New York, N.Y.

6. International Conference of Building Officials.
    1979. Uniform building code. Whittier, Calif.

7. National Forest Products Association.
    1977. National design specification for wood construction. Washington, D.C.

8. National Research Council of Canada.
    1977. National building code of Canada. Ottawa, Ontario.

2.2   Duration of Load*

2.2.1   Introduction

Wood can withstand higher loads for a short period of time than it can for a long time period. This effect is referred to as "duration-of-load". A satisfactory scientific explanation for this behavior has not been postulated, but the observation that beams which have carried a high load for an extended period of time tend to sag may have lent credence to the acceptance of the duration-of-load concept, even though what is observed is a creep phenomenon.

In the late 1940's, work was in progress at the Forest Products Laboratory, Madison, Wisconsin, aimed at quantifying the duration-of-load effect in bending. It is reported by Lyman Wood (4). Pairs of matched specimens of clear Douglas-fir 1 x 1 x 16 inches in size were tested. One specimen of the pair was tested using a standard bending test where a concentrated load at the center of the span was increased to cause failure in about 5 minutes. The observed strength was used to infer the strength of the pair. The second specimen was subjected to a constant load placed at the center of the span at a preselected percentage of the strength of the pair and the time to failure was observed. Eight stress ratios were used with 16 replications for each. The results (4) are shown in Figure 2.2.1. The ordinate is stress ratio and the abscissa is the logarithm of time to failure.

A mean curve was fitted to the data points and extrapolated to give a prediction of the strength after 50 years, equal to 9/16 of the short-term strength.

2.2.2   Design Procedure

The above information has been incorporated into the design procedure for timber structures used in most North American building codes.

---

*Author: Borg Madsen, Professor, Dep. of Civ. Eng., University of British Columbia, Vancouver, B.C.

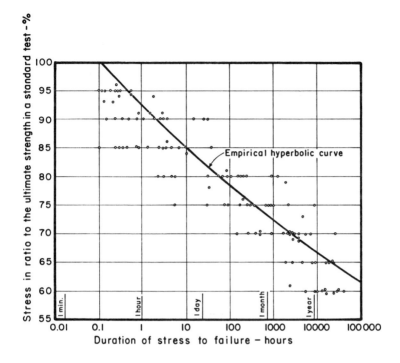

Figure 2.2.1--Madison test data, small clear.

The strength properties are stated as allowable stresses valid for loads acting for a period of 10 years. Should the actual loads act for a longer or shorter period than 10 years, the designer should apply the adjustment factors shown in Table 2.2.1 to the allowable stresses.

Table 2.2.1--Duration-of-load adjustment factors used in structural design

Loading conditions	Adjustment factor
Continuous	0.90
10 years	1.00
2 months	1.15
1 week	1.25
1 minute	1.33
Impact	2.00

The time period to be used is the time that the full design load acts on the structure either continuously or as the aggregate of shorter periods of full design load throughout the life of the structure.

The basis for the above procedure was the bending experiment described earlier. Similar experiments for other properties (tension, compression, etc.) do not exist but nevertheless the concept derived from bending data was applied to the other strength properties and is used, as well, for connections. The values for modulus of elasticity are not adjusted for long-term loading.

2.2.3 Recent Work

The appropriateness of applying results obtained from experiments with small clear wood specimens to commercial lumber or timber has been questioned recently due to the obvious differences in failure modes. The failure mode observed in small clear bending specimens is one of compression followed by tension, while the failure mode in commercial lumber most often is a tension failure associated with localized slope of grain caused by knots.

Duration-of-load experiments (2) conducted with commercial lumber have indicated that the duration-of-load effect is less severe than for clear wood and that the effect may be dependent upon the strength of the material (i.e. smaller effects for the lower-strength material).

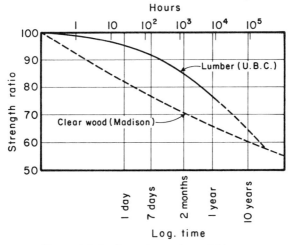

Figure 2.2.2--Comparison of trend lines.

Figure 2.2.2 shows the overall trend for commercial lumber compared to the concept derived from small clear wood specimens. The abscissa is the logarithm of time to failure and the reader should realize that most of the change in strength takes place during the first 5 months.

The logarithmic presentation may distort that fact unless properly interpreted. Figure 2.2.3 represents the trend line for commercial lumber loaded to different load levels. The spread in the data is not shown here, but it is large, making it difficult to establish accurate values of duration-of-load effects. Plans for developing a more extensive experiment are presently being made.

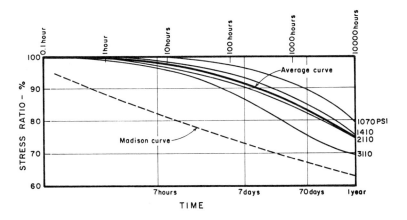

Figure 2.2.3--Preliminary data for commercial lumber.

Experiments dealing with the effect of rate of loading (3) have also shown the abovementioned trends--i.e., little or no effect for the weak material while some effect was observed for the very strong material. That experiment indicates that it may be on the unsafe side to use a duration-of-load factor of 2.0 for impact loads.

2.2.4 Application

The above information pertains to timber which has not been impaired by decay or other deleterious effects. It is extremely difficult to generalize on cases where decay or mechanical damage has occurred. Each case has to be examined carefully. The load history of the structural member should be established (estimated) to find out to what extent it might have been overloaded and for how long a time period the excessive loading might have lasted. The member should be examined closely for evidence of newly developed cracks which may indicate a change in the condition of the member. Such cracks can usually be detected by the appearance of "fresh wood" while older cracks will normally have darkened with time. Of specific concern is the highly stressed zone of the structural member.

Splits--openings going through from one side of the timber to the other--occurring where high shear stresses are present should be given specific attention since the phenomenon "slow crack growth" may be present. The extent of the cracks should be marked and dated so that a re-examination at a later date could reveal if slow crack growth had indeed taken place. Splits can severely reduce the carrying capacity and an analysis based upon fracture mechanics such as is outlined in (1) ought to be performed.

The effect decay has on the long term strength is not well known so that specific guidelines cannot be provided. Any remedial action necessitated by decay would depend on the extent of the decay, whether or not it affects highly stressed zones, the particular load history, etc. Expert advice should be sought so that the interplay of the above conditions can be fully evaluated.

Heavily loaded beams subjected to high-humidity conditions or frequently changing moisture conditions may exhibit "pronounced sagging". This by itself will not indicate that the beam is in distress since wood subjected to the conditions described can creep considerably without a similar loss of strength. The creep can be two to three times the elastic deformation in humid conditions while under dry conditions it can be about 1.6 times the elastic deformation if the beam is loaded continuously.

2.2.5 Summary

The duration-of-load effect relates to the full design load lasting for the specified length of time. Practically speaking, this is a relatively rare condition for wood structures. The recent experiments with commercial lumber indicate that the duration-of-load concept used in our present design codes is very conservative indeed except for impact loads. In evaluating the effect of duration of load on existing structures, particular attention should be paid to the development of splits. Sagging of beams does not necessarily indicate a loss of strength.

The whole subject of the duration-of-load effect is undergoing intensive study. The data on commercial lumber presented herein should, therefore, be considered as preliminary.

2.2.6 References for Section 2.2

1. Barrett, J. D., and R. O. Foschi.
    1977. Shear strength of uniformly loaded dimension lumber.
    Can. J. of Civ. Eng., Vol. 4, No. 1, pp. 86-95.

2. Madsen, Borg, and J. D. Barrett.
    1976. Time strength relationship for lumber. Struc. Res. Ser.
    Rep. No. 13. The University of British Columbia, Dep. of Civ.
    Eng.

3. Spencer, Richard.
    1978. Rate of loading effect in bending for Douglas-fir lumber.
    Proc. of First Int. Conf. on Wood Fracture, Banff, Alberta.
    ISBN 0-86488-014-6.

4. Wood, Lyman W.
    1951. Relation of strength of wood to duration of load. For.
    Prod. Lab., U.S. Dep. of Agric., Madison, Wisconsin, Rep.
    No. 1916.

2.3   Temperature*

2.3.1. Introduction

As with all structural materials, changes in the temperature of wood
will alter its mechanical properties. However, unlike inorganic materials, this organic polymer is, in addition, significantly affected by
duration of the temperature change if the temperature exceeds a critical level. The properties are strongly correlated to the degree of
thermal degradation which, in turn, is related to the temperature of
the wood and the period during which it is maintained. Thermal degradation is reflected largely by the loss of wood mass, at elevated temperatures, hence mass loss is a sensitive indicator of mechanical property change.

The degree to which overall properties of a wood section undergo change
further depend upon the temperature history of each point within the
section and location of the wood fibers critical to the given loading
condition. For example, the top and bottom of a beam contain wood
fibers that respond rapidly to change in environmental temperature
thereby altering bending strength. On the other hand, a compression or
tension member relies upon all wood fibers to carry load, and heating
or cooling takes longer to alter the load-carrying capacity. The
analysis of temperature profiles within sections of wood is beyond the
scope of this discussion, but several references covering this subject
are available (5,11,14). If however, the environmental temperature is
maintained for long periods of time (greater than 1 day), many common
sizes of timber and plates will approach equilibrium with the environment. For brief periods of heating or cooling, only fibers near the
surface change. As a result, brief or long term heating or cooling
effects are often analyzed in a simpler fashion than that for intermediate periods.

It is convenient to discuss the thermally induced changes in mechanical
properties, for application to timber structural elements, by de-
scribing the response within two topic headings (see fig. 2.3.1):

1. The immediate effect on properties when the wood is subjected
   to a given temperature level, and
2. The time-dependent effect on properties as the wood is exposed
   to heat or cold for increasing time periods.

The properties at a given temperature are seen to exhibit recovery of,
or permanent damage to, levels achieved at room temperature (68° F or
20° C). The ratio of permanent to recoverable change generally
increases with increasing temperature and duration of exposure. As

---
*Author: E. L. Schaffer, Supervisory Research Engineer, Forest Products
   Laboratory, Madison, Wisconsin.

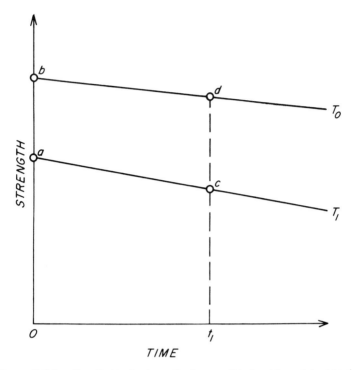

Figure 2.3.1.--Hypothetical strength change with duration of heating/cooling at temperature, T_1. ($T_1 > T_o$)

I. Immediate effect (t = 0)

    a. Property at temperature T_1.

    b. Property at reference temperature, T_o, after pulse type exposure to temperature, T_1.

II. Time-dependent effect (t = t_1)

    c. Property at temperature, T_1.

    d. Property at reference temperature, T_o, after duration of exposure, t_1, to temperature T_1. (Permanent effect or residual strength.)

temperatures decrease below comfortable room conditions (68° F or 20° C), the inverse is true; most properties increase and the levels at room temperature are fully recoverable.

2.3.2 Immediate Effect

2.3.2.1 Total Effect at Elevated and Low Temperatures*

Considerable information is available on the mechanical and thermal properties of wood at temperature levels commonly encountered in structures (e.g., -50° C to +50° C, or -58° F to 122° F). A summary of the effects of temperature in this range on mechanical properties is given in table 2.3.1. The properties are observed to decrease with an increase in temperature and to increase with a decrease in temperature. Increasing the moisture content at the same time magnifies the response. Tensile strength parallel to the grain is the only property found to decrease with lower temperature. Other results (25) show that work to maximum load increases until 0° C (32° F) is reached and then decreases with a further drop in temperature.

Heating wood above given temperature levels causes a more precipitous decrease in some mechanical properties than in others. Tensile strength and modulus of elasticity (MOE) exhibit this behavior above 200° C (392° F) (22). Other properties uniformly decrease with increased temperature.

The thermal properties, with the exception of thermal expansion, are functions of temperature.

In both hardwoods and softwoods the parallel-to-the-grain thermal expansion coefficient ranges only from 0.9×10^{-6} to 1.4×10^{-6} per °C (1.7×10^{-6} to 2.5×10^{-6} per °F). This is about one-tenth that for steel (12×10^{-6} per °C) and concrete (10×10^{-6} per °C) and explains the thermal stability of long timber beams. Across the grain, however, thermal expansion is greater than that along the grain and is also a function of wood specific gravity (6). For softwoods of structural quality, specific gravity can be assumed to be 0.5, and the resulting coefficients are between 14×10^{-6} per °C and 19.4×10^{-6} per °C (26×10^{-6} to 35×10^{-6} per °F). Hence across-the-grain thermal expansion is similar to that of steel or concrete.

Of importance to analyzing heat flow in, or insulative capability of, wood are the thermal conductivity, k, heat capacity, c, and thermal diffusivity, α. The thermal conductivity, k, is a function of moisture content (in percent), M, and dry specific gravity, ρ_o:

$$k = \rho_o(1.39 + 0.028M) + 0.165 \quad \{\text{Btu in./ft}^2 \text{ hr °F}\}$$

For a softwood with dry specific gravity of 0.5 and moisture content of 12 percent the thermal conductivity is about 1.0.

---

*Information for this section is largely derived from the interpretive survey by C. C. Gerhards (7).

Table 2.3.1.—Approximate middle trend effects of temperature on mechanical properties of clear wood at various moisture conditions (7)

Property	Moisture condition	Relative change in mechanical[2] property from 20° C (68° F)	
		At -50° C (-58° F)	At +50° C (122° F)
	Pct	Pct	Pct
MOE parallel to grain	0	+11	-6
	12	+17	-7
	>FSP[1]	+50	--
MOE perpendicular to grain	6	--	-20
	12	--	-35
	≥20	--	-38
Modulus of rigidity	>FSP	--	-25
Bending strength	≤4	+18	-10
	11 to 15	+35	-20
	18 to 20	+60	-25
	>>FSP	+110	-25
Tensile strength parallel to grain	0 to 12	>-10	>-4
Compressive strength parallel to grain	0	+20	-10
	12 to 45	+50	-25
Shear strength parallel to grain	>FSP	--	-25
Tensile strength perpendicular to the grain	4 to 6	--	-10
	11 to 16	--	-20
	≥18	--	-30
Compressive strength perpendicular to the grain at the proportional limit	0 to 6	--	-20
	≥10	--	-35

[1] FSP: Fiber saturation point. Wood above 30 pct moisture content.
[2] Dashes reflect no information on the property.

SERVICEABILITY FACTORS 73

The specific heat, c, depends upon temperature and moisture content of the wood, but is practically independent of specific gravity. For dry wood, it is given by:

$$c_o = 0.25 + 0.0006\ T \qquad \{Btu/lb\ °F\}$$

where T is the temperature in °F. The dry specific heat, c_o, is corrected to other fractional moisture contents, M, using the expression:

$$c = [(M + c_o)/(1 + M)] + A$$

Here A varies from 0.03 to 0.04 over the range of 85° F to 140° F at a fractional moisture content of 0.10.

The thermal diffusivity, α, is defined as the function:

$$\alpha = k/\rho c$$

and is about 2.5×10^{-4} {in.^{2}/sec} for a softwood of dry specific gravity of 0.5 and moisture content of 12 percent.

2.3.2.2 Recoverable Effect

This section will discuss what is known about the recovery of mechanical properties after momentary (i.e., lasting not more than a few minutes) changes in temperature and return to room temperature.

Theoretically, momentary exposure to intermediate heating and cooling regimes (-300° F to 550° F; -184° C to 288° C) should result in complete recovery of initial property levels. Short periods (minutes) of heating or cooling do occur which can have a practical effect. It is this which is of interest. Longer term (hours, days, ...) behavior will be covered in Section 2.3.3.

Because momentary cooling does not result in thermal degradation, it can be safely assumed that all mechanical properties completely recover after such exposure.

Reversibility results are quite limited in the literature. Available are the effect on tensile and compressive strengths, and on modulus of elasticity parallel to grain for Douglas-fir (12,21). Compressive strength is recoverable for brief exposures up to 250° C as shown in figure 2.3.2 (10). Tensile strength appears fully recoverable up to 100° C with decreasing recoverability above this temperature level (fig. 2.3.3) (10). No appreciable change in MOE reversibility is detectable up to temperatures of 150° C in dry (0 pct MC) Douglas-fir (17).

2.3.3 Time-Dependent Effects

A most commonly encountered design problem is the need to have a structural component continue to perform satisfactorily while subjected to heating or cooling for extended periods. Examples are beams and columns in proximity to boilers, beams in the roofs of process mills that

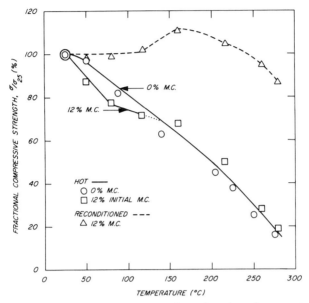

Figure 2.3.2.--Compressive strength as function of temperature while hot as well as after cooling (21,12).

generate heat, or even members employed in frigid environments. One needs information on expected changes in the strength of wood, of glued joints, and of other fasteners to design for such heated or frigid environments as these.

As in the previous section, time-dependent effects can best be treated by showing how a property changes after various heating or cooling periods (1) while the wood is at the given temperature levels, and (2) upon returning to a reference temperature level of 68° F (20° C). Case (2) is normally termed the "permanent" effect on strength properties in many references.

It is well recognized that strength properties correlate strongly with the thermal degradation (or mass loss) of the wood (see for example ref. (23)). Though such degradation may occur at temperatures of 20° C or less, it is not evident until time spans of hundreds of years have passed. At these temperatures, the time-dependency effects are considered fully reversible; that is, though a change in properties occurs upon a change in temperature, they do not change with time of exposure and will return to initial levels when reversed. As a result, the property changes as given in Section 2.3.2 on immediate effects are assumed to apply at low temperature exposures independent of exposure time. When, however, wood is heated above room temperature (20° C) for

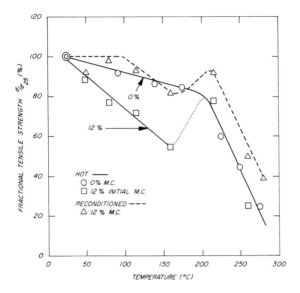

Figure 2.3.3.--Tensile strength as function of temperature while hot as well as after cooling (21,12).

long periods of time, one must be on guard for permanent or irreversible changes that can occur. This is more likely to occur in steam-heated than in oven-heated type exposures, due to higher rates of degradation in hot moist environments. Figure 2.3.4 illustrates weight (mass) loss for wood heated in an oven or beneath a molten metal (23). Figure 2.3.5 is a similar plot for steamed wood.

2.3.3.1 Strength After Heating for Various Durations and Tested in the Heated Condition

Information on changes in properties evaluated at the exposure temperature after various durations is very limited.

Tension (parallel to grain)--It is known that tensile strength parallel to grain is well correlated to cellulose crystallinity in wood. Hence until temperatures are generated that cause depolymerization of cellulose to occur, loss in tensile strength is not evident (e.g., 10). Such loss in cellulose is not observed until temperatures in excess of 400° F (205° C) are reached. This effect can be seen in the decrease in tensile strength above 205° C (fig. 2.3.3). Hence, below this level tensile strength can be expected to be reversible. After ovendry Douglas-fir has been above 200° C, for two hours, however, tensile strength decreases about 30 percent more than that found after immediate heating (20). No similar information is available at higher moisture content levels or for longer durations of heating.

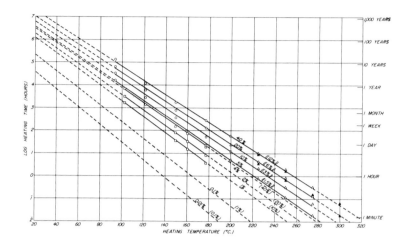

Figure 2.3.4.--Logarithm of the heating time versus the temperature to attain various degrees of degradation of wood (23).
  Open symbols--heating in oven.
  Shaded symbols--heating beneath surface of molten metal.
  Circles and Triangles--weight loss data.
  Squares--modulus of rupture loss data.
  Percentages
    No parentheses, weight loss on oven heating.
    Double parentheses, weight loss on heating beneath the surface of molten metal.
    Single parentheses, modulus of rupture loss on oven heating.

Compression (parallel to grain)--Compressive strength, unlike tensile strength, appears to increase with duration of heating at temperatures that cause significant thermal degradation to occur (fig. 2.3.2). When tested after 2 hours of heating, the compressive strength of ovendry Douglas-fir is increased as much as 10 percent when exposure temperatures exceed 93° C (20). Below this temperature level for this duration of heating, strength on return to room temperature is the same regardless of strength level at the elevated temperature. Information for longer periods of heating is unavailable.

Modulus of elasticity (parallel to grain)--The MOE is relatively insensitive to durations of heating for temperatures below 140° C. However, short periods (2 h) above 140° C do result in some reduction in the MOE of ovendry wood (20).

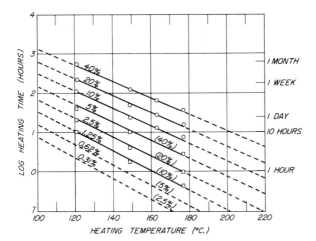

Figure 2.3.5.--Logarithm of heating time in steam versus the temperature to attain various degrees of degradation of wood (23).
 Open circles--data for the loss of weight of softwood specimens.
 Open squares--data for the loss of modulus of rupture of Sitka spruce specimens.
Percentages
 No parentheses, weight loss.
 Parentheses, modulus of rupture loss.

Creep (parallel to grain)--Though long-duration creep has been examined at temperatures of 25° C and at several moisture contents, no similar long-term creep information is available at higher temperature. Increasing the exposure temperature increases the rate of creep deformation (1,19,20). As moisture content is increased as well, the creep rate is increased proportionately (2). Hence, hot moist conditions are conducive to high creep deformation.

Total creep strain (ε_c) can be described as a function of temperature, θ, by a single exponential function (1):

$$\varepsilon_c = \sigma A t^N \exp(\alpha\theta)$$

with

$A = 0.27 \times 10^{-4}$
$N = 0.25$
$\alpha = 0.042$

and σ is the applied stress, t is time in minutes, and θ, the temperature in °C. Such a form is employed to predict the total deformation with time in a small column loaded parallel to grain using a finite difference technique (1).

To partition the creep into recoverable (elastic) and irrecoverable (permanent or plastic) deformation at elevated temperature has proven difficult to fully quantify (19), but the ideal model has the form:

$$\varepsilon_c = g_1(\sigma) \int_{-\infty}^{t} D_1(\xi - \xi') \frac{dg_2(\sigma)}{d\xi'} d\xi' + \int_{-\infty}^{t} \alpha_1(\xi - \xi') \frac{d\phi_1(\zeta)}{d\xi'} d\xi'$$

where

σ: stress

$g_1(\sigma)$, $g_2(\sigma)$: functions of stress

θ = absolute temperature {°K},

D_1 = creep compliance,

α_1 = thermal expansion or shrinkage (time dependent),

ϕ_1 = function of the temperature difference,

$\zeta = \theta - \theta_0$, and

$$\xi = \int_0^t \frac{dt'}{a_T}$$

ξ' = integration variable for ξ

a_T = shift factor = $a_T(\theta)$.

Experiments with ovendry wood have shown that one may break the model into the sum of a delayed term, $\underline{\varepsilon_t}$, and a steady-state rate term, $\underline{\varepsilon_s}$:

$$\varepsilon_c = \varepsilon_t + \varepsilon_s$$

where

$$\varepsilon_t = \Sigma_i \left\{ g(\sigma_1) D_1^{(i)}/a_T + \zeta_1 \alpha_1^{(i)} \right\} [1 - \exp(-t/a_T \tau_i)]$$

$$\varepsilon_s = g(\sigma_1) D_1^s \frac{t^n}{a_T}$$

and

σ_1: constant stress level

ζ_1: constant temperature difference

Of these terms, ε_s and $\Sigma\zeta_1\alpha_1^{(i)}[1 - \exp(-t/a_T\tau i)]$ are irrecoverable strains respectively associated with plastic flow and thermal-degrade-induced shrinkage. The remaining term is recoverable creep strain. The exponent, n, in the plastic flow term is less than 1. Though not exact, with n equal to 1, the parameters for the above equation are:

$$g(\sigma_1) = (\sigma_1/\sigma_o)^m$$

m = 1.07 tension, 1.54 compression,

σ_0 = 16,300 pounds per square inch (tensile strength at 25° C),

$D_1^{(1)}$ = 0.0³,192,

$D_1^{(2)}$ = 0.0³,078,

D_1^s = 10^{-8},

$a_T\tau_1$ = 446 seconds,

$a_T\tau_2$ = 25 seconds,

$\alpha_1^{(1)}$ = -0.0³,001,28 {°K^{-1}},

ζ_1 = T_1 - 298 {°F},

With i = 1 only:

a_T = exp $\{-\Delta E_{avg}/R(\frac{1}{T_o} - \frac{1}{T})\}$,

$\Delta E_{Avg} \cong$ 5,400 cal/mole/°K.

Figures 2.3.6 and 2.3.7 (as derived from Bach ([2])) may be used to estimate total creep strain for 100-minute and 17-hour periods at intermediate temperatures (to 70° C) and MC (to 12 pct). The coincidence between that obtained by Bach ([2]) and information on ovendry wood ([20]) is shown in table 2.3.2.

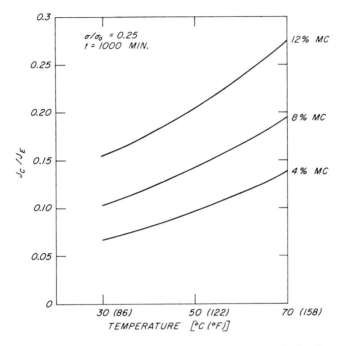

Figure 2.3.6.--The ratio of parallel-to-grain creep and elastic compliance as a function of temperature and moisture content (2).

Table 2.3.2.--Ratio of parallel-to-grain tensile creep compliance, J_C, to elastic compliance (J_E) at 100 minutes as developed by two authors (2,20)

Temperature (°C)	Stress level	J_C/J_E	
		Maple (4 pct MC)	Douglas-fir (0 pct MC)
	Pct		
25-30	40	0.048	0.038
25-30	80	.060	.045
50	40	.072	.073

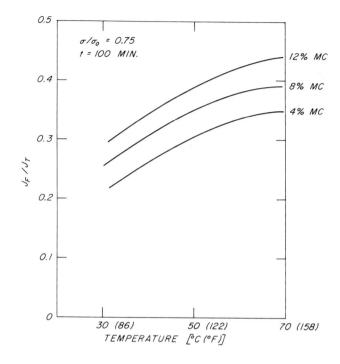

Figure 2.3.7.--Ratio of creep flow to total observed creep for wood parallel to grain as a function of temperature and moisture content (2).

2.3.3.2 Strength at Cool (20° C) Condition After Heating for Various Durations

Information in this area is more extensive. It allows one to assess the residual strength of members after heating and to determine what "permanent" damage may have been done. Because of the limited information available on strength change after heating or cooling for extended periods, it is normal to assume that the residual strength (as discussed in this section) can be added to the immediate effects (discussed in Section 2.3.2.1) to estimate total change at the given temperature level.

It was previously mentioned that the heating medium has a strong bearing on residual strength properties. Freezing apparently has a negligible effect, but increasingly greater effects with medium are found in the order oven, steam, and water heating. The change in strength properties is also well correlated with wood weight loss.

WOOD STRUCTURES

Tension and compression (parallel to grain)--Similar to the observations stated in the section on recoverable strength, tensile strength is more sensitive to duration of oven heating than is compressive strength (17). As shown in figure 2.3.8, a nonrecoverable effect on tensile strength is significant at weight loss levels of 0.1 percent, whereas weight losses of greater than 2 percent are necessary to reduce compressive strength in hardwoods and softwoods. A weight loss of 20 percent or more results in very large losses in residual strength. Oven temperatures ranged from 100° to 200° C.

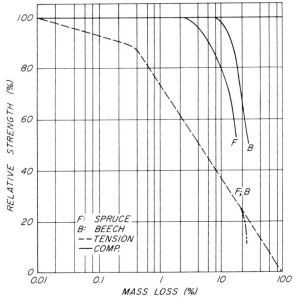

Figure 2.3.8.--The relative strength of dry beech (B) and spruce (F) in tension and compression as dependent upon the degree of thermal degradation (loss of mass) (17).

Modulus of elasticity (parallel to grain)--The residual MOE is affected less than 5 percent for durations of oven heating resulting in less than 8 percent weight loss for a hardwood and softwood (fig. 2.3.9) (17). The effect can be seen with duration of oven heating in figure 2.3.10 (15). The residual MOE is not affected for periods of oven heating up to 1-1/2 years at temperatures at or below 150° F (65.5° C) (11).

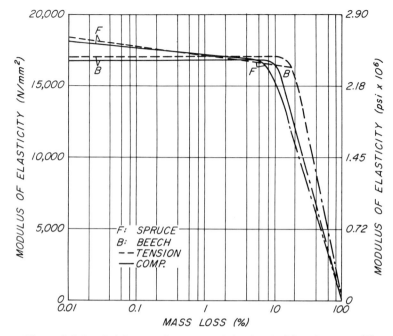

Figure 2.3.9.--Modulus of elasticity of dry beech (B) and spruce (F) in tension and compression as dependent upon degree of thermal degradation (the loss in mass) (17).

Modulus of rupture (parallel to grain)--The MOR has received more attention as a strength index than have other properties. Figures 2.3.11, 2.3.12, and 2.3.13 show the effect of various heating periods in water and steam, and in an oven (6). Note in figure 2.3.13 that the MOR has a residual of about 90 percent of initial after heating in water at 150° F (65.6° C) for 300 days. No similar significant decrease in residual MOR has been found when heating in an oven at 150° F for up to 1.5 years in both softwood and hardwood species (13).

Work to maximum load (bending parallel to grain)--Work to maximum load is a property most sensitive to heat exposure. The effect on residual work capacity after heating in water at two temperature levels is shown in figure 2.3.13 (6). Oven heating produces less effect, but periods of heating to 1-1/2 years at 150° F result in an average reduction of 22 percent in Douglas-fir, Sitka spruce, and yellow birch (13). At higher temperatures, the effect is exemplified by that for Douglas-fir shown in figure 2.3.14 (13).

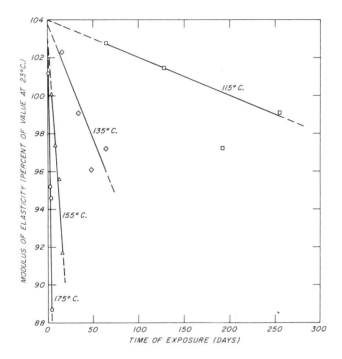

Figure 2.3.10.--Permanent effect of oven heating on MOE as based upon four softwood and two hardwood species (15).

2.3.4 Effect on Plywood and Adhesive Bonded Products

The satisfactory functioning of a structural laminated product under exposures to elevated or frigid temperature has, as a required condition, that an adhesive acceptable for structural components has been used in its fabrication. This implies that the adhesive is relatively durable under wet exposures. As a result, such components are commonly bonded with thermosetting resins such as: resorcinol, phenol, phenol-resorcinol, melamine, and melamine-urea. These resin-based adhesives have been found to provide such durability.

Though wood properties generally increase with a decrease in temperature, plywood exhibits some losses due to the response of the adhesive.

The tensile strength of a balsa joint using resorcinol adhesive at -300° F was between 63 and 97 percent that at room temperature (6). Impact shear strength at -70° F is also reduced to between 30 and 50 percent of that at room temperature (4). Initial moisture contents of the wood were 12 percent.

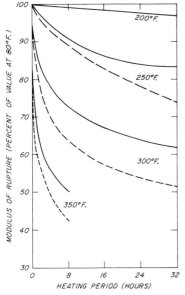

Figure 2.3.11.--Permanent effect of heating in water (solid line) and in steam (dashed line) on the MOR. Data based on tests of Douglas-fir and Sitka spruce (6).

In frigid-temperature (-12° and -65° C) exposures of water-saturated wood laminates (24), urea- and phenol-resorcinol-based adhesives exhibit greater residual shear strengths than do melamine-urea, phenol, and cross-linked polyvinyl-acetate adhesives. Solid wood exhibits no significant degrade in shear strength. Such strength loss increases with number of cycles of freeze-thaw to which a laminate is subjected. As a result, adhesives with high rigidity (such as urea and phenol-resorcinol) compared to the wood substrate should be avoided in frigid exposure conditions.

The residual shear and internal bond strength of plywood joints heated wet and dry has been investigated (8,9,18). As a result of these investigations, it can be concluded that, in products bonded with the phenol and resorcinol structural adhesives, the deterioration of strength occurs in the wood and not the adhesive. Hence, the information previously presented on wood strength change can be applied to wood structural products laminated with these adhesives. Melamine and melamine-urea bonded plywoods were less resistant to water soaking than phenol- or resorcinol-based systems, but exhibited equivalent performance under dry heating (8).

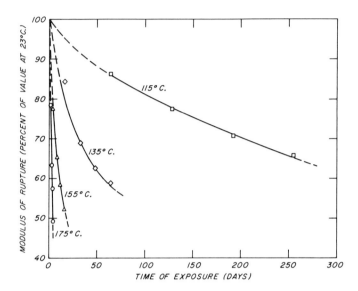

Figure 2.3.12.--Permanent effect of oven heating at four temperatures on MOR. Results based upon four softwood and two hardwood species (6).

2.3.5 Application Notes

The information presented can generally be interpreted to conclude that as wood is cooled below normal temperatures its strength increases. When heated its strength decreases. The effect occurs immediately upon exposure and up to 150° F (65.5° C) is reversible. That is, a member can be expected to recover essentially all of its strength when temperature is reduced to normal. The effect is intensified with increasing moisture content of the wood. Prolonged heating above 150° F (65.5° C) can cause a permanent reduction in strength.

In some applications, structural members can be periodically exposed to elevated temperatures. Normally, in buildings, the accompanying relative humidity is lowered and results in low wood moisture contents. For example, wood moisture content is about 12 percent (on ovendry weight basis) in 70° F--65 percent relative humidity. If the temperature is raised to 150° F, the relative humidity decreases to 5 percent and wood eventually attains a moisture content of less than 1 percent (24). The response for other temperatures compared to when wood is at

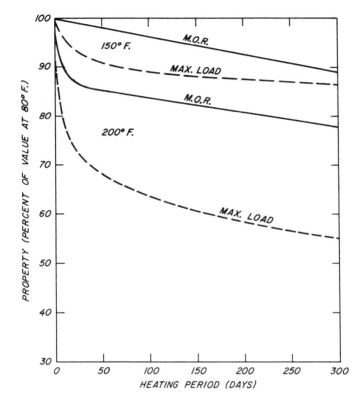

Figure 2.3.13.--Permanent effect of heating in water on MOR and work to maximum load. Data based upon tests of Douglas-fir and Sitka spruce (6).

equilibrium at 70° F is shown in figure 2.3.15. Note that, if a wood product is initially at 11 percent equilibrium moisture content (EMC) at 70° F (a common level in some regions of the U.S.), and the temperature is then raised without a change in atmospheric moisture level, wood EMC is reduced to 6 percent at 90° F and less than 1 percent at 120° F. Hence, for long-term elevated temperature exposures greater than 120° F for wood products normally at 12 percent EMC, the wood is heated in a nearly dry state. In addition, strength properties increase with decreasing moisture content. As a result of considering these offsetting factors, the National Forest Products Association recommended that allowable stresses specified in the National Design Specification for Wood Construction (16) be used as given for ordinary temperature fluctuations and occasional short-term heating to air temperatures as high as 150° F.

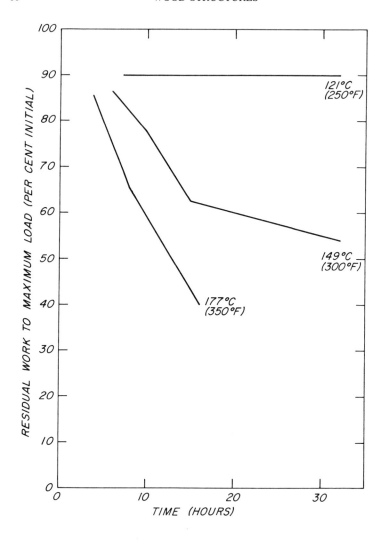

Figure 2.3.14.--Work to maximum load (as a pct of initial levels) as function of oven heating for Douglas-fir (13).

For purposes of designing for extended exposures of wood structural members to very low temperatures at high moisture contents or heated at temperatures to 150° F, adjustment of the design values is specified as follows (16) (table 2.3.3).

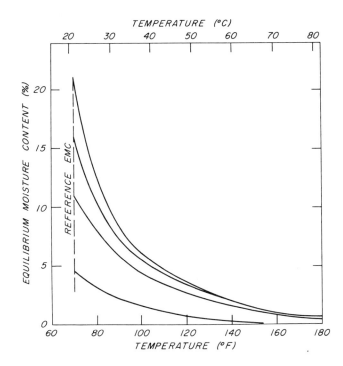

Figure 2.3.15.--Change in wood equilibrium moisture content when temperature is changed in air held at constant absolute moisture content. Reference moisture contents at 70° F are 4.5, 11, 16, and 21 percent for each of four curves.

Table 2.3.3.--Percent increase or decrease in design values for each 1° F decrease or increase in temperature (16)

Property	Moisture content	Cooling below 68° F (min. -300° F)	Heating above 68° F (max. 150° F)
	Pct	Pct	Pct
Modulus of elasticity	0	+0.04	-0.04
	12	+0.14	-0.19
Other properties	0	+0.17	-0.17
	12	+0.32	-0.49

2.3.5 References for Section 2.3

1. Arima, T.
   1973. Creep during temperature changes. III Prediction of creep at elevated temperature. J. Jap. Wood Res. Soc. 19(2):75-79.

2. Bach, L.
   1965. Nonlinear mechanical behavior of wood in longitudinal tension. Ph. D. Thesis, Syracuse Univ., Syracuse, N.Y.

3. Beall, F. C.
   1980. "Effect of temperature on the structural uses of wood and wood products" in "Structural use of wood in adverse environments." Van Nostrand, New York, N.Y.

4. Boller, K. H.
   1954. Wood at low temperature. Mod. Packag., Sept.

5. Carslaw, H. S., and J. C. Jaeger.
   1947. Conduction of heat in solids. Oxford Univ. Press, N.Y.

6. Forest Products Laboratory.
   1974. Wood handbook: Wood as an engineering material. Mech. Prop. of Wood, chapter 4, USDA Handb. No. 72 Rev., USDA For. Serv.

7. Gerhards, C. C.
   1980. Effect of moisture content and temperature on the mechanical properties of wood: an analysis of immediate effects. Submitted for publ. to Wood and Fiber.

8. Gillespie, R. H.
   1965. Accelerated aging of adhesives in plywood-type joints. For. Prod. J. 15(9):369-378.

9. Gillespie, R. H., and B. H. River.
   1975. Durability of adhesives in plywood: Dry-heat effects by rate-process analysis. For. Prod. J. 25(7):26-31.

10. Ifju, G.
    1964. Tensile strength behavior as a function of cellulose in wood. For. Prod. J. 8:366-372.

11. Kern, D. Q.
    1950. Process heat transfer. McGraw-Hill Book Co. Inc., N.Y., p. 639-656.

12. Knudsen, R. M., and A. P. Schniewind.
    1975. Performance of structural wood members exposed to fire. For. Prod. J. 25(2):23-32.

13. MacLean, J. D.
    1955. Effect of oven heating and hot pressing on strength properties of wood. Am. Wood Pres. Assoc. Proc.
14. McAdams, W. H.
    1942. Heat transmission. 2d Ed. McGraw-Hill Book Co. Inc., N.Y.
15. Millett, M. A., and C. C. Gerhards.
    1972. Accelerated aging: Residual weight and flexural properties of wood heated in air at 115° to 175° C. Wood Sci. 4(4).
16. NFPA.
    1977. National Design Specifications. Nat. For. Prod. Assoc., Wash., D.C.
17. Rusche, H.
    1975. Strength properties of dry wood after heat treatments. Verein Deutscher Ingenieure, Zeitschrift 11:87-92. (From German by D. Pronin FPL Translation.)
18. Sasaki, H., H. Kaneda, and T. Makee.
    1976. Dry-heat degradation of plywood-type joint. Jap. J. of Wood Res. 59/60:58-66.
19. Sawabe, O.
    1974. Studies on the thermal softening of wood III effects of temperature on the bending creep of dry Hinoki wood. J. of Jap. Wood Res. Soc. 20(11):517-522.
20. Schaffer, E. L.
    1970. Elevated temperature effect on the longitudinal mechanical properties of wood. Ph. D. Thesis, Univ. of Wisconsin, 217 pp.
21. Schaffer, E. L.
    1973. Effect of pyrolytic temperatures on longitudinal strength of dry Douglas-fir. ASTM J. Test. Eval. 1(4):319-329.
22. Schaffer, E. L.
    1978. Temperature-time dependency of longitudinal mechanical behavior of dry Douglas-fir. NSF Workshop Proc. "Gen. Const. Relat. for Wood and Wood-Based Mat." Syracuse Univ., Syracuse, N.Y., July.
23. Stamm, A. J.
    1956. Thermal degradation of wood and cellulose. Ind. and Eng. Chem. 48:413-417.
24. Steiner, P. R., and S. Chow.
    1975. Low-temperature durability of common wood adhesives. For. Prod. J. 25(8):26-29.

25. Thunell, B.
1942. Quality and strength in wood. Federal Research Lab,
Report 89, Stockholm, Sweden, p. 4.

2.4 Moisture*

2.4.1 Introduction

Wood in the growing tree contains varying amounts of moisture, ranging
from about 30 percent of the weight of wood substance to more than
200 percent, with the moisture content of the sapwood generally being
substantially higher than that of the heartwood. Since wood is a
hygroscopic material, a piece of wood fresh from the tree will immediately start to lose moisture to the surrounding air (unless the air is
saturated) until it reaches equilibrium with the conditions surrounding
it. Since a number of changes take place during the time the wood is
attempting to reach the equilibrium moisture content, it is common
practice to dry (season) lumber less than about 3 inches in nominal
thickness. The most common practice is to dry it in a kiln at elevated
temperature to hasten the operation, while drying in the open air or in
a shed may also be used if time is not a factor. Veneer is always
dried to relatively low moisture content to permit proper bonding of
the adhesive in the manufacture of plywood or of parallel-laminated-
veneer products. Material from about 3 inches on up in thickness is
commonly not dried because of the long time required and the difficulty
of seasoning it without undesirable changes such as the development of
deep seasoning checks. Long-term air drying can be used to achieve a
degree of seasoning with the larger timbers. It is generally considered best practice to have the moisture content of a piece of wood
as it goes into place at about the moisture content to be expected in
service. Table 2.4.1 indicates desirable moisture contents, and
figure 2.4.1 indicates the areas referred to in the table.

Following sections will discuss a number of factors related to moisture
content of wood which affect performance in service.

2.4.2 Dimensional Change

Moisture in wood can exist as water or as water vapor in the cell
cavities and as water "bound" chemically within the cell walls. As
green wood dries, the point at which the cell cavities no longer contain water, but the cell walls are completely saturated (all "bound"
water) is called the "fiber saturation point." This point, generally
at about 30 percent moisture content, is important in that, as the
moisture content drops below the fiber saturation point, physical and
mechanical properties begin to change as a function of moisture
content.

One of the properties affected is dimension. That is, as moisture content decreases below the fiber saturation point, wood shrinks; conversely, in going from a lower to a higher moisture content below the

---

*Author: Alan D. Freas, formerly Assistant Director, Forest Products
Laboratory, Madison, Wisconsin.

SERVICEABILITY FACTORS

Table 2.4.1.--Recommended moisture content values for various wood items at time of installation

Use of wood	Moisture content for					
	Most areas of United States		Dry south-western area[1]		Damp, warm coastal areas[1]	
	Average[2]	Individual pieces	Average[2]	Individual pieces	Average[2]	Individual pieces
	Pct	Pct	Pct	Pct	Pct	Pct
Interior: Woodwork, flooring, furniture, wood trim, laminated timbers, cold-press plywood	8	6-10	6	4-9	11	8-13
Exterior: Siding, wood trim, framing, sheathing, laminated timbers	12	9-14	9	7-12	12	9-14

[1] Major areas are indicated in figure 2.4.1.
[2] To obtain a realistic average, test at least 10 pct of each item. If the amount of a given item is small, several tests should be made. For example, in an ordinary dwelling having about 60 floor joists, at least 10 tests should be made on joists selected at random.

fiber saturation point, wood swells. Wood shrinks most in the direction parallel to the annual rings (tangentially), about half as much perpendicular to the annual rings (radially), and only slightly along the grain (longitudinally). These differences in shrinkage with direction can distort the shape of a piece of wood as is indicated in figure 2.4.2.

As a piece of wood dries, the moisture leaves it through its surfaces, the remaining moisture migrating toward the surface on its road to escape. In a piece having a relatively large cross section, the interior may still retain large amounts of moisture and be above the fiber saturation point (and thus stay constant in dimension), while the outer layers drop below the fiber saturation point and attempt to shrink as the moisture content is reduced. The inner core, however, keeps the shell from changing dimension, and tensile stresses perpendicular to grain develop. When these stresses exceed the tensile strength of wood perpendicular to grain, failure results and an opening (termed a seasoning check) develops. As the moisture content continues to drop, the opening widens and deepens, and it is this phenomenon which explains why many timber structures built from large timbers have deep checks.

94 WOOD STRUCTURES

Figure 2.4.1.--Recommended average moisture content for interior use of wood products in various areas of the United States.

Similar separations can occur if two points in a member are held in position and the member tries to shrink. For example, if a steel plate is held to a piece of wood at a connection by bolts near top and bottom of the wood member, at a moisture content above equilibrium, the bolts will prevent the member from changing dimension as it dries and a separation will occur. If shear in a beam, for example, happens to be critical, such a separation can seriously reduce the load-carrying capacity of the member.

In some cases, moisture can be reintroduced into a member after it has been initially dried, as for example if it is pressure treated with waterborne salts to impart fire or decay resistance. In such a case, large amounts of water will be introduced through the end grain but will not escape as easily. As the water tries to escape, the central core will prevent shrinkage as described earlier, and longitudinal separations will occur. Depending upon the size of the member, the results may be large enough to reduce load-carrying capacity. It is for this reason that treatment of large laminated members with waterborne preservatives or fire retardants after gluing is not permitted by industry standards. If treatment is required, the individual laminations will be treated and redried prior to bonding.

In the case just described, water was deliberately injected into a wood member for a specific reason. Similarly large amounts of water can

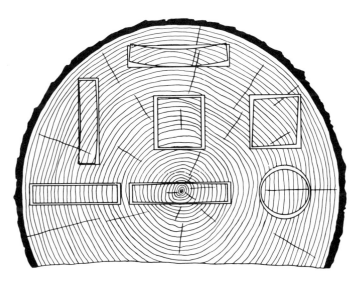

Figure 2.4.2.--Characteristic shrinkage and distortion of flats, squares, and rounds as affected by the direction of the annual rings.

enter a member accidentally as from a roof leak, hosing down a food preparation area, or from rain or lawn sprinklers in the case of exterior members. This unintentional introduction of moisture can cause excessive swelling, can reduce mechanical properties to a level below original design values, or can result in decay.

The moisture content of wood changes seasonally as the ambient relative humidity changes, and wood subjected to these changes will attempt to shrink and swell accordingly. If a piece of wood is attempting to swell and is restrained from increasing its dimension, as for example by plates or washers beneath the head and nut of a bolt, it will be compressed and, if this compression is maintained over a long period, will tend to "set" in the restrained size. Then, as the humidity drops and the wood shrinks, it will lose contact with the washers or plates, and the connection will no longer be tight. Such loosening can continue over a series of cycles of relative humidity and, if long-continued, can seriously reduce the load-carrying capacity of the connection. Checking the tightness of nuts in bolted connections is, therefore, a standard item of periodic maintenance for timber structures.

96   WOOD STRUCTURES

2.4.3   Fungus Attack

Decay and other fungi require a certain combination of circumstances in order to flourish. One such circumstance is the presence of moisture at or above the fiber saturation point. Thus, a wood member which has been installed at a high moisture content or which has been subject to wetting, whether by leakage, by spraying, by rain, or whatever, may if it carries a high moisture content for a long period be subject to decay and thus to serious reduction in strength. Installation of dry lumber, design to prevent wetting by rain or other sources, and careful maintenance are essential to prevent decay. This is covered in detail in section 2.8.

2.4.4   Finishes

Finishes on wood must, for best results, generally be applied to dry wood. The entry of water after application of finishes may, however, have various deleterious effects. Water-staining or the development of mildew can, at the very least, spoil the appearance of a finished wood component. The entry of water into wood below a film-forming finish such as paint or varnish can, however, result in the total failure of the finish. A familiar example of this is the peeling of painted wood outside of high-humidity areas such as baths or kitchens or at junctures between elements such as where two pieces of siding are butted end-to-end or where siding butts against window or door frames. The latter occurrences can be alleviated in most cases by application of water-repellent preservatives to the end grain.

The matter of finishes and their performance is covered in detail in Chapter 6.

2.4.5   Insects

A number of insects require moisture to prosper. Some are found only where the moisture content of the wood is relatively high. Thus, for example, carpenter ants will be found only where the wood is saturated, commonly from leaks or entry of rainwater through end grain. See section 2.8 for more detail.

2.4.6   Mechanical Properties

Mechanical properties such as bending strength, tensile strength, or compressive strength or modulus of elasticity vary with moisture content--again, only below the fiber saturation point as is true of physical properties such as changes in dimension. Thus, a structural member which will be subject to high moisture contents in service, such as one which is under water or one which is subject to exposure to the elements and thus periodic wetting, will have less capacity to carry load than one which is kept continually dry as in the interior of a building. Design stresses for wood are, therefore, keyed to service moisture content. That is, design stresses are commonly given for "dry" conditions of use or for "wet" conditions. Methods for adjustment of design stresses for moisture content are given in handbooks. If, however, investigation of a structure discloses that a member or a group of members is at high moisture content for whatever reason, the investigator should check the design carefully to be sure that the member has the strength to support the imposed loads at this moisture content. And, of course, the member should be checked also for evidence of decay, as well as to determine if any of the other effects discussed herein have been at work.

2.4.7 Summary

It is evident from the preceding discussion that high moisture content in wood in service can be deleterious in a variety of ways. Thus it would seem that the watchword for good service of a timber structure would be "Keep it dry!" This can be accomplished by care in design and by careful maintenance. In further summary, the investigator of the effectiveness of an existing structure should pay particular attention to the points discussed herein if there is evidence of high moisture content at some point in the present or past life of the structure.

2.5. Chemicals*

2.5.1 Introduction

Wood offers many advantages as a material of construction for equipment and structures for use in corrosive environments. Knowledge of the chemical resistance and durability of wood is important to persons responsible for selecting materials of construction for use under such conditions.

The most important advantage of wood in corrosive environments is its ability to withstand mild acid conditions. Wood is much more resistant than mild steel and cast iron in this respect.

Wood is composed of lignin, cellulose, hemicellulose, and extractive materials. The lignin and cellulose impart strength. The extractives provide both resistance to penetration of liquids and natural resistance to decay. The resistance to chemical attack depends to some extent upon the inability of the chemical to permeate the wood structure.

A species is usually chosen for chemical resistance on the basis of past experience and availability. In general, softwood species are more resistant than hardwood species, and heartwood is more resistant than sapwood. Experience has shown that lumber cut from the following species is suitable for water tanks: baldcypress (Taxodium distichum), southern yellow pine (Pinus spp.), Douglas-fir (Pseudotsuga menziesii), redwood (Sequoia sempervirens), northern white-cedar (Thuja occidentalis), and western redcedar (Thuja plicata). Heartwood from the first three species is well suited for tanks where resistance to chemicals in appreciable concentrations is an important factor. These six species combine moderate-to-high resistance to water penetration with moderate-to-high resistance to decay. Some imported species also

---

*Author: Andrew J. Baker, Chemical Engineer, Forest Products Laboratory, Madison, Wisconsin.

provide resistance to chemical attack. For structural members where exposure is usually limited to chemical fumes and dusts, the common structural species such as Douglas-fir, southern pines, true firs, and hemlock in the form of lumber and plywood provide adequate resistance to chemical attack.

2.5.2   Chemical Effects on Strength of Wood

Chemicals affect the strength of wood by two general types of action. One is an almost completely reversible effect involving swelling of the wood substance. The other type of action is nonreversible and involves permanent changes in the wood substance as a result of a reaction between the chemical and the wood.

In the first type, chemicals such as alcohols, other polar solvents, and water will swell air-dry wood. The swelling causes no permanent effect on the strength properties, and removal of the swelling liquid returns the wood to nearly the original dimensions and strength. The loss in strength per unit area upon wetting wood in a polar solvent is in proportion to the amount of swelling. Nonpolar liquids such as petroleum hydrocarbons, which are often carriers for preservatives, have a negligible swelling effect and, thus, do not appreciably change the strength properties of wood.

The second type of action which causes permanent changes is caused by acids, alkalis, and salts that attack the wood substance. For wood structures, chemical vapors and dusts are of concern. For instance, acid vapors and ammonia gas, which is alkaline, and dust from stored salts, especially those that are hygroscopic, can affect structural members. The chemical attack reduces the strength properties in proportion to the amount of cross-sectional area attacked and the degree of attack.

Chemical substances can be grouped into classes of chemicals such as inorganic acids, organic acids, alkalis, salts, and oxidants. Each of these classes of chemicals reacts with wood in a different manner. A prediction as to whether loss of strength will be fast or slow can be made by knowing the concentration of the chemical, the pH, and the temperature. A brief description of the effects of a water solution of these classes of chemicals on wood follows:

Inorganic acids.--Acids cause the hemicellulose and cellulose polymers of wood to hydrolyze to water-soluble sugars. The rate of hydrolysis depends upon acid concentration and temperature. Acid concentration can be conveniently measured in terms of pH of the solution. For wood such as in wood tanks, in direct contact with acid solutions, it is recommended that the pH of the solution be above 2 when at room temperature. At conditions of pH 2 and room temperature, a wood tank will have a service life of at least 5 to 8 years. With acid solutions of higher pH, the service life is usually several times longer. It is important to remember that the rate of hydrolysis is highly dependent on temperature, and as a general rule, the rate of hydrolysis will double for each 10° C temperature rise. The usual temperature limit of wood in direct contact with an acid solution is 50° C for long durations such as years at pH 4 to 6, but it is common to go as high as 65° C for short durations such as a couple of hours per day. This exposure to the higher temperature will decrease service life, however. The combination of high temperature and low pH should not be considered for wood in direct contact with inorganic acids.

Organic acids.--Acids such as formic, acetic, propionic, and lactic usually do not hydrolyze wood rapidly because the pH of these acids in solution is about the same as the pH naturally occurring in moist wood (pH 3 to 6). The same temperature restrictions as mentioned under inorganic acids should be observed.

Alkalis.--Alkaline chemicals such as sodium, calcium, and magnesium hydroxide will swell the wood structure, react with the hemicellulose and cellulose, and dissolve the lignin. Wood is seldom used in contact with chemicals more than mildly alkaline. For wood in direct contact with these solutions, a pH of more than 11 at room temperature is considered harmful and can cause rapid loss of strength properties to wood tanks and chemical equipment. At temperatures of 40° C to 50° C, solutions of about pH 9.5 or higher will result in swelling and rather rapid strength loss. Because wood is seldom used in alkaline conditions there is no reliable experience available to estimate service life.

Salts.--As with acids and alkalis, the effect of these chemicals can usually be predicted on the basis of pH. Some salts are acid, some neutral, and some are alkaline. The acid salts can be considered as weak acids and will not have a rapid effect on the strength of wood unless the temperature is high. Most neutral salts will not react with wood to cause a strength loss. Similarly, alkaline salts are harmful to wood and can be considered as weak alkalis. The effect these salts have can be estimated on the basis of pH and temperature as for alkalis. That is, pH above 9.5 will result in rather rapid strength loss.

Although sodium chloride does not react with wood, it can cause the wood structure to rupture. This is a surface effect but, after a number of years of exposure, rupturing may reach to a depth of 1 inch or more. This salt-rupturing effect is noticed only when wood is exposed to alternate wetting with saltwater and rapid drying such as on the top and undersides of wood decks on boats, nonpreservative-treated wood exposed to saltwater spray on docks, and spillage of salt solutions in chemical plants. The rupture of the wood structure is due to the growth of salt crystals that grow large enough to force openings in the fibrous structure. Figure 2.5.1 shows part of a piece of wood piling that was exposed to saltwater spray. Warehouses to store road salt are often constructed with oil-type preservative-treated side walls, doors, and roof structural members because the oil-type wood preservatives do not allow the salt solution to penetrate the wood structure, and the wood is not affected by the corrosiveness of the salt.

Oxidizing chemicals.--Oxidizing chemicals such as sodium and calcium hypochlorite swell and react with wood. They will cause rather rapid

Figure 2.5.1.--Example of rupturing of wood structure due to exposure to saltwater.

loss of strength. Nitric acid also reacts rapidly with wood. The reaction rate and rate of strength loss are more rapid than would be predicted on the basis of pH and temperature. Wood is seldom used in contact with these types of chemicals and there is no reliable experience available to estimate service life.

2.5.3   Estimation of Conditions for Long Service Life

Because usually even a small amount of chemical reaction has a great effect on the strength properties, it is not possible to chemically analyze partially degraded wood to estimate the amount of strength loss. Usually, before a chemical change can be detected by routine chemical analysis, the wood has lost a substantial proportion of its strength.

A number of investigators have attempted to determine the effect of various chemicals on the strength of wood in terms of percentage of strength loss versus concentration, time, and temperature. Generally, small wood samples, measuring about 1/4 by 1/2 by 8 inches, were soaked in various chemicals at room temperature and at some elevated temperature. Samples were removed after various times and tested in the wet condition for bending strength. Control samples were soaked in water for the same time and at the same temperature and similarly tested. The results were then reported as percent strength retention of wood soaked in various chemicals compared to the retention of samples soaked in water.

The results in Table 2.5.1 are typical of what is found in the literature (1). The data indicate that the softwood species have better retention of strength properties than do the hardwood species but the data do not indicate service life of wood in use when in contact with these or similar chemicals. The data indicate only the relative effects of the chemicals on the species. Literature data should be compared and used with caution because of the different experimental procedures used to obtain the results.

Table 2.5.1.--Strength retention after chemical exposure as a percent of water-soaked control

Concentration	Nitric	Sulfuric	Acetic	Caustic
Pct	Pct	Pct	Pct	Pct
		DOUGLAS-FIR		
2	80	92	90	22
6	60	89	88	0
		WHITE OAK		
2	44	80	101	20
6	13	60	101	0

Sample size - 1/4 by 1/2 by 6 inches.

Soaking time - 1 week; temp. 120° F.

Data are not available to permit calculation of service life or amount of strength loss associated with the length of time a wood member is in contact with a chemical as "corrosion rates" for wood in various environments are not available for wood as they are for metals. The factors that do affect the rate and amount of chemical reaction are known, however, and it is possible to describe upper limits of the factors.

Service conditions suitable for use of wood in contact with chemicals, then, are:

        (a) When the pH of the solution is between 2 and 11,
        (b) When the temperature is usually less than 50° C,
        (c) When there is no contact with oxidizing chemicals.

At the extremes of the pH range, the service life is shortened, especially at elevated temperatures.

The waterproof phenolic adhesives used in glued wood products for wet conditions are usually considered more resistant to any chemicals than is wood substance. For severe exposures, plywood can be obtained with a phenolic-impregnated or other plastic film overlay that is also more resistant to chemicals than is wood. Use of this type of plywood usually requires construction such that the unprotected edges are not exposed to the chemicals.

2.5.4   Chemical Effects on Structural Wood

Wood structural elements are seldom in constant direct contact with chemicals, but can be exposed to chemical spills, corrosive gases, and hygroscopic dusts. This type of chemical exposure is seldom detrimental to the service life of wood structural members themselves, but the metal fasteners or other metal hardware used with them are subject to corrosion.

Occasional chemical spills will not be a hazard to the service life of structural wood if the chemicals are washed away. The method used to wash the chemical from the wood will depend upon the chemical and site conditions, but usually flushing with large volumes of water will be sufficient. Since wood structural members are usually rather large in cross section, washing away of the chemical within a few hours will be sufficient to eliminate any significant hazard to the wood. Chemical reactions will occur only on the surface of the wood during such short exposure. Of immediate concern however, after chemical contact with wood structural members, is the effect of the chemical on the metal fasteners and other connecting hardware. Chemicals should not be allowed to penetrate around metal fasteners because the chemical is difficult to wash away or neutralize. In addition, a small amount of chemical could severely affect a metal fastener and thus the strength of the joint. If repeated exposure to liquid chemicals occurs, the structural members should be washed well and then protected from additional contact with the liquids. Any physical barrier installed should leave an airspace between the barrier and the structural member. This airspace will allow drying and also will provide for visual inspection to note damage.

Corrosive gases and water vapor usually have more effect on the fasteners and other connecting hardware than they have on the wood structural members. This is because gases and vapors have only a surface effect on the wood since their concentrations are kept low within a building, due to normal requirements for human occupancy and the need to control the gases to minimize air pollution. Also, wet conditions due to condensing water vapor are seldom tolerated. It should be remembered that nearly all corrosive gases are acidic, and it is these conditions that wood can tolerate best. It is these acidic conditions, however, that are corrosive to commonly used metal or metallic structural and fastener materials.

Common sense can be used to evaluate the severity of exposure of wood to corrosive gases and to judge if corrective action should be taken. First, it should be determined if the enclosed structure or portion of the structure is suitable for human occupancy. If it is, then chances are that conditions are not severe enough to affect the wood. There could possibly be enough corrosive materials present, however, to cause corrosion of metal fasteners. If there are conditions to which the wood is exposed that are not suitable for human occupancy due to corrosive gases, then the gases should be collected by an exhaust hood and either vented or otherwise disposed of. Where neither of the foregoing are possible, then a simple test such as pH or other chemical analysis of the moist surface of the wood or of the water squeezed from a wet cloth exposed to the gases should indicate how serious the problem is. It should be mentioned again, however, that the most serious problem will most likely be corrosion of the fasteners.

Structural failure of some laminated wood beams in warm, moist conditions above log heating chambers in a veneer mill was traced to stress corrosion cracking of tempered carbon steel fasteners (2). As a result, differently tempered fasteners with a galvanized finish are now recommended.

In warehouses for the bulk storage of chemicals such as fertilizer and salts, the corrosive conditions are created by dust of the chemical stored. The chemicals are sometimes hygroscopic, and the layer of dust can adsorb moisture from the air and create corrosive conditions. These conditions are not considered corrosive to wood, however, since they are only mildly acidic or alkaline. The hygroscopic dust can cause corrosion to fasteners and hardware. Glued laminated wood fertilizer warehouses frequently have been constructed with stainless steel type 304 base shoes, connecting hardware, and nails. The metal selected for the hardware depends upon the material stored. For less hygroscopic but yet corrosive chemicals, epoxy-painted or hot-dip galvanized fasteners have been specified for bulk chemical storage warehouses.

2.5.5 Inspection and Maintenance of Wood Structures for Chemical Damage

Wood members should be inspected for color change as a first sign of chemical damage. If a color change is noted, then a probe such as a knife blade should be used to determine the depth of the chemical damage. Estimations of the strength loss can then be made, based on the dimensions of the sound wood remaining.

The actual degree of chemical attack, sometimes measured in terms of marked color change or weight loss, need not be extensive to cause significant changes in certain strength properties. Thus, wood showing a marked color change and easily probed with a knife blade should be considered to contribute no strength to the wood member. It is particularly important to inspect the wood around the fasteners and the fasteners themselves. Fasteners should be checked for general corrosion, pitting, and reduced cross section.

A maintenance program should consist of protecting the wood structural members from direct contact with corrosive chemicals. For liquids this could be a physical barrier placed so that an inspection and ventilation space is provided between the structural members and the barrier.

Moist, corrosive gases should be collected near the source and properly exhausted. In cold climates, insulation should protect the inside surfaces from temperatures low enough to allow condensation of moist, corrosive gases for extended periods of time. Interior structural wood exposed to corrosive chemicals is usually not surface coated except for appearance. This allows continual visual inspection and prevents entrapment of moisture and corrosive chemicals under film-forming surface coatings.

2.5.6 Metal Corrosion in Wood

The corrosion of metal in wood, preservative-treated wood, and fire-retardant-treated wood products has been discussed in two reports (3,4). The subject matter of these reports is highlighted here to describe briefly the problems and precautions.

Corrosion of metal fasteners in wood is serious because: (1) the fasteners can lose thickness or cross section, and (2) the chemical byproducts of the corrosion can attack the wood adjacent to the fastener. The result is that the fastener becomes smaller and the hole in the wood becomes larger, and the joint becomes weakened by both effects. Fortunately, however, corrosion in wood is not usually a problem, because if protected from weather, the moisture content is usually too low to allow corrosion. That is, the electrical resistance of wood is high enough to suppress current flow and inhibit corrosion so that wood exposed to equilibrium moisture content conditions that result in about 20 percent moisture content or less does not cause a serious corrosion problem. It is only when wood is exposed to high humidity, in contact with moisture or soil, or treated with hygroscopic chemicals that corrosion is a problem.

The most commonly observed corrosion problem involves fasteners exposed to exterior conditions. Iron staining is common on exterior siding and on timbers with steel fasteners or galvanized steel fasteners after the zinc has corroded away. When a steel fastener corrodes, the ions of ferrous iron leaving the steel are oxidized to ferric ions. This oxidation catalyzes a chemical reaction that results in loss of strength of the adjacent wood. This accounts for the enlarged holes around nails in old house and barn siding and loosening of the spikes and plates on ties on railroad tracks. If water that contains salt (chloride ions) is present, a steel fastener can undergo crevice corrosion that results in the formation of acidic conditions around the steel embedded in the wet wood. In these conditions not only does the iron-catalyzed wood degradation occur, but the accumulated acid condition results in hydrolysis of the wood cellulose as well as increased corrosion of the steel to liberate more ferrous ions. Thus, the corrosion and wood degradation proceed at ever-increasing rates.

If two different metals are used as a fastener (such as a steel bolt and a brass washer) in wet wood, galvanic corrosion will occur. The chemical reactions depend upon the metals present and can be predicted by the galvanic series. In this case, the steel will corrode and degrade the wood adjacent to the bolt. The washer will not corrode, but alkaline corrosion byproducts from the washer will accumulate

SERVICEABILITY FACTORS

between the washer and the wood. Eventually, the alkaline conditions will swell and soften the wood under the washer. The bolt will become loose and the joint will weaken.

When wood is used in contact with soil or in moist conditions, it is treated with preservatives to inhibit decay. Fastener corrosion has not been reported to be a problem (except in rail ties) in wood treated with oil-type preservatives. The oil evidently increases the electrical resistance at the wood-metal interface enough to inhibit corrosion. Wood treated with waterborne salt preservatives, however, can cause corrosion to many common fastener materials. Since many such preservatives contain copper compounds, this corrosion is apparently the result of free copper ions depositing on certain metals and alloys; these conditions establish numerous galvanic corrosion cells. For use with chromated copper arsenate and ammoniacal copper arsenate-treated wood in wet conditions where long service life is required, stainless steel type 304 or 316, silicon bronze, or copper fasteners are recommended.

The fire performance of wood products is improved by treating with fire-retardant chemicals such as boric acid, borax, ammonium sulfate, aluminum sulfate, and mono- and di-ammonium phosphates. These compounds are corrosive to many metals in humid or moist environments. Because some of them are also hygroscopic, they can be corrosive at low relative humidities, too. The corrosiveness of a fire-retardant-treated product varies with the formulation used. Some fire-retardant formulations include corrosion inhibitors such as sodium dichromate and ammonium thiocyanate.

There have been no reports of corrosion with the nonleachable type of fire-retardant treatment that is recommended for wood exposed to weather. This is probably due to the fact that the fire-retardant chemicals are of an organic nature rather than in salt form.

The general recommendations for protection and use of fire-retardant-treated wood products in contact with metals is to keep them protected from moisture in storage, and to use them where relative humidity is maintained below about 80 percent. Wood products that are treated with fire-retardant chemicals in solution should be redried prior to use with metals. If high relative humidities are expected, plastic-coated or stainless steel fasteners should be considered. However, the long-term effect of the fire-retardant chemicals on the stainless steel has not been determined.

2.5.7 References for Section 2.5

2.5.7.1 Literature Cited

1. Baechler, R. H.
   1954. Wood in chemical engineering construction. J. For. Prod. Res. Soc. 4:332-336.

2. Baker, A. J.
   1974. Degradation of wood byproducts of metal corrosion. USDA For. Serv. Res. Pap. FPL 229, For. Prod. Lab., Madison, Wis.

3. Baker, A. J.
    1980. Corrosion of metal in wood products. In Durability of
    Build. Mater. and Components. ASTM STP 691, P. J. Sereda and
    G. G. Litvan, Eds., Am. Soc. for Testing and Mater.,
    pp. 981-993.

4. Fox, S. P., G. M. Barton, and E. B. Howbolt.
    1977. Environmental cracking of tempered carbon steel nails in
    Douglas-fir wood. Mater. Performance 16(10):36-40. Oct.

2.5.7.2 Additional References

5. Thompson, W. S.
    1969. Effect of chemicals, chemical atmospheres, and contact
    with metals on southern pine wood: a review. Mississippi
    State Univ. For. Prod. Util. Lab. Res. Rep. 6, 33 pp.

6. Wangaard, F. F.
    1966. Resistance of wood to chemical degradation. For. Prod. J.
    16(2):53-54.

7. Ross, J. D.
    1956. Chemical resistance of western woods. For. Prod. J.
    6:34-37.

2.6   Weathering*

2.6.1   Introduction

Wood is an extremely durable material even under adverse conditions, but the durability depends upon the environment. Buried deep under ground, fully exposed to the weather, submerged under water, or hidden in an ancient tomb, it can last for centuries (fig. 2.6.1 and 2.6.2). Examples of structures, ships, and other wooden objects which have survived centuries of use are many (4,15). The same type of wood exposed to an unfavorable environment, however, may vanish almost without a trace within a year or two. In the tropics, a wooden house may under certain conditions disintegrate in a few years.

Stalker (40) conveniently divided the environmental agencies that brought about wood degradation into categories. "Physical" forms of energy were used to describe all factors other than fungi, insects, or animals. In Table 2.6.1, the importance of the various destructive agents on wood can best be considered by comparing two situations, inside and outside wood structures. The most serious risk to wood indoors comes from the intense heat of an accidental fire. Outdoors the factor most deserving of attention is weathering--a complex combination of chemical, mechanical, and light energies.

Weathering is not to be confused with decay, which results from decay organisms (fungi) acting in the presence of excess moisture and air for

---

*Author: William C. Feist, Chemist, U.S. Forest Products Laboratory, Madison, Wisconsin.

Figure 2.6.1.--Old Fairbanks house at Dedham, Massachusetts. Built in 1637, most of the white pine clapboard siding was replaced in 1903 and has stood 75 years without paint.

an extended period of time (15). Under conditions suitable for the development of decay, wood can deteriorate rapidly and the resulting phenomenon is far different than that observed for natural outdoor weathering. (See section 2.8.)

An excellent annotated bibliography was compiled by the California Redwood Association (8), which provides a source of historical references which pertain to wood weathering, applied surface treatments and/or coatings, as well as to wood substrate and/or extractive modification, materials, and methods. The bibliography covers references to significant published material and to pertinent unpublished material as well. A recent review of weathering was compiled by Feist (18).

2.6.2 The Weathering Process

In outdoor weathering of smooth wood, original surfaces become rough as grain raises and the wood checks, and the checks grow into large cracks; grain may loosen, boards cup and warp and pull away from fasteners. The roughened surface changes color, gathers dirt and mildew, and may become unsightly; the wood loses its surface coherence and becomes friable, splinters, and fragments come off. All these effects, brought about by a combination of light, water, and heat, are comprehended in one word: weathering (fig. 2.6.3).

108   WOOD STRUCTURES

Figure 2.6.2.--White oak log cabin near Middleton, Wis., construction around 1845 and never painted or finished.

2.6.2.1 Weathering Factors

   Action of water--The principal cause of weathering is frequent exposure of the wood surface to rapid changes in moisture content (43,44). Rain or dew falling upon unprotected wood is quickly absorbed by capillary action on the surface layer of the wood followed by adsorption within wood cell walls. Water vapor is taken up directly by adsorption under increased relative humidities. Adsorbed water has been shown to virtually add its volume to that of the cell walls, resulting in swelling (41,42). Stresses are set up in the wood as it swells and shrinks due to moisture gradients being set up between the surface and the interior (46). These induced stresses are greater the steeper the moisture gradient, and are usually concentrated near the surface of the wood. When unbalanced, they may result in warping, cupping, and face checking. Grain raising results from differential swelling and shrinking of summerwood and springwood.

   Action of light--The photochemical degradation of wood or wood-related materials has been reviewed in several publications (14,27,37). It was recognized quite early that the initial color change of wood exposed to sunlight was a yellowing or browning (8,18). The graying of wood occurs after browning and was at one time thought to be related to iron salts. Sunlight, particularly the ultraviolet (UV) end of the spectrum, degrades the organic materials in wood; lignin decomposes

Table 2.6.1.--Relative effect of various energy forms on wood (Stalker)

Form of energy	Indoor		Outdoor	
	Result	Degree of effect	Result	Degree of effect
Thermal				
Intense	Fire	Severe	Fire	Severe
Slight	Darkening of color	Slight	Darkening of color	Slight
Light				
Visible and UV	Color change	Slight	Large color changes	Severe
			Chemical degradation (especially lignin)	Severe
Mechanical	Wear and tear	Slight	Wear and tear	Slight
			Wind erosion	Slight
			Surface roughening	Severe
			Defiberization	Severe
Chemical	Staining	Slight	Surface roughening	Severe
	Discoloration	Slight	Defiberization	Severe
	Color changes	Slight	Selective leaching	Severe
			Color changes	Severe
			Strength loss	Severe

preferentially to a relatively shallow depth of 0.05 to 0.5 mm (27,37). Photo-degradation by UV light induces changes in chemical composition, particularly in the lignin (27,29).

It is important to note here that the two most important elements of weathering--light irradiation and water--tend to operate at different times. Exposed wood can be irradiated after having been wet by rain or when surface moisture content is high from overnight high humidity or from dew. Time of wetness, therefore, is an important parameter in relating climatic conditions to exterior degradation. The action of the combined elements can follow different degradation paths, with irradiation accelerating the effect of water or the converse.

Action of heat--The role of temperature in the natural weathering process is generally felt to be of less importance than those of light and water.

2.6.2.2 Property Changes

There is a great deal of literature on the chemical changes occurring during the outdoor weathering of wood and it can be concluded that

110   WOOD STRUCTURES

Figure 2.6.3.--Close-up view of weathered
white oak logs in figure 2.6.2.

absorption of UV light by lignin on the wood surface results in preferential lignin degradation. In the graying of wood, most of the solubilized lignin degradation products are washed out by rain. Fibers, high in cellulose content and whitish to gray in color, remain on the wood surface and are resistant to UV degradation (8,18,27).

Color changes--The color of wood exposed outdoors is affected very rapidly. Generally, all woods change toward a yellow to brown due to the breakdown of lignin and extractives (47,48). This yellowing or browning occurs after only several months of exposure in sunny, warm climates (fig. 2.6.4). Woods rich in extractives, like redwood and cedar, may first become bleached before the browning becomes observable. In the absence of micro-organisms, wood weathers to a soft, silver gray as a result of the leaching of decomposition products of wood lignin.

Changes in wood color reveal chemical changes in wood during weathering (27). Only those parts of the wood close to the exposed surface are affected. Initial browning penetrates only 0.01 to 2 mm into the wood (7). As rain leaches the brown decomposition products of lignin, a silver gray layer of 0.08 to 0.2 mm thick consisting of a disorderly arrangement of loosely matted fibers develops over the brown layer. The gray layer is composed chiefly of the most leach-resistant parts of the wood cellulose. This surface color change to gray is observed when the wood is exposed to intense radiation of the sun in cooler climates with little precipitation in the form of rain. This would also occur

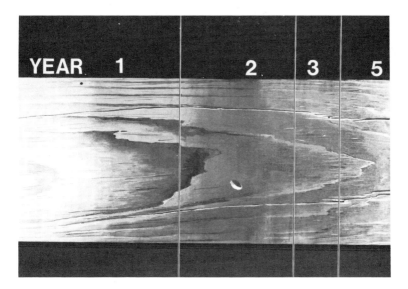

Figure 2.6.4.--Artist's monochromatic rendition of color changes and surface wood change during the outdoor weathering process of a typical softwood.

on structures with large roof overhang, which protects the wood underneath from rain and sun. However, another mechanism of graying of weathered wood usually predominates, particularly in the presence of moisture.

All modern studies on the weathering of wood conclude that the discoloration (graying) of woods in the presence of moisture is practically always due to growth of fungi on the surface of the wood (16,33,35,39, 48). The most frequently observed fungus species is Aureobasidium pullulans (Pullularia pullulans) (6,34) which under favorable conditions, grows not only on wood surfaces, but also on the surface of coatings and on various organic and inorganic materials (6,47). The ecological requirements of this fungus are modest, the most important condition for its growth being the sporadic supply of bulk water. The fungus is otherwise relatively resistant and adaptable.

Sell (35) found that A. pullulans infested finished as well as untreated wood surfaces after only 1 year of outdoor weathering. He concluded that discoloration of wood by mildew is more widely distributed than commonly believed. Fungal infection was the result of a temporary or lasting wetting of the wood surface with liquid water. Sell and Leukens (39) subjected 20 European and non-European wood species of widely differing density and mechanical properties to unprotected outdoor weathering of 45° inclined to the south in Switzerland. The initially distinctly different behavior among the different species gradually changed, and photo-chemical and mechanical deterioration as well as intensity of attack by the blue stain fungi were observed to even out. After only 1 year's weathering, all wood surfaces had a uniformly weathered and gray appearance.

Physical changes--In addition to chemical and color changes of wood on outdoor weathering, damage that is mechanical in nature occurs on the exposed wood surface (fig. 2.6.4). Decomposition of a wood surface due to the combined action of light and water causes surface darkening and leads to formation of macroscopic to microscopic inter- and intracellular cracks or checks. There is a loss of strength in cell wall bonds near the wood surface. As the weathering process continues, rain water washes out degraded portions and further erosion takes place. Because of different types of wood tissue on the surface, erosion and checking differ in intensity and the wood surface becomes increasingly uneven (fig. 2.6.5).

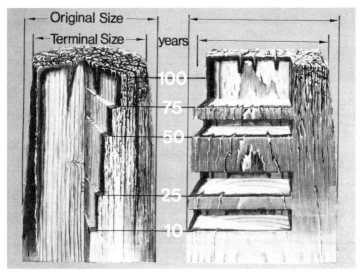

Figure 2.6.5.--Artist's rendition showing the weathering process of round and square timbers. Cutaway shows interior wood below the surface is relatively unchanged.

Browne (7) reports that the weathering process is so slow that "only 1/4 inch (6.4 mm) of thickness is lost in a century." A value of 1 mm per century has been reported for wood exposed in northern climates. Feist and Mraz (19) report an erosion value of 13 mm per century for western redcedar. This value was based on exposure data from 8 years outdoor weathering at 90° facing south. Erosion data obtained on controlled accelerated weathering of redwood, Douglas-fir, Engelmann spruce, and ponderosa pine were used to estimate outdoor weathering and showed that these species would erode at a rate of approximately 6 mm per 100 years, a value similar to Browne's. Borgin (4) reports on erosion of wall cladding on stave churches in Norway and estimates that 10-mm-thick cladding has been reduced by half over a few hundred years' weathering.

From the above, it is obvious that the physical loss of wood substance from the wood surface during weathering is dependent upon the amount of irradiation, rain action, wind, degree of exposure and, generally, climate. In addition, defiberization, or loss of surface wood cells, occurs most rapidly in thin-walled fibers of springwood and, at a slower rate, in the denser summerwood (7,19,48). Because of this, weathered wood assumes a corrugated or serrated appearance. Accompanying this loss of wood substance are the swelling and shrinking stresses set up by fluctuations in moisture content. All these result in surface roughening, grain raising, differential swelling of springwood and summerwood bands, and the formation of small checks or cracks. Larger and deeper cracks may develop and warping may result (fig. 2.6.5).

Borgin (4) reported on the appearance of broad planks fully exposed to weather for at least 500 years without cracking. The depth of weathering of the surface was proportional to the density across annual rings. Round timbers developed deep cracks due to internal stresses, but surface erosion was still slow without any deterioration of the interior of the timber. All structures investigated were of Norwegian pine.

Strength changes--One area receiving relatively little attention is that of change in strength properties of solid wood with outdoor weathering. Modified wood products such as plywood and particleboard have received considerable attention and will be discussed in the next section. Concerning solid wood, Stamm (41) reports that the various weathering effects (physical, chemical, etc.) have little influence upon modulus of rupture, modulus of elasticity, and compressive strength of wood. Only toughness and abrasion resistance are materially reduced as a result of penetrating thermal effects. This apparent lack of change in strength properties is undoubtedly due to the fact that weathering is an essentially surface effect (fig. 2.6.5). In the absence of decay, little change occurs in the overall physical properties of wood. Borgin et al. (5) indicated that, under weathering conditions free from decay, mechanical and structural properties appear to be only slightly affected. Sell (37) also indicates that weathering does not markedly affect the strength properties of wood.

2.6.2.3 Weathering of Wood-Based Materials

The weathering process described thus far has been that observed for solid wood. The introduction of another variable, the adhesive, in the weathering of wood-based materials such as plywood and particleboard creates additional complications. Wood substance is still exposed to the elements in products such as plywood and particleboard, and deteriorates in a manner similar to that for solid wood. The wood-adhesive bond is the new element in exposure.

Plywood--The weathering of plywood is directly related to the quality of the veneer exposed and to the adhesives used. Water resistance of the adhesive is of great importance. Because of its tendency to check, most exterior plywood is protected with a finish or with overlay material. Finished or overlaid plywood weathers and performs similarly to solid wood (1,2,3).

Strength change studies on both naturally and artificially weathered plywood have been reported (1,13,25,32). For unfinished plywood, rapid surface checking occurs in the early weathering stages. Water absorption increases and percent shrinkage and swelling parallel to the fiber direction decrease slightly. Kaneda and Maku (28) investigated lauan plywood loaded in bending and tension under artificial weathering, and found that shear strengths varied greatly with the type of adhesive.

Particleboard--As in the case of plywood, durability of particleboard in outdoor weathering depends very much on wood species, and on the amount and nature of resin (binder, adhesive) used in preparing the board (9,10,12). In outdoor weathering of unprotected particleboards, outer layers are subjected to greater wear and tear. As long as the outer cover layers are intact, inner layers are protected from the elements of weathering. When outer layers of exposed boards deteriorate and loosen, shrinking and swelling of inner layers result due to exposure to changes in moisture content. Accelerated deterioration of inner layers generally results, cohesion is lost, and boards may fail under mechanical loads (21). Only 1 or 2 years of weathering can cause significant strength loss and increased swelling (20,23,26,30). Deterioration of particleboard during outdoor weathering takes place because of the combined effects of springback from compression set, deterioration of resin, and differential shrinkage of adjacent wood particles during moisture content change (24). Phenolic resins appear to give the best overall performance (31). Additional related studies have reported on the effect of natural outdoor and artificial accelerated weathering on durability and strength properties of particleboard and related materials (11,20,22).

Protection of exterior particleboard by surface coatings, treatments and overlays, and subsequent weathering performance, has received considerable attention (20,23,26,30,31). Painted and overlaid boards are much more durable than unfinished boards. Addition of wax as a water repellent is of benefit to the performance of the boards, but does not help protect against water vapor (24,36,38,39).

Results from outdoor weathering experiments with different surface treatments indicate that lasting protective effects of surface treatment, and protection of board edges against moisture, are the most important prerequisites for long-term functioning of coating systems. Complementary construction measures are necessary for good performance (36). Meierhofer and Sell (31) conclude that waterproofing and sealing of the wood surface on particleboards are imperative. Above a minimum value, the gradient of water vapor diffusion of the coating is only of secondary importance. They also concluded that the visual state of the board and thickness swelling are considered the best criteria for evaluating the protective effect of surface treatments. Surface stability is recognized as one of the most important criteria in retaining finish and maintaining satisfactory protection and appearance (30).

2.6.3 Protection Against Weathering

Problems associated with outdoor weathering, in the absence of wood decay, are inclined to be aesthetic more than anything else, as strength properties for solid wood are not markedly affected. Surface coatings for wood are required to provide protection against weathering for mainly cosmetic reasons, by reflecting light or by absorbing it (17,45). In the case of wood products such as plywood and particleboard (including waferboard), however, surface coatings play an important role in substrate protection.

On the other hand, however, light and water protection provided to a wood surface by the surface treatment will be affected by the weather resistance of the bonding agents of the finish (drying oils, synthetic resins, latexes, etc.). These bonding agents are subject to photolytic degradation to some degree. Protection of wood exposed outdoors by various finishes and other factors, to compensate for effects of weather, is discussed in Chapter 6.

2.6.4 Summary

Wood exposed outdoors without protection undergoes: (1) Photodegradation by ultraviolet light, (2) leaching, hydrolysis, and shrinking and swelling by water, and (3) discoloration and degradation by microorganisms. Unfinished wood surfaces change color when exposed to weather, are roughened by photodegradation and surface checking, and erode. The appearance of unprotected wood exposed outdoors changes markedly in a few months; then the wood remains almost unaltered for years. In the absence of decay, wood exposed to the weather can and does last for centuries. Although, physical as well as chemical changes occur due to weathering, these changes affect only the surface of the exposed wood. Wood only a few millimeters under the surface is essentially unchanged and unaffected.

2.6.5 References for Section 2.6

1. Adhesives Age.
    1977. Outdoor weathering of plywood and composites. Adhes. Age 20(2):15-18.

2. American Plywood Association.
    1978. Stains and paints on plywood, APA Pamphlet B407. Am. Plywood Assoc., Tacoma, Wash.

3. Black, J. M., Mraz, E. A., and Lutz, J. F.
    1976. Performance of softwood plywood during 10 years' exposure to weather. For. Prod. J. 26(4):24-27.

4. Borgin, K.
   1969. The stability, durability, and weather resistance of wooden houses under cold climatic conditions. Archit. Builder, June-July.

5. Borgin, K., Parameswaran, N., and Liese, W.
   1975. The effect of aging on the ultrastructure of wood. Wood Sci. Tech. 9(2):87-98.

6. Brand, B. G., and Kemp, H. T.
   1973. Mildew defacement of organic coatings. The Paint Res. Inst., Fed. of Soc. for Paint Technol., Philadelphia, Pa.

7. Browne, F. L.
   1960. Wood siding left to weather naturally. So. Lbrmn., pp. 141-143.

8. California Redwood Association.
   1962. Wood and wood finishes in exterior use--An annotated bibliography of significant studies of coatings, treatments, and substrates modifications. CRA, San Francisco, Calif.

9. Clad, W., and Pommer E. H.
   1971. Open air weathering tests on chipboard. Holz-Zentralbl. 97(28):397-398; (32):453-454; (35):505-506.

10. Clad, W., and Schmidt-Hellerau, C.
    1976. Tests on wood-based materials bonded with melamine resin. 1. Open-air weathering tests on particleboards bonded with melamine and phenolic resins. Holz-Zentralbl. 102(24):313-314.

11. Clad, W., and Schmidt-Hellerau, C.
    1976. Tests on wood-based materials bonded with melamine resin. 2. The behavior of particleboards in stability tests. Holz-Zentralbl. 102(37):505.

12. Clad, W., and Schmidt-Hellerau, C.
    1976. Tests on wood-based materials bonded with melamine resin. 3. The effect of the wood species used on the durability of particleboard. Holz-Zentralbl. 102(40):543.

13. Clad, W., and Schmidt-Hellerau, C.
    1976. Tests on wood-based materials bonded with melamine resin. 4. Open-air weathering tests on plywood. Holz-Zentralbl. 102(43):583.

14. Coupe, C., and Watson, R. W.
    1967. Fundamental aspects of weathering. Proc. Ann. Conv. Brit. Wood Preserv. Assoc. 37-49.

15. DeGroot, R. C.
    1976. Your wood can last for centuries. USDA For. Serv. So. For. Exp. Stn., New Orleans, La.

16. Duncan, C. G.
    1963. Role of microorganisms in weathering of wood. Off. Dig. J. Paint Tech. Eng. 35(465):1003-1012.

17. Feist, W. C.
    1977. Finishing wood for exterior applications--Paints, stains and pretreatments. In I. S. Goldstein, Ed., Wood Technology: Chemical aspects. ACS Symp. Series No. 43. Am. Chem. Soc., Washington, D.C.

18. Feist, W. C.
    1980. Weathering of wood in structural uses. Proceedings of symposium on structural use of wood in adverse environments, Vancouver, B.C. 1978. Society of Wood Science and Technology.

19. Feist, W. C., Mraz, E. A.
    1978. Comparison of outdoor and accelerated weathering of unprotected softwoods. For. Prod. J. 28(3):38-42.

20. Geimer, R. L., Heebink, B. G., and Hefty, F. V.
    1973. Weathering characteristics of particleboard. USDA For. Serv. Res. Pap. FPL 212. For. Prod. Lab., Madison, Wis.

21. Gressel, P.
    1968. Investigations on weather aging of particleboard. A comparison between 2-year weathering and three different accelerated test methods. Holz Roh- Werkst. 26(4):140-148.

22. Hall, H., and Haygreen, J. G.
    1975. The effect of short periods of simulated weathering on the impact performance of particleboard. Wood Fiber 7(2):91-103.

23. Hann, R. A., Black, J. M., and Blomquist, R. F.
    1962. Three years of tests help answer this pertinent question...How durable is particleboard? For. Prod. J. 12(12):577-584.

24. Hann, R. A., Black, J. M., and Blomquist, R. F.
    1963. How durable is particleboard? Part II. The effect of temperature and humidity. For. Prod. J. 13(5):169-174.

25. Hunt, M. O., and Matterson, D. A., Jr.
    1976. Structural characteristics of weathered plywood. J. Struct. Div. Proc. Am. Soc. Civ. Eng. 102(4):759-768.

26. Jokerst, R. W.
    1968. Long term durability of laboratory-made Douglas-fir flakeboard. USDA For. Serv. Res. Note FPL-0199. For. Prod. Lab., Madison, Wis.

27. Kalnins, M. A.
    1966. Surface characteristics of wood as they affect durability of finishes. Part II. Photochemical degradation of wood. USDA For. Serv. Res. Pap. FPL 57. For. Prod. Lab., Madison, Wis.

28. Kaneda, H., and Maku, T.
    1976. Studies on the weatherability of composite wood. V. Weatherability and durability of adhesive joint of plywood under loaded conditions. J. Jap. Wood Res. Soc. 22(3):173-183.

29. Kleinert, Th. N.
    1970. Physical and chemical changes occurring in naturally exposed wood. Holzforsch. Holzverwert. 22:21-24.

30. Lehman, W. F.
    1968. Durability of exterior particleboard. Proc. 2nd WSU Particleboard Symp., Washington State Univ., Pullman, Wash. pp. 275-306.

31. Meierhofer, U. A., and Sell, J.
    1975. Investigations on weathered particleboards. 3. Results of weathering test over several years with boards having different surface treatments. Holz Roh- Werkst. 33(12):443-450.

32. Nakamura, F.
    1975. Changes in the surface properties of plywood exposed in open air. J. Hokkaido For. Prod. Res. Inst. 2:11-15.

33. Schmidt, E. L., and French, D. W.
    1976. *Aureobasidium pullulans* on wood shingles. For. Prod. J. 26(7):34-37.

34. Schmitt, J. A.
    1974. Mildew-induced defacement of organic coatings (MIDOC) and related phenomena. J. Paint Technol. 46(596):53-59.

35. Sell, J.
    1968. Investigation of the infestation of untreated and surface-treated wood by blue-stain fungi. Holz Roh- Werkst. 26(6):215-222.

36. Sell, J.
    1973. Surface treatment of particleboards for external use: Experiments to elucidate principal requirements. Holz-Zentralbl. 99(147):2337-2338.

37. Sell, J.
    1975. Basic requirements for the surface treatment of exterior wood. A review of pertinent literature. Holz Roh- Werkst. 33(9):336-340.

38. Sell, J., and Krebs, U.
    1975. Studies on weathered particle and chip boards. Part II. Combating moisture with water repellents and surface coatings Holz Roh- Werkst. 33(6):215-221.

39. Sell, J., and Leukens, U.
    1971. Investigations of weathered wood surfaces. Part II. Weathering phenomena of unprotected wood species. Holz Roh-Werkst 29(1):23-31.

40. Stalker, I. N.
    1971. Protection of timber from fire and weathering. Chem. Ind. 50:1427-1431.

41. Stamm, A. J.
    1963. Effect of dimensional stabilization on weathering properties of wood. Am. Paint J. 48(25):72-88.

42. Stamm, A. J.
    1964. Wood and cellulose science. The Ronald Press Co., New York.

43. Stamm, A. J.
    1965. Wood and cellulose-water relationships and their effect upon finishes. Off. Dig. J. Paint Technol. Eng. 37(485):654-669.

44. Stamm, A. J., and Loughborough, W. K.
    1942. Variation in the shrinkage and swelling of wood. Trans. Am. Soc. Mech. Eng. 64:379-385.

45. Timber Research and Development Association.
    1965. Maintaining timber exposed to the weather. Timber Res. Devel. Assoc., High Wycombe, Buckinghamshire, England.

46. U.S. Department of Agriculture, Forest Service, Forest Products Laboratory.
    1974. Wood Handbook: Wood as an engineering material. USDA Agri. Handb. No. 72. Washington, D.C.

47. U.S. Department of Agriculture, Forest Service, Forest Products Laboratory.
    1975. Wood finishing: Mildew on house paints. USDA For. Serv. Res. Note FPL-0128. For. Prod. Lab., Madison, Wis.

48. U.S. Department of Agriculture, Forest Service, Forest Products Laboratory.
    1975. Wood finishing: Weathering of wood. USDA For. Serv. Res. Note FPL-0135. For. Prod. Lab., Madison, Wis.

2.7 Fire*

2.7.1 Introduction

Heavy timber beams, columns, and decking are not only excellent performers during fire, they also provide the potential for rather simple restoration both functionally and esthetically after a fire. A common restoration procedure after fire damage is to remove the surface char through sandblasting and then to calculate the residual load-carrying capacity. Structural modification can then be predicated upon analytic results obtained. In many cases, where a long, severe fire has not occurred, so that char depth is small, removing the char leaves a section capable of continuing in service with no additional modification. It is the purpose of this section to discuss specifics of the influence of fire on wood members. Application of this information to calculation of the residual load-carrying capacity of the restored members will be covered in detail in chapter 4.

2.7.2 Fire and Wood

There are several parameters that require quantification in order to estimate residual strength of timber members. Most important are the temperatures achieved in the residual (uncharred) wood, and the level of strength and stiffness retained by the uncharred wood.

2.7.2.1 Temperatures Achieved

As wood burns, it is progressively converted to char. Though surface temperatures may be on the order of 1800° F (982° C), the temperature at the interface between uncharred and charred wood is generally about 550° F (288° C). This becomes the highest temperature to which wood is exposed, then, before conversion to a char having negligible strength. After about 20 minutes of fire exposure (or development of about a 1/2-inch layer of char for softwood species), the temperature gradient further into the wood (with respect to the char-wood interface) reaches a pseudo steady-state, as illustrated in figure 2.7.1 for a Douglas-fir section. Temperatures do not exceed that at the char-wood interface. For example, a half inch into the wood, the temperature, T, is 0.4 the difference between the interface temperature, T_c, of 550° F and the initial temperature, T_o, of the wood. At 2 inches into the wood from the char-wood interface, the temperature of the wood is close to its initial level. To assess the temperature influence on mechanical properties, only an examination of what occurs in the 2-inch zone of wood beyond the char-wood interface is needed. It should be noted that this steady temperature gradient can be practically described (after about 20 min. of fire exposure) by the equation (6):

$$(T - T_o)/(T_c - T_o) = \exp[-\beta x/\alpha]$$

---

*Author: E. L. Schaffer, Project Leader, Fire Design Engineering, U.S. Forest Products Laboratory, Madison, Wisconsin.

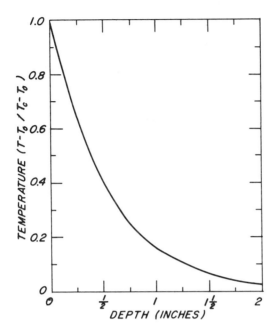

Figure 2.7.1.--Temperature gradient beyond char-wood interface in Douglas-fir during fire exposure.

where

β is the wood charring rate
α is the thermal diffusivity of the wood
x is the distance into wood <u>beyond</u> char-wood interface.

For exposure periods less than 20 minutes, the temperatures achieved are less and can be approximated, but with more analytical difficulty (5). The mean charring rates, β, for Douglas-fir and southern pine at 12 percent moisture content are 1-1/2 inches per hour (0.64 mm/min) and 1.8 inch per hour (0.76 mm/min) respectively (7). Thermal diffusivities can be assumed to be 0.000 230 square inch per second (1.5×10^{-3} cm^2/sec). Thermal diffusivity is defined by the equation (3):

$$\alpha = 0.000\ 3752 - 0.000\ 292\ \rho \quad (in.^2/sec)$$

where

ρ is the specific gravity of the wood (at in-use moisture content).

2.7.2.2 Mechanical Properties

In the previous section we examined what was available to describe temperatures achieved in wood during fire. We now examine what we know about the influence the short-time exposure to heat has on the stiffness and strength of wood after it has been cooled again to room temperature. Though longer-term exposures to elevated temperature have a degrading influence on strength and stiffness (see Section 2.3), cooling after exposure to momentary high temperature (maximum of 550° F) decreases this effect (9). Modulus of elasticity is least influenced by any period of elevated temperature as compared to other mechanical properties. There appears to be no evidence that the modulus of elasticity decreases after an exposure of a few minutes to elevated temperatures as high as 550° F (288° C). It has been observed to increase as much as 5 percent! It may be safely assumed, therefore, that modulus of elasticity is largely unaltered in uncharred wood.

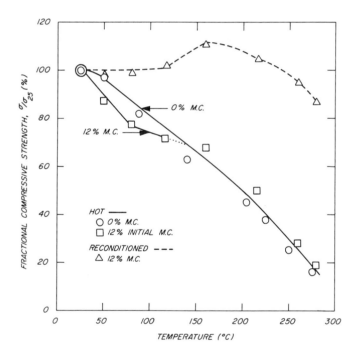

Figure 2.7.2.--Compressive strength of Douglas-fir in heated condition (solid line) and after cooling (dashed line).

The available data show that tensile, compressive, and bending strength after a short time at elevated temperature and evaluated after cooling to room temperature are influenced more significantly than is modulus of elasticity. The compressive strength of Douglas-fir is unchanged for temperatures up to about 100° C, then increases between 100° C and 160° C; it then decreases to about 85 percent of original strength at 288° C (2) (fig. 2.7.2). Tensile strength, on the other hand, decreases between 100° and 200° C, and decreases more rapidly above 220° C. At 288° C, or the incipient char temperature, it retains only about 40 percent of initial strength after cooling (2) (fig. 2.7.3).

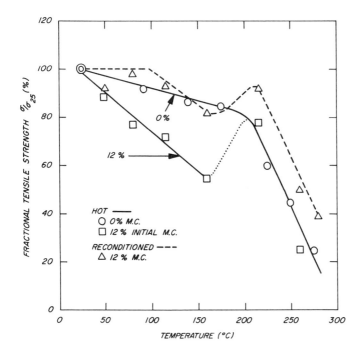

Figure 2.7.3.--Tensile strength of Douglas-fir in heated condition (solid line) and after cooling (dashed line).

The room-temperature modulus of rupture of Douglas-fir after short periods at temperatures as high as 177° C have been reported (4,1) to be unchanged as compared with wood which has not been at high temperature. Temperatures higher than 177° C, however, do decrease room-temperature modulus of rupture.

2.7.3 Adhesive Behavior

The adhesives commonly used in the United States to fabricate laminated beams, columns, arches, decking, and other structural members are thermosetting types. Included are phenolic, resorcinol, and melamine formulations. The available information shows that panels and members bonded with these adhesives produce a thermal response equivalent to solid wood. That is, the gluelines maintain their integrity (until the wood is well charred) during and after fire exposure.

Casein-glued laminated members are not common now, but casein was once used in their manufacture. Structural plywood and particleboard panels currently employ only phenolic resin adhesives in their manufacture. It is known (8) that the casein glueline is affected to about one-fourth inch below the char-wood interface in members laminated from structural lumber. Separation and weakening of the joint must be compensated for in this case.

This is accomplished by deducting one-fourth inch from glueline length beyond the char-wood interface for gluelines perpendicular to a fire-exposed surface. A glueline extending through a member exposed to fire on both sides (or more) requires a 1/2-inch deduction. The new glueline length in the member cross section is the reduced shear width for calculating residual shear strength. A casein glueline parallel to a fire-exposed surface, such as in the bottom laminate of a fire exposed beam, would be assumed to be fully effective until the char-wood interface is within one-fourth inch of the line. Thereafter, the glueline is ineffective and may delaminate.

2.7.4 Properties of Timbers Exposed to Fire

The modulus of elasticity is apparently unchanged in a cool timber member after fire exposure, so member stiffness based upon the residual dimensions is easily calculated.

The strength of tensile and compressive zones of columns and beams may be ascertained by employing the temperature distribution information in the previous section. If this is done, the resulting tensile and compressive strength in Douglas-fir are shown, with respect to distance from the char-wood interface, in figure 2.7.4. Note that compressive strength can be considered insignificantly altered, but that at depths less than 0.18 inch (4.6 mm) wood is reduced significantly in tensile strength. It is also reduced on the order of 10 percent to 15 percent for much of the zone between 0.18 and 0.65 inch. It is clear, therefore, that the lower edge of beams exposed to fire must be analyzed to account for this effect.

The evaluation of members following a fire is discussed further in Chapter 4. However, it can be concluded that, because of the short time that wood just beyond the charline has been at its maximum temperature, the overall strength loss in heavy sections will be small and the residual load-carrying capacity will be closely approximated by using the initial strength properties of the uncharred residual cross section as a base.

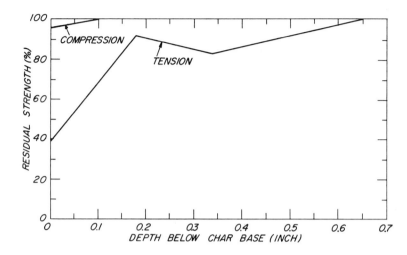

Figure 2.7.4.--Residual strength of fire-exposed and cooled Douglas-fir section as function of depth beyond the char layer (strength at 25° C is 100 pct).

2.7.5 References for Section 2.7

1. Fushitani, M.
    1968. Effect of heat treatment on the static visco-elasticity of wood. J. of Jap. Wood Res. Soc., 14(4):208-213.

2. Knudsen, R. M., and A. P. Schniewind.
    1975. Performance of structural wood members exposed to fire. For. Prod. J. 25(2):23-32.

3. MacLean, J. D.
    1940. Relation of wood density to rate of temperature change in wood in different heating mediums. Proc. Am. Wood-Preserv. Assoc., 36:220-248.

4. MacLean, J. D.
    1945. Effect of heat on the properties and serviceability of wood. U.S. FPL Rep. 1471, 12 p.

5. Rattner, F., and E. L. Schaffer.
    1967. Determination of temperature distribution in wood with variable surface temperature by numerical integration of Duhamel's integral. USDA, For. Serv. Res. Note FPL-0169.

6. Schaffer, E. L.
    1965. An approach to the mathematical prediction of temperature rise within a semi-infinite wood slab subjected to high-temperature conditions. Pyrodynamica 2:117-132.

7. Schaffer, E. L.
    1967. Charring rate of selected woods--transverse to grain. USDA, For. Serv. Res. Pap. FPL 69.

8. Schaffer, E. L.
    1968. A simple test for adhesive behavior in wood sections exposed to fire. USDA, For. Serv. Res. Note FPL-0175. For. Prod. Lab., Madison, Wis.

9. U.S. Forest Products Laboratory.
    1974. Wood Handbook: Wood as an engineering material. USDA, Agric. Handb. No. 72.

2.8   Insects, Fungi, and Other Organisms That Attack Wood

2.8.1   Introduction

Most wood degrading organisms require oxygen, a favorable temperature, moisture, and food. Elimination of one essential component from the wood environment protects wood from biodeterioration. Oxygen is difficult to eliminate from wood used above ground; but air-tight wraps have been successfully used to protect marine pilings from borers. Temperature also is difficult to control; however, lethal heat treatments provide temporary sterilization. Moisture is the environmental variable that is most easily controlled by structural design, i.e., through designs that keep wood dry. Where moisture cannot be controlled, biocidal treatments are needed to protect wood. Because moisture control is a fundamental concern in the design of durable wood structures, this section will emphasize the moisture requirements of wood-destroying pests.

Bacteria and decay fungi require free water to induce biochemical deterioration of wood tissues. The animal pests, depending on their type, have moisture requirements that vary from saturation to air dry wood; and they destroy wood by mechanically biting off bits of tissue.

2.8.2   Bacteria

Bacteria are mostly single-celled organisms, passively dispersed within the environment. They grow within or upon a food substrate. Bacteria require free water to grow and are common in saturated, water-soaked, or moist wood.

Wood in cooling towers, logs stored under continuous sprays of water or floated in rivers, piles under buildings and bridges, and wood used or maintained elsewhere under conditions of water saturation are subject

---

*Authors: Rodney C. DeGroot, Plant Pathologist, and Glenn R. Esenther, Entomologist, Forest Products Laboratory, Madison, Wisconsin.

to bacterial attack (7). Sapwood of all wood species is susceptible to attack. Heartwood is generally not affected in short-term exposures, and in long-term exposures of decades or centuries is less affected than is sapwood.

The initial consequence of bacterial infestation of wood is a marked increase in permeability, a slight decrease in toughness, but no loss in specific gravity. Degradation of the secondary cell wall usually does not occur with short-term bacterial attack. Pulpwood bolts, for example, can be stored up to 1 year under water sprays without loss of specific gravity (5).

Long-term bacterial action may cause important alteration of strength properties. A significant reduction in bending strength, stiffness, and specific gravity occurred in untreated red pine (Pinus resinosa Ait.) piles during 85 years of use in the Milwaukee River (2). Lumber cut from those piles could not be recommended for structural use, according to published allowable design stresses for the species. Similarly, significant bacterial attack in cell walls of untreated piles of Scots pine (Pinus sylvestris L.) and Norway spruce (Picea abies (L.) Karst) has also been observed under a 75-year-old building in Stockholm. The compressive strength, bending strength, and modulus of elasticity (MOE) of the sapwood were distinctly reduced. Strength properties of heartwood were affected, but less severely than those of the sapwood (4).

2.8.3 Fungi

2.8.3.1 Wood-Decay Fungi

Mushrooms, brackets and other fleshy to "woody" fruiting bodies that grow on wood exposed to excessive wetting are commonly-recognized evidences of fungi. But it is the filamentous, vegetative portion of these fungi, growing within wood, that causes rot or decay. Individual strands (hyphae) that grow into and through wood cells induce a biochemical attack within individual cells of wood tissue.

Wood-decay fungi are customarily grouped into three categories: White rots, which tend to bleach the affected wood; brown rots, which produce a brown, crumbling type of decay; and soft rots, a term originally coined to describe a progressive softening of the surfaces of wood.

For typical decay fungi to grow, at least a film of moisture has to be present on the surface of wood cell walls. For practical working limits, it is generally assumed that wood with a moisture content below 20 percent is safe from decay and that wood with a moisture content above 30 percent will be susceptible to decay unless otherwise protected. The upper limit of wood moisture content that permits decay is dependent upon the specific gravity and relative volume of void space within the wood. As the cell lumens become filled with water, aeration decreases and becomes the limiting factor for microbial growth.

Soft-rot fungi have a greater tolerance for low oxygen levels than do the brown- and white-rot fungi. This partly explains the prevalence of soft-rot decay fungi to the virtual exclusion of other types of decay fungi (6) in saturated woods of cooling towers.

The soil constitutes a major potential source of moisture, and the decay hazard is always great for untreated wood in contact with the ground. This is one of the reasons that building codes and standards specify minimum distances between untreated wood products and ground level.

At any given location, the severity of decay hazard for above-ground components is directly related to their exposure to water. Concentrated flowages of rain over the roof edge, against and down vertical walls, against supporting structural framework, and splashing or flowing against the bases of columns or walls contribute to a significant decay hazard in the structural components they wet. Equally important are unrepaired plumbing leaks, unchecked escapes of water from poorly sealed showers, and condensate in walls or under floors of houses that lack adequate protection against water vapor movement.

The moisture content of exterior woodwork that doesn't receive an extraordinary exposure to a concentrated flowage of water is determined by rain-wetting, seepage, and the rate at which drying occurs.

Seepage can affect the potentials for decay at the ends of wood members. Water is absorbed by wood more rapidly and to a greater depth through transverse surfaces than it is through either the radial or tangential surfaces. This moisture will contribute to decay. Seepage-associated decay is a potential problem in joints of exterior millwork, docks, porches, steps, and fences unless appropriate procedures are taken.

Symptoms of rain seepage in exterior woodwork are paint failure that appears first at joints, buckling of siding, nail pulling caused by the swelling and shrinkage of intermittently wetted wood, and finally fungus fruiting bodies where wetting has continued sufficiently long for decay to occur (11).

In contact with the soil, both hardwoods and softwoods are attacked by brown- and white-rot fungi. Hardwood species are more readily attacked by soft-rot fungi than are softwoods.

In wood used above ground, white-rot fungi are associated more frequently with hardwood than with softwood lumber and products. Brown-rot fungi predominate in softwood used above ground. Hence, brown rot is the predominant type of decay in above-ground construction.

Brown-rot fungi cause a more rapid loss of mechanical properties per unit of weight loss due to decay than do white-rot fungi (Table 2.8.1). This is because cellulose-degrading mechanisms of brown-rot fungi penetrate wood cell walls and rapidly depolymerize the bulk of the cellulose molecules in wood cell walls faster than the degraded cellulose can be metabolized. In contrast, the cellulolytic enzymes of white-rot fungi appear to be restricted to wood cell wall surfaces; hence, unconsumed molecules in cell walls are substantially intact, so that the

Table 2.8.1.--Reduction in strength properties of southern pine sapwood by a brown-rot fungus, Lenzites saepiaria, and a white-rot fungus, Peniophora gigantea[1]

Property	Loblolly pine		Slash pine	
	L. saepiaria (brown rot)	P. gigantea (white rot)	L. saepiaria (brown rot)	P. gigantea (white rot)
	- - - - - - - - - - - - - Pct - - - - - - - - - - - - -			
Static bending tests				
Specific gravity	6	8	7	7
Fiber stress at proportional limit	49	17	60	17
Modulus of rupture	46	15	57	12
Modulus of elasticity	14	8	33	7
Work to proportional limit	69	26	74	25
Work to maximum load	57	13	73	16
Compression parallel to grain				
Specific gravity	5	9	6	7
Maximum crushing strength	18	15	22	11

[1] Richards, C. A., and M. S. Chidester. 1940. The effect of Peniophora gigantea and Schizophyllum commune on strength of southern yellow-pine sapwood. Proc. AWPA 36:24-32.

loss in mechanical properties due to white rot is nearly proportional to the percent weight loss.

With brown rot in some woods, modulus of rupture (MOR) and work to maximum load in static bending are often drastically reduced even though little or no weight loss is observed. In practical terms, presence of a brown, crumbling type of decay is visible evidence that mechanical properties of the affected wood have been seriously reduced. As indicators of incipient decay by the white-rot fungus, Polyporus versicolor, work to maximum load and total work beyond maximum load are outstanding. For indicating the amount of decay, MOR and maximum crushing strength appear to be most dependable (8).

2.8.3.2 Stain Fungi

Stain fungi produce bluish-black to steel-gray or brownish discolorations in sapwood of recently cut trees and lumber and in unprotected sapwood of rewetted, previously dried wood. Dark-colored hyphae of stain fungi initially colonize the ray tissues of softwoods, but in the early stages of infection in hardwoods, hyphae are more abundant in the vessels and fibers without much growth in the rays, except where the stain infection is severe. Stain fungi attack sapwood soon after it is cut and may continue to grow in softwoods until the wood moisture content drops below 24 percent and in hardwoods until the wood moisture content drops below 55 percent.

Fully developed blue stain in naturally infected sapwood of southern pine may reduce specific gravity by 1 to 2 percent, strength in compression parallel to the grain and MOR from 1 to 5 percent, and toughness by 15 to 30 percent. In laboratory studies, sapstain in hardwoods had no effect on specific gravity, but reduced work to maximum load by 6 to 15 percent and reduced MOR usually by less than 5 percent.

Stain fungi also increase the permeability of wood. Fungus-stained wood may also have incipient decay. Kiln drying will kill fungi in wood, but the permeability of stained wood remains and, when exposed to rain, it will absorb more water than will unstained wood. This can promote decay. In the Gulf States of the United States, decay has been observed to develop soon after construction in southern pine siding suspected as being stained or having incipient decay at time of installation. In contrast, siding produced from kiln-dried wood had essentially no decay after 5 to 10 years of service on comparable buildings (9,10).

2.8.3.3 Mold Fungi

Whereas stain fungi produce discoloration as a result of dark-colored hyphae in wood, mold fungi discolor the surface of wood with an abundance of colored vegetative growth or spores. Mold fungi can grow on air-dry wood exposed in environments with continuously high relative humidity (3).

The ability of mold fungi to penetrate wood varies with species, but generally the potential for molds to attack the cellulosic component in wood increases as the ratio of surface area to volume of individual wood units increases. Thus, molds are important deteriogens in fiberboards and they can attack particleboards, but their presence on lumber or veneer has usually been regarded as merely a blemish to the aesthetic appearance of wood.

2.8.4 Insects

2.8.4.1 Termites

Termites are by far the most economically important wood-destroying insects. Depending on their type, termites vary in moisture requirements from very high to very low. The dampwood termites of the western

United States require damp wood for successful colonization. Subterranean termites also have a high moisture requirement, but as long as they have access to a source of water, they can attack very dry wood. Normally, the ground is the source of their water; however, the Formosan termite is notorious in Hawaii for its ability to use aboveground water sources without any ground contact by the colony. The range of moisture tolerance by drywood termites varies greatly depending on the species. They are able to fulfill much of their moisture requirement with metabolic water, but the tolerance to low moisture conditions is related to the degree to which they have evolved a shell impervious to loss of water by evaporation.

Evidence of attack by termites is oftentimes first noticed as flights of winged adults swarming from an established colony. Earthen shelter tubes on surfaces, blisters or dark stains in flooring, or ruptured, paper-thin surface coverings over termite galleries in wood might also be detected. Subterranean termites feed principally upon earlywood, thereby forming irregular tunnels confined within annual rings. These cavities have mud flecked surfaces and may be packed with soil.

2.8.4.2 Beetles, Ants, and Bees

Some insects require high moisture content in wood to initiate attack. The large wood borers (beetles and wood wasps) and bostrychid powder-post beetles begin attacks on dying trees, green logs, or wet wood. Normally they require the presence of bark to begin an attack. The larvae of several species of large wood borers can survive after the wood air dries (but kiln drying is normally lethal) and they can complete development into adult insects in dimension stock cut from infected logs. Emergence of adults terminates the infestation, usually within 1 or 2 years (1).

The "old house borer" (a large wood-boring beetle) is an exception to the typical large wood borer. It can initiate infestations of finished wood products in humid environments, including the reinfestation of such wood. The larvae and emerging adults of large wood boring beetles produce tunnels that are elliptical in cross section.

Powder-post beetles are so named for the way in which larvae reduce wood to a mass of floury sawdust. The anobiid and lyctid families of powder-post beetles infest dry wood. In fact, wood does not become susceptible to anobiid or lyctid powder-post beetle attack until it dries to near 30 percent MC; they can tolerate moisture contents below 10 percent. Different species of anobiids attack hardwoods and softwoods. Anobiids are a major problem in European buildings that do not have central heating. In the United States, their infestations are similarly associated with higher moisture content of wood in poorly heated basements and unheated crawl spaces. In the humid southern States, anobiids are becoming more prevalent. The anobiids benefit nutritionally from fungally decayed wood which indicates that they favor wood whose ambient moisture content rises above 20 percent. The lyctids attack only the nutrient-rich sapwood of large-pored hardwoods. Initial lyctid infestations in buildings are associated primarily with the usage of preinfested wood such as flooring, molding, or veneers.

Damage to wood by the large wood borers or by powder-post beetles can usually be detected by the presence of emergence holes. Damage by powder-post beetles can be identified by presence of powder within their cavities. Emergence holes, alone, cannot be used to identify the insects which were, or may still be present. Thus, if evidence of large wood borers or of powder-post beetles is detected, a specialist should be consulted for identification of the causal insects.

Carpenter ants select, as nesting sites, naturally soft wood or wood softened by fungal decay. Because immature ants are susceptible to dessication, the wood also must be moist or in a high-humidity location. To form nests, ants excavate large cavities that have clean interior surfaces and cross several annual rings. Cavities formed by subterranean termites, by contrast, usually have mud-flecked walls, and commonly occur within individual annual rings, i.e., in the springwood, but not in the more dense summerwood. Ants do not eat wood; therefore, they are not deterred by distasteful chemicals such as occur naturally in some softwoods, notably redwood.

Carpenter bees that burrow into structural timbers are large, robust insects that resemble bumble bees. They produce elongate, usually unbranched galleries that average 10 to 15 cm in length, but sometimes may be 2 to 3 m in length. These galleries will oftentimes show evidence of being subdivided into a series of chambers, in which individual offspring develop.

2.8.5   Marine Borers

Wood exposed in marine waters may be attacked by small, invertebrate animals known collectively as marine borers. Wood attacked by _Limnoria_ or "gribble" (crustaceans) acquire the familiar "hourglass" shape in the intertidal zone. The _Limnoria_ produce a lacework of burrows just beneath the surface of the wood. The action of waves in the intertidal zone removes the thin covering of the burrows as the animals go progressively deeper into the wood. Teredo (borers within the phylum Mollusca) burrow deep within infested wood and may completely riddle wood within 6 months with little visible evidence of attack on the surface of the wood. Hence, physical tests of suspect construction need to be made to ascertain the extent of attack by these marine mollusks known as "woodworms." Marine borers cause their most severe damage in tropical waters.

2.8.6   Miscellaneous Animals

Rodents and woodpeckers will damage wood in their efforts to reach food or to establish a nesting site. This is particularly annoying when woodpeckers select wood siding of buildings for their destructive activity. If the siding is insect-infested, elimination of the insects should stop feeding by woodpeckers on the siding. Nesting activities by both woodpeckers and rodents, however, are more difficult to control. Physical barriers may be required to prevent nesting activity.

SERVICEABILITY FACTORS

2.8.7 Measures to Provide Protection Against Attack

2.8.7.1 Wood Preservatives

Treatment of wood with preservative chemicals contributes to resistance against a broad spectrum of bacteria, fungi, and insects. The degree of protection rendered is dependent upon the treating process, level of treatment, and the chemicals employed. Treatment processes provide either a superficial or an indepth protection to wood.

Superficial protection of wood with water-repellent preservatives is achieved with short dips of precut lumber in treating solutions, and inplace, brush-on application at the construction site. Such treatments do not provide protection for wood used in contact with the ground or in other uses with equally severe decay hazard. They are very effective in preventing decay above ground associated with end-grain absorption of water. Millwork, for example, commonly receives a factory treatment with water-repellent preservatives.

Deeper protection of wood products is provided by pressure treatment with a wood preservative. Those preservative treatments that have been proven to provide adequate protection of wood from biological attack when exposed under adverse conditions have been incorporated into various governmental and industrial standards--cf. Fed. Spec. TT-W-571 or American Wood-Preservers' Association Standards.

2.8.7.2 Site Sanitation and Insecticide Treatments

Site sanitation and soil treatments are the two principal means presently available for protecting residential structures against subterranean termites. Wood debris and other cellulosic materials buried at the building site provide the foodstuffs which termites require for the establishment of large colonies. Removal of such materials from around the house is a recognized "first step" in protection against subterranean termites.

Soil treatments with persistent chlorinated hydrocarbons have been remarkably effective for controlling subterranean termites. Some chlorinated hydrocarbons, when used as soil treatments, have remained effective for over 25 years. To be effective, the insecticide must be thoroughly distributed throughout a continuous layer of soil under and around the foundation. Where soil treatments are needed, materials used for backfill should be free of wood scraps, paper, other cellulosic materials that could be used as food by termites. It is also desirable that backfill be free of hard substances such as bricks, surplus concrete, or rocks which would interrupt even distribution of the soil treatment chemical around the foundation. Pressure-treated wood is resistant to termite attack, but does not deter attack to other untreated parts of structures, because termites can build shelter tubes over treated wood.

2.8.7.3 Designs

Design details that protect wood above ground from decay and termites either shelter wood from exposure to water or employ decay-resistant woods where high moisture contents are anticipated. Because of the prevalence of brown-rot type of decay in above-ground construction, particular attention should be given to elimination of potential water entrapment and end-grain absorption of water by wood at intersurface (wood-wood, wood-metal, and wood-concrete) contacts. In architecture, roof overhangs should be designed to protect critical areas, and buildings should be oriented to provide maximum protection to exposed wood members. Exposed timbers should be preservative treated. The special design criterion to alleviate termite problems is the elimination of hidden access routes. For example, the use of a 4-inch-thick reinforced concrete cap on top of a hollow block foundation wall forces termites that may be within the wall to emerge and build their shelter tubes over the exterior of the solid cap where they can be easily detected. Care also is required to avoid hidden access routes behind attached steps or flower planters. The presence of visible shelter tubes indicates that remedial treatments are needed to suppress an active infestation of termites.

2.8.8 References for Section 2.8

1. Baker, W. L.
   1972. Eastern Forest Insects. U.S. Dep. Agric., Misc. Publ. 1175. Washington, D.C., p. 642.

2. Bendtsen, B. A.
   1974. Bending strength and stiffness of bridge piles after 85 years in the Milwaukee River. U.S. Dep. Agric. For. Serv. Res., Note FPL-0229. For. Prod. Lab., Madison, Wis., 5 p.

3. Block, S. S.
   1953. Humidity requirements for mold growth. Appl. Micro. 1:287-293.

4. Boutelje, J. B., and A. F. Bravery.
   1968. Observations on the bacterial attack of piles supporting a Stockholm building. J. Inst. Wood Sci. 4(2):47-57.

5. Djerf, A. C., and D. A. Volkman.
   1969. Experiences with water spray wood storage. Tappi 52:1861-1864.

6. Duncan, C. G.
   1961. Relative aeration requirements by soft rot and basidiomycete wood-destroying fungi. For. Prod. Lab. Report No. 2218. U.S. Dep. Agric. For. Serv., For. Prod. Lab., Madison, Wis., 6 p.

7. Rossell, S. E., E.G.M. Abbot, and J. F. Levy.
    1973. Bacteria and wood. A review of the literature relating to the presence, action and interaction of bacteria in wood. J. Inst. Wood Sci. 6(2):28-35.

8. Scheffer, T. C.
    1936. Progressive effects of Polyporus versicolor on the physical and chemical properties of red gum sapwood. U.S. Dep. Agric. Tech. Bull. 527. Washington, D.C. 46 p.

9. Verrall, A. F.
    1951. Permanence for wood siding. For. Path. Spec. Rel. N. 36. U.S. Dep. Agric. Bur. Plant Ind., Soils, and Agric. Eng., Div. For Pathol., Washington, D.C., 7 p.

10. Verrall, A. F.
    1960. Preventing moisture problems in wood siding in the South. For Prod. J. 10(3):148-151.

11. Verrall, A. F.
    1966. Building decay associated with rain seepage. U.S. Dep. Agric. Tech. Bull. 1356. Washington, D.C., 58 p.

CHAPTER 3   INSPECTION*

3.1   Introduction**

Careful and thorough inspection of a structure is a vital first step prior to its repair, modification, or upgrading. The inspection provides knowledge of the structure and its components, the loadings to which it is or has been subjected, the quality of the materials involved and their condition, and the effectiveness of the fastenings and other appurtenances. All must be known in order to evaluate the ability of the structure to perform its intended function, to establish the procedures necessary to restore its original function, or to prepare it for a new function. This chapter provides guidelines for that inspection. It will not provide all the details of inspection techniques and tools, but rather a guide which must be supplemented by reference to specific publications and by the inspector's knowledge and experience.

Establishment of the cause of damage or structural disability plays an important part in developing methods of correction or repair. Sound inspection methods, in turn, can be vital to the establishment of reasons for structural disability.

Inspections may be made for several purposes and the inspection procedures are likely to differ depending upon the purpose. Three of the most common purposes are discussed below.

---

* Chapter coordinator: Alan D. Freas, Former Assistant Director, Forest Products Laboratory, Madison, Wis.

**Authors: A number of people have prepared material for this chapter with respect to inspection of specific types of structures. Obviously, many aspects of inspection are valid regardless of the structural type so that there was, inevitably, some duplication among the sections submitted. To avoid undue repetition, many of the common aspects are presented in sections 3.2 and 3.3, much of which was adapted from the original material. Removal of such material from the original manuscripts necessitated adaptation of the remainder for inclusion in section 3.4. This is the reason for the wording of the author identification in section 3.4. Every effort has been made to obtain agreement between the original and the final wording. In some cases, the subsection has been prepared from general information or appropriate literature; in such cases, no author has been shown. The chapter coordinator has been responsible for the adaptation and presentation of material in this chapter.

Routine preventive maintenance

The structure should be examined for obvious problems such as fractured members, obvious distress short of fracture, and loadings in excess of those provided for in design.

Dimensional changes resulting from moisture changes combined with vibration and other factors can cause a general loosening of bolted connections. This can become a serious matter if significant reduction of bearing area or disengagement of connectors occurs. Bolts should be tightened when members approach moisture equilibrium at the end of 4 or 5 months of operating conditions and at approximately 5-year intervals thereafter.

All members should be inspected for decay or insect infestation, with special attention to areas where there is evidence of high moisture content. Checks, splits and delamination should be examined to determine whether the openings are beyond the limits imposed by the lumber grade; whether there is evidence, in laminated structures, that the glue bond has deteriorated; or whether the openings have permitted entrance of sufficient moisture to result in interior decay.

Evaluation of revised loadings or usage

Original architectural and structural documents (section 3.2.1) should be located and considered in the evaluation. Requirements of the code under which the structure was designed and built should also be considered in the evaluation. Not only will it provide the requirements for the original structure, but it will, in addition, provide information on the extent of upgrading required. In some instances, remodeling may require bringing the structure into full conformance with the current code; in others, the requirements of the existing code may need to be met only if the cost of remodeling exceeds a certain percentage of the value of the structure (section 1.6). Under some codes, only some parts of the structure must be brought as close as possible to the current code.

Special attention must be given to current or planned loadings and to their points of application in comparison with the design assumptions. Member sizes and grades must be established to determine their suitability under the new conditions of use and loading.

Evaluation of distressed structure

Inspection of a distressed structure should determine first whether the distress posses an immediate threat to life or property. This is necessarily an engineering judgment for which no general guidelines can be established. The following points must be considered and appropriate decisions made promptly:

--Is evacuation of the structure necessary?

--Are shoring or other emergency procedures justified?

--If shoring is required, may the structure be reoccupied upon completion of the shoring?

--Can repairs be completed without disruption of normal use?

--Is the presumed distress structurally significant?

The second step is to provide protection of property and to establish security against vandalism or removal of evidence relating to the cause of the distress.

After these steps have been completed, determination of the cause of the distress may begin. Potential causes are discussed in detail in succeeding sections, but among the more important sources of distress are:

--Overloading, whether due to snow, ponding, or unplanned addition of equipment such as hoists, conveyors, and the like.

--Loads applied to the structure at points not planned for.

--Splits, delamination, or other openings, particularly at connections.

--Fractures, distortions, or misalignment of members.

--Failure or distress in connections.

--Deterioration of materials.

3.2 Where to Look and What to look For

3.2.1 Documents on Design and Construction

The use of a structure may change one or more times over a period of years. A warehouse may have been converted to an office building; a barn may have been converted to a store; a home may have been converted to offices. Each use imposes its own special requirements on the structure. Thus changes in use may have created problems for the structural frame as a result of changes in loading or in the environment. In addition to changes in requirements as a result of changes in use, the structure may have been subjected to an earthquake, a hurricane or tornado, or other extreme condition which may have caused distress or partial or complete failure. Whatever the circumstances, historical documents relating to the use or uses of the structure, design documents, specifications and drawings reflecting both the design and the as-built conditions will be of great value as a basis for analysis. Records of structural modifications, if any, should be examined. Sadly, many such documents have a way of disappearing, but, if available, they can be extremely helpful.

Design calculations, if available, provide the assumptions and loadings considered in the original design, a means for comparing the loadings actually found with those anticipated by the designer, and a basis for evaluating the possibility that the structure has been overloaded. Conformance to currently accepted design procedures for timber structures should be evaluated. Specifications, if available, should indicate the grades of lumber, plywood, and other wood products used in the structure, the wood moisture content anticipated at the time of construction, and whether preservative, fire-retardant, or other treatments were applied.

Engineering and construction drawings provide a basis for comparing the structure as actually built with the design details. From them it is possible to detect differences between design and construction and, perhaps, clues to the reasons for difficulties.

Memoranda and informal notes in the file may give important clues to changes made in the structure that were not translated into drawings. Memories of old-timers, while perhaps not classifiable as documents, may lead to information about additions, changes, accidents, or other circumstances which can affect the utility of a structure.

3.2.2 The Structure in General

In buildings, sloping floors, cracks in walls, or other evidence of settlement may suggest either foundation failure or local crushing at bearing points such as columns, girders, or bearing walls. Water stains or other discolorations on wood or finishes may indicate roof or wall leaks or defective plumbing. Evidence of excessive water should lead to careful examination for decay as evidenced by fruiting bodies (fig. 3.1), softening of the surface of the wood, or abnormal surface shrinkage (13). Flat or nearly flat roofs should be examined carefully to be sure that clogged drains or scuppers have not been the cause of leaks, and sizes of drains and scuppers should be checked against rainfall records to be sure that they are adequate. Because reroofed areas may indicate points at which a leak had occurred in the past, the interior of a building at such locations should be examined for evidence of damage. Special attention should be paid to enclosed spaces such as attics and crawlspaces where condensation from high humidity can lead to extensive decay. Excessive humidity can be generated in attics by inadequate vapor barriers, by inadequate or closed vents, or by exhausting bathroom, kitchen, or clothes-dryer vents into the space. Excessive humidity in crawlspaces can result from lack of or improper installation of soil covers, from inadequate or improper venting, or from inadequate drainage. Even when adequate and properly installed, vents may become blocked.

Irregularities in the surface of a flat or nearly flat roof which can collect water should be viewed with concern. For example, in a roof without adequate pitch, water may collect in irregularities and increase in depth as the roof deflects until the load from such ponding becomes dangerous or even catastrophic. Evidence of water accumulation or lack of drainage on a flat or nearly flat roof should lead to analysis of the roof system for ponding; analytical methods are available (2,11). Ponding may be a particular problem, with high-intensity storms, where long members are parallel to the outer wall of the building.

140   WOOD STRUCTURES

Figure 3.1--Examples of decay fungus fruiting bodies in a
   growing or fresh condition. They soon darken and
   deteriorate if not eaten by insects or other animals.

The general appearance of the structural frame may give clues to loss
of effectiveness. Examples are misalignment of members or excessive
deflection of trusses or of individual members. Evidence that a structural member or truss component has been replaced may signal the
existence of problems.

Evidence that the structure has been exposed to a fire calls for
examinations to determine the degree of damage and what steps, if any,
have been taken to offset the damage (see sections 2.7, 4.5.4 and 7.4).
Such damage may occur not only in the wooden elements of the structure
but, as well, to attachments and fastenings. Thus hangers, bolts,
split rings, and other fastenings should be examined.

Long-term exposure to high temperature can lead to degradation of the
wood even though the temperatures have not been sufficiently high to
cause a fire (section 2.3). Areas around ovens, boilers, and processing equipment involving high temperatures should be inspected.
Discoloration (browning) of the wood indicates degradation and the

load-carrying capacity of the affected part of the structure should be carefully considered.

Evidence of the use or storage of chemicals (section 2.5) should prompt examination of affected members and connections. While many chemicals have little or no adverse effect on wood, some will reduce strength, particularly resistance to impact. Metal fastenings are subject to corrosion, particularly by chemicals that absorb moisture from the air (7).

Exterior structures are subject to many of the same influences which have been disucssed above for buildings. Because of their exposure to wetting by rain, however, they are very vulnerable to damage from decay and insects. Waterfront structures in sea water are exposed to decay above the water surface (10) and to marine borers below (1,9). Special attention should be paid to closely fitted surfaces where water enters the joint but drying is impeded and slow. Moisture may be a problem even in relatively dry areas. For example, dirt accumulating on the tops of bridge abutments may absorb and hold moisture from dew and occasional rains and subject the bearing ends of bridge girders to decay.

Nailed laminated bridge decks have a tendency to loosen under traffic loads and thus to open the joints between the laminations (4). Water stains on the lower surface of laminated bridge decks and on supporting structure indicate the need for inspection.

Fastenings which enter the structure from surfaces subject to wetting may be suspect. Water may enter around the fastening and follow it into the member and into other members. Deck nails, drift pins, and the like represent areas where water may have entered the structure, causing wet areas within the members which favor internal decay if there is no preservative protection (fig. 3.2).

3.2.3 Individual Members

Individual members should be checked for grade stamps or other evidence of quality to be sure that the grade of the material conforms with drawings and specifications. If there are no drawings or specifications with which to compare, the information is still needed as a basis for analysis of the load-carrying capacity of the members. Lacking grade marks, members should be examined for knots, cross grain, checks and shakes, and other strength-determining factors to determine whether they are cause for concern. It may be desirable to have this examination made by a certified grader from the appropriate rules-writing or grading agency. He will be able to relate each member to a current grade. Knowing the grade and the species or species groups involved, an appropriate design stress can be evolved as a basis for analysis of the load-carrying capacity of the member. Depending upon the nature of the structure, the grader may not be able to examine the ends of members to determine rate of growth or percentage of summerwood if these are grade criteria. However, this information can be obtained by removing cores or plugs.

142   WOOD STRUCTURES

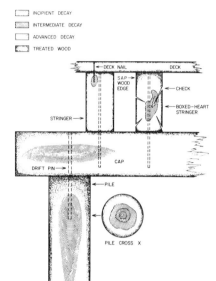

Figure 3.2--Schematic diagram of portion of pile bent, stringers, and deck showing locations where decay may occur.

Seasoning checks are to be expected in solid sawn members of large cross section because of dimensional changes during drying since it is impractical to dry them to service moisture content before installation. If deep, they may seriously reduce shear strength. If completely through the thickness of a member, they may cause excessive deflection. If deep they may extend below the treated shell of a timber and expose the untreated core to decay (fig. 3.3). Checks should always be related to those permitted in a grade because their presence does not necessarily indicate a defective beam. Checks may occur in glued laminated members as well as in solid members, but generally they are not large.

Longitudinal separations at the glue lines in laminated members should be examined carefully to determine whether the separation is in the glue line or in the wood. Failure in the wood should be treated as equivalent to a seasoning check in sawn timbers. Glue failure, indicated by the presence of glue on one or both faces of the opening and the absence of wood failure should be viewed with concern. Extensive delamination will be obvious, but even a small area of glue failure could portend serious delamination in the future. Shear tests of the glue line may be desirable to determine the extent of the problem (14).

Figure 3.3--The rot pocket in this Douglas-fir pole section is associated with a narrow but deep seasoning check.

Beams which are framed into masonry walls should be carefully examined for evidence of decay in the portion extending into the wall. Lack of ventilation in the recess may have led to high moisture contents at the end of the beam. Similarly, columns which are supported on concrete floors at grade or are embedded in concrete may have decayed as moisture entered the wood from the concrete.

Beams or arch legs extending outside the walls of a structure are subject to wetting from rain, roof overflow, or lawn sprinklers. Beam ends should be examined for evidence of decay. Arch legs which are exposed should be examined with special care where they are attached to concrete buttresses or other support and for surface checks that may trap water. Some fittings at the leg-buttress location may collect and hold water so that the end-grain face at the end of the leg absorbs substantial amounts of water and raises the hazard of decay. Older laminated members which have been bonded with casein adhesive (which is not waterproof) should be examined for separations at the glue line.

Damage to a member warrants an effort to determine the type of stress that caused the damage and the fracture, if present. Did the damage occur during shipment or erection? Did it result from misalignment or miscutting during fabrication? Was the member altered (perhaps by mechanical trades) after it was put in place? Did the supports settle? Were loads applied which were not anticipated in design? Did connections loosen or fail?

3.2.4 Connections

Fastenings can loosen if timbers dry in service and they should be tightened as a matter of regular maintenance. Thus during inspections they should be checked to be sure that they are tight. Excessive deformation at a connection--a chord splice, for example--may indicate loose hardware, undersize hardware, omission of shear plates or split rings during assembly, or teeth bent during installation of toothed truss plates in light-frame trusses. In connections involving stresses perpendicular to grain, failure to use stitch bolts or other methods of reinforcement may cause members to split. Splits may occur, as well, where a load-bearing fitting is attached to the supporting member by bolts or lag screws too close to the bottom edge of the member.

Fastenings should be checked for corrosion when they are exposed to wetting or to high humidity, to chemicals or their vapors, or are in contact with wood treated with salt-type fire retardants or water-borne preservatives (see section 2.5 and ref. ($\underline{7}$)). Holes for fastenings can trap water which can lead to internal decay.

3.2.5 Loads on the Structure

Loads on a structure should be checked against those assumed in design. Conveyors, unit heaters, or other appurtenances never considered in the design may have been attached to members, possibly at points other than those assumed in design. For example, appurtenances may be suspended from a lower truss chord at a position between, rather than at, panel points, resulting in unplanned bending stresses in the lower chord and greatly increased or reversed web stresses. In addition, a ceiling may be attached to the lower chord and used for storage, imposing large, unplanned bending stresses in the chord. On occasion, the end bay of the roof between the end wall and the first truss may be framed flat, with the joists supported on the lower chord of the truss. While trusses may be designed for this condition, frequently trusses used in this way were designed for top chord loading only and thus the lower chord is subject to joint and member forces not used in design. It is reported that this situation has been the cause of numerous bowstring truss distress situations.

In structures having flat or minimum-pitch roofs, special attention should be paid to the possibility of ponding which, when combined with a flexible roof system, can lead to catastrophic overloads. Thus special attention should be paid to the slope, to roof irregularities, to blocked drains or scuppers, and to the stiffness of the roof-support system ($\underline{2},\underline{11}$). The roof drainage system, particularly on large buildings, should be checked for the effect of high-intensity storms.

Snow loads on roofs can be nonuniform and cause loadings not anticipated in design. The effect of roof form and adjacent buildings on drifting of snow and the effect of snow removal practices should be examined carefully to determine if nonuniform loading could have caused distress in a structure (sections 2.1 and 4.3).

Additions to a structure warrant careful attention. Addition of conveyors, crane beams and hoists, heaters, or other appurtenances has already been mentioned. More insidious additions include new layers of roofing, unforeseen mechanical units for heating, ventilation, or air-conditioning, or added thicknesses of flooring. Marks indicating that chain hoists or other equipment may once have been attached to the structure should be looked into. In some instances, the addition of equipment may not greatly increase the calculated stress. If, however, the equipment induces vibration in the member to which it is attached (an air-conditioning unit, for example), the number of cycles may become sufficient, over a long period, to raise questions of fatigue; this possibility should be considered.

3.2.6 Harsh Environments

The function of the structure may result in environments harmful to wood and connections. Food storage and preparation areas, for example, may involve high humidities, sometimes at high temperatures if steam is generated during cooking, and will require frequent washdowns to insure cleanliness. Decay can be expected when wood is wetted repeatedly. Long exposure to acidic or alkaline atmospheres generated in chemical processes or during chemical storage may be detrimental to both wood and fastenings. Long exposure to temperatures much above 80°C or perhaps even lower is detrimental to wood strength. Tan, brown, and black discolorations of wood near a heat source indicate progressive stages of reduced strength. In a fire-charred timber, the charred zone has little or no residual strength, but normal-appearing wood beyond the char line may have retained all or nearly all of its original strength because of the insulating qualities of the char (sections 2.7, 4.5.4, and 7.4).

Untreated timbers exposed to wetting, whether by rain, sprinklers, condensation, or other means will decay whenever water remains in contact with wood or whenever water which has entered the wood cannot escape and its moisture content becomes and remains high (above about 20 percent). Typically, decay may be found where seasoning checks or bolt holes permit water to enter, where end grain absorbs water, or where adjoining surfaces trap water etween them. Paint films, glass-fiber coatings, and tight flashing too often become incubation chambers for decay fungi once water penetrates inside them. Flashing or other mechanical barrier to prevent contact of water with wood should be installed with an air space between the wood and the barrier to provide ventilation.

Preservative-treated timbers may decay when openings permit water to penetrate beyond the treated shell. Openings may result from deep seasoning checks, from gouging by pointed tools, from loosened fastenings such as nails or drift pins (fig. 3.2), or when cuts or holes made after treatment are left unprotected. Laminated timbers treated with pentachlorophenol in light solvents or with waterborne preservatives check more in service than do those treated with pentachlorophenol in heavy oils or with creosote.

3.3 Techniques and Equipment

3.3.1 Decay Detection (4,8)

3.3.1.1 Visual

The color of wood may or may not indicate whether it has become decayed. As wood approaches advanced stages of decay it loses luster and may experience notable changes in color. In the early stages, however, the wood may appear unchanged although it may have lost substantial percentages of its strength, particularly in shock resistance (15).

The presence of fruiting bodies (fig. 3.1) indicates that a decay fungus is present in the member where the bodies occur; some fungi produce fruiting bodies after little or moderate decay while others do not produce fruiting bodies until after extensive decay has occurred.

Another visible clue to the presence of decay is the localized depression or sunken faces over decay pockets which extend close to the surface of the member. Termites, carpenter ants, and beetles are associated with decayed wood and evidence of infestations by these insects may be evidence of decay.

A number of conditions provide visual evidence of conditions conducive to decay and areas exhibiting these conditions should be inspected carefully. Evidence of water, such as watermarks, indicate areas which may be or may have been at a high moisture content and thus should be investigated. Rust stains on wood surfaces show possible excessive wetting, particularly if the source of the iron is a wood-penetrating fastener. Appreciable growth of moss or other vegetation on wood surfaces or in checks or cracks is evidence of potentially hazardous wetting. Special attention should be paid to wood adjacent to water-trapping areas such as within joints where end-grain surfaces occur and at interfaces between members.

3.3.1.2 Sounding

In the hands of an experienced person, a hammer is a simple, rapid, and effective tool for detecting extensive internal advanced decay. A sharp ring when a wood member is struck indicates sound wood, whereas a hollow sound or a dull thud may indicate rot. Because shakes and checks can affect the sound, suspicious areas should be drilled or cored to confirm the indication.

3.3.1.3 Drilling

Drilling with a brace and bit or with an electric drill indicates rot by reduced resistance which may be abrupt when voids are present. Wet wood may give a false indication of decay. Discolored chips and fines indicate decay. Shell thickness may be determined by inserting a thin metal rod with a hook on the end into the hole and pulling the rod back until the hook catches on the edge of the rot pocket. This procedure works best in cedar, where transition from sound to rotted heartwood is abrupt, but can be misleading in nondurable heartwoods such as Douglas-fir and pine.

3.3.1.4 Coring

An increment borer (fig. 3.4) or plug cutter may be used to extract cores which can be examined for rot, measured for preservative penetration and shell thickness and cultured for positive evidence of decay fungi.

Figure 3.4--Cores extracted with an increment borer permit detection of rot, as well as measurement of shell thickness and depth of preservative penetration. Cores can be retained and cultured for fungi.

3.3.1.5 Special Devices

The Shigometer measures electrical resistance to detect internal rot. A probe--two twisted insulated wires with the insulation removed near the tip--is inserted to various depths in a hole 3/32 inch in diameter. A marked change in electrical resistance as the probe goes deeper indicates moisture, rot, or another defect. While the device effectively detects rot, it can also give "bad" readings on apparently sound wood.

A device proposed for decay appraisal involves measurement of the resistance to penetration of a needle (3). The usefulness of the device appeared to be limited by its lack of sensitivity to intermediate decay and by the relatively short depth of effective penetration. It was found, however, to be effective in identifying serious decay and in indicating the depth of a shell of sound wood.

The extent of the sound shell is important because it provides a measure of the remaining strength of infected poles. The amount of intact sound shell can be used to decide whether a pole must be replaced or whether it can be retained with periodic inspection.

A sonic testing device, the Pol-Tec (8), utilizes two probes pressed against opposite sides of a pole to detect internal voids which may or may not be decay. Tripping a hammer sends a sound wave down one probe, through the pole, and up the other probe to an indicator dial. A low reading compared with that for a sound pole of similar diameter indicates decay or a void such as a ring shake. The device is said to work well on Douglas-fir and western redcedar but, apparently, not as well on southern pine.

Another ultrasonic device used for detection of internal decay (6) and damage from insects or marine organisms (1) is the James "V" meter. The manufacturer (James Electronics, Inc., 4050 North Rockwell St., Chicago, Illinois 60618) has prepared a number of publications relating to its use. The ultrasonic equipment consists of a transmitter, receiver and instrumentation to measure the transmission time in microseconds (10^{-6} sec). The equipment generates ultrasonic pulses at regular intervals of time and measures the time of flight between transducers. The presence of voids, decay, soft pockets of wood and other discontinuities within the cross-section are readily detected by a higher required sonic transmission time. The test method requires a grid to be drawn on opposite sides of the questionable member and readings taken at the grid intersections (fig. 3.5). Typically, a 4- or 6-inch square grid is used. The device is nondestructive and proven to be effective on Douglas-fir and southern pine. When used in conjunction with an increment boring tool, this equipment can provide valuable information on the internal condition of a wood member.

Indications of internal rot by these devices should be verified by coring to determine the nature of the defect.

INSPECTION 149

Figure 3.5--The "V" meter being used to detect decay in a 12- by 12-inch column.

3.3.1.6 Scrapers

A shovel, scraper with triangular blade, or dull probe can be used to detect below-ground surface rot of poles.

3.3.1.7 Pick Test

When a sliver of wood is lifted by a pointed tool, a splintering break indicates sound wood. A brash break suggests decay.

3.3.2 Moisture Content Determination

Resistance-type moisture meters (fig. 3.6) are very useful for determining whether wood contains insufficient moisture to support decay-- less than about 20 percent. A sliding hammer is used to drive two insulated electrodes with exposed tips into the timber to depths of about 3 inches. By noting the moisture percentage at 1/2-inch intervals into the wood (that is, the moisture gradient), one can estimate the moisture content at depths beyond 3 inches.

Although moisture meter dials are calibrated from 7 to 65 percent, the accuracy of the readings decreases above 20 percent. It should be

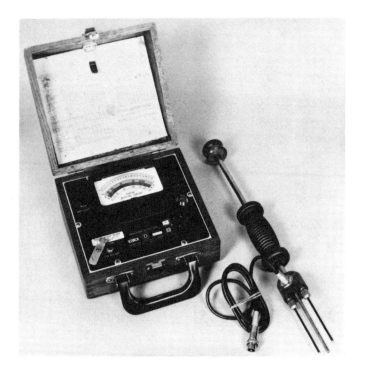

Figure 3.6--Resistance-type moisture meter.

noted that, where members have been treated with waterborne chemicals (preservatives or fire retardants), the salt in the wood will cause erroneous readings. High moisture content may, however, be indicated by corrosion of truss plates or other connections or by the presence of mold or decay.

3.3.3 Glue Joint Soundness

Open joints in laminated members may be probed with feeler gages to estimate their <u>minimum</u> depths. To more accurately determine glue joint soundness, a plug reaching through several laminations and crossing several glue lines may be removed with a plug cutter. Open joints in the plugs may be examined to determine whether there is a high degreee of wood failure, which indicates good adhesion, or little or no wood failure, which indicates poor adhesion. To obtain a quantitative measure of bond quality, shear strength of the glue line can be measured by means of a tool adapted to hold the cylindrical plugs (<u>14</u>).

3.4 Inspection of Specific Structural Types

Many aspects of inspection are common to several types of structures. Many of these have been covered in section 3.2. The discussions in section 3.4, on the other hand, generally deal with aspects related rather specifically to the structural type being considered. In planning for inspection of a particular type of structure, therefore, the applicable items of section 3.2 should be considered together with the points covered in section 3.4. For example, unusual loadings, whether they be on a bridge, a heavy truss, or on a factory floor should be considered carefully with respect to the original design. Likewise, any structure which has been exposed to fire or to high temperature should be examined for the effects of such exposure even though it is not mentioned specifically in connection with the discussion of inspection of a specific structural type.

Consideration should be given also, in planning, to the purpose of the inspection as discussed in the closing paragraphs of section 3.1.

3.4.1 Heavy Timber Buildings*

3.4.1.1 Checking

Many older timber buildings were constructed with large solid timbers for beams, girders, columns, and truss members. Because such timbers characteristically have dried in place, large checks are common. Their location and depth should be measured to determine whether they might have serious effects on load-carrying capacity and whether or not they exceed the limits permitted by current grading rules. In beams, for example, are they severe enough to seriously reduce shear strength? In columns, are they sufficiently severe to cause the column to act as if it were made of two pieces, with sharply increased values of length/least dimension and sharply reduced load-carrying capacity? Laminated members, which are fabricated from seasoned lumber, are not likely to check as severely as are solid sawn timbers. Nonetheless, checking or delamination may occur as a result of unusual or unexpected circumstances. In one instance, serious checking occurred near the ends of laminated beams, apparently as the result of treatment of the glued members with a water-soluble fire-retardant. In another case, water collecting in checks and then freezing caused checks to deepen and lengthen. The cause of unusual checks or other separations such as delaminations should be sought in order to reduce or eliminate the probability of recurrence or extension.

3.4.1.2 Evidence of Water

Watermarking or other evidence of the presence (current or past) of liquid water from condensation, roofing or plumbing leaks or periodic flooding should be sought. Special attention should be paid to areas

---

*Author: Adapted from material furnished by Dick W. Ebeling, Dick W. Ebeling, Inc., Consulting Structural Engineers, Portland, Oregon.

of the structure where drainage is poor or nonexistent, where ventilation is poor, or where condensation can occur (section 3.2.2). Wood members supported in masonry walls (fig. 7.5.10) are especially vulnerable to these conditions. Moisture meter readings should be taken and evidence of decay looked for in such locations.

3.4.1.3 Mechanical Damage

All members near traffic aisles should be checked for damage resulting from impact by forklifts or other mechanical equipment. Damage may range from split-off corners or fractures to columns knocked off their base anchorage and seriously out of plumb. Members in areas adjacent to vibrating machinery may have checks or other damage accentuated and extended by vibrations and, in addition, fastenings may have loosened.

Mechanical damage may have occurred during shipment or erection, from cutting by mechanical trades, and from other causes (section 3.2.3). The damage should be identified and its effects evaluated.

3.4.2 Heavy Trusses*

Older trusses were commonly constructed of solid sawn timber with either the chords or the webs made up of two or more members with the webs and chords occupying separate planes to permit web-chord connections to be made at the point of member overlap. Inspection for bolt tightness is particularly important in such trusses because of the potential for shrinkage and thus for loosening of bolts. Newer trusses may be made of glued-laminated members with chord and web in the same plane so that shrinkage is not so important.

Member damage may occur as a result of improper handling during erection. On occasion, truss members may be removed or partially cut through to accommodate wiring, plumbing, or ductwork. Members should be inspected for this type of damage.

Recent research has shown that allowable tensile stresses used in design for some years were not justified and the current allowables are somewhat lower. Lower truss chords in older structures, therefore, should be examined for tensile fractures which may have resulted from the use of high allowable stresses.

Web members exert force perpendicular to the grain of chords. The chords should be examined at panel points for splits resulting from such forces.

3.4.2.2 Truss Bracing

Trusses are inherently rather unstable when standing alone and must be properly braced to be structurally viable. Erection bracing, commonly X-bracing between trusses, lower-chord struts in line with the X-braces, and top-chord joists or roof sheathing comprise the primary

---

*Author: Adapted from material furnished by Donald W. Neal, Neal Engineering Associates, Consulting Engineers, Portland, Oregon.

bracing elements. Stability of the compression chord is particularly important and should be a prime concern in the inspection, with particular attention given to areas where the roof diaphragm is not in the plane of the compression chord as in the case of a raised skylight or dormer. Unless otherwise braced in such areas, there is a potential for buckling of the compression chord.

3.4.2.3 Truss Connections

Older trusses may contain connections with little or no hardware, such as compressive web members notched into chords, leaving the potential for dislocation. In such connections, in addition, shear area may be inadequate and evidence of excessive deformation as a result of shear failure should be looked for.

Bolted or connectored joints at the areas of overlap between members commonly involve eccentricity when the axes of webs and chords do not intersect at a common point. Web-chord connections and the members entering such connections should be examined for evidence of distress as a result of eccentricity.

Splices in the lower chord and connections at the heel joint are critical since the entire lower-chord tensile force must be taken at these connections. They thus need careful examination to be sure that the bolts are tight, that there is no unusual deformation, and that there is no fracture at bolt holes.

Allowable connector loads (as well as allowable design stresses for wood members) were more or less arbitrarily increased as a material conservation measure during World War II (section 1.4). As a consequence, some connections may be overstressed and show distress. Connections, particularly lower-chord splices and heel joints should be examined for evidence of excessive deformation.

If end distances in bolted and connectored joints are inadequate, the bolts or connectors may cause splits in the ends of members.

If the gage of straps connecting webs to chords is inadequate, the straps may buckle when the web stress is compressive.

Connections should be examined for disengagement of connectors or loss of bearing area because of loose bolts.

There have been reports that connectors were omitted occasionally from joints in wartime structures. Where there is evidence of unusual deformation at a joint, it may be desirable to probe the interfaces between members with a feeler gage or similar device to be sure that connectors are in place. Where shear plates have been used, feeler gages may not be suitable, since the shear plates are set into the faces of the member. In such cases, a magnetic device such as a stud finder may be useful.

3.4.3 Light Trusses*

3.4.3.1 General

Design of light trusses employing metal truss plates as connections between members at a joint is somewhat specialized. It is desirable that efforts be made to contact the engineer who originally designed the trusses or the manufacturer of the truss plates used in the trusses to insure that all background information related to design and use is known. The importance of knowledge of original design assumptions was emphasized in section 3.2.1. Information on standard designs may be obtained from the Truss Plate Institute.

3.4.3.2 Lumber Grade

The grade of lumber in the various members may be determined from grade stamps on the lumber. Lacking a grade stamp, or if there is a question about the accuracy of the grade, each member may be regraded according to the appropriate grading rules. As indicated in section 3.2.3 it may be desirable to have the regrading done by a certified grader.

3.4.3.3 Moisture Condition

Evidence of current or past moisture problems resulting from roof or plumbing leaks, from inadequate vapor barriers, or other source should prompt careful examination of truss members for signs of decay and a check of the moisture content by means of a moisture meter.

3.4.3.4 Connections

Connections between truss members may be of any of a number of types--split rings or shear plates in conjunction with bolts; plywood gusset plates bonded to the members or, perhaps more commonly, metal truss plates. If split rings or shear plates have been used, the tightness of the bolts should be checked, and the joint should be examined to be sure that the connectors are actually present, particularly if there is evidence of unusual deformation in the joint (section 3.4.2). Glued gussets should be checked for integrity of glue bond both in the plywood and in the gusset-member bond; older trusses may have had the gussets bonded to the members with an adhesive which is water-resistant but not waterproof, so that unusually moist conditions may have degraded the bond. Metal truss plates should be checked to be sure that all teeth are effective. All connections involving metal--nails, bolts, or plates--should be carefully checked for corrosion, especially when there is evidence of unusually moist conditions or when the wood is known to have been or thought to have been treated with a salt-based fire retardant or preservative.

---

*Author: Adapted from material furnished by Prof. Donald H. Percival, Small Homes Council-Building Research Council, University of Illinois at Urbana-Champaign, Champaign, Ill.

3.4.3.5 Damage

Members and connections should be checked for evidence of mechanical damage during fabrication or erection or in use. Although it is probable that most damage occurs during fabrication and site handling, it is possible that members have been cut to accommodate some unplanned feature such as a fireplace or an attic stairway. Ends of members should be checked for splits which may have occurred in handling but which may also be the result of moisture change.

3.4.3.6 Deformation

Warped or bowed members may cause movement of the truss. This may create a phenomenon called ceiling-floor-partition separation (5) which manifests itself as a separation either at the junction of the ceiling and the partition or at the junction of the partition and the floor in light-frame construction. Such separation may occur as well from other causes: (a) improper construction of walls, floors, partitions, and foundations which can result in differential settlement; (b) use of wet wood in the floor system, which can result in excessive shrinkage; (c) frost heaving or expansive soils which raise the exterior walls; and (d) attachment of bottom chords of trusses to interior partitions.

3.4.4 Glued Laminated Structures

Glued laminated members may be used in any application where solid sawn members might be used. The preceding comments are applicable to either type of material. In addition the integrity of the glued bonds between laminations should be examined. If delamination is found, its extent should be determined as described in section 3.3.3.

3.4.5 Bridges (4)

3.4.5.1 General

Bridges, like other heavy timber structures, should be checked for quality and soundness of members and connections. Because of their exposure to the weather, however, they are especially subject to prolonged exposure to water and thus to the hazard of decay. The members are usually pressure-treated with preservative.

3.4.5.2 Moisture and Decay

The structure should be examined for visual evidence of conditions conducive to decay and for evidence of decay as described in section 3.3.1 and reference (4). Special attention should be paid to nail-laminated decks and to the structures supporting them since such decks commonly leak badly. Areas around fastenings such as deck nails should be carefully examined since loosened fastenings, bolt holes, and drift pins are locations where water may readily enter (fig. 3.2). Members more or less tight against each other, as in laminated decks or in multi-leaved trusses are commonly areas of decay because the water which gets between them is slow to leave. Bearing areas between deck and girders, between girders and pile caps or abutments, or pile caps and piles should be carefully examined. Even in arid or semi-arid areas, there is risk of decay at the ends of girders because of moisture from dew or occasional rain in the dirt which inevitably collects on top of the abutment and partially surrounds the end of the girder.

3.4.5.3 Mechanical Damage

All portions of the bridge, both superstructure and substructure, should be examined for mechanical damage. On highway bridges built with through trusses, for example, all members, but especially the end posts, are subject to damage from impact by vehicles. Pile bents are subject to impact by ice and debris and, depending on the nature of traffic in the river, by boats or ships.

Members are, in addition, subject to damage from poor handling during shipment or erection. Therefore evidence of impact such as abrasions or gouging, misalignment or warping of members, deformed connections, and the like should be watched for as clues to areas of possible damage.

3.4.6 Pole Structures (8)

3.4.6.1 Condition Above Ground

The pole should be examined for general condition, unusual damage (including mechanical damage as from impact by vehicles), and the size and location of seasoning checks. Even narrow checks may penetrate deeply into the poles, exposing untreated heartwood to the hazard of decay (fig. 3.3). Look for round or elliptical holes made by beetles, for mounds of sawdust as well as the carpenter ants that make them, for woodpecker holes, and for evidence of attack by termites. Cedar poles should be examined for surface and shell rot typical of untreated sapwood above the treated butt.

3.4.6.2 Sounding

The pole should be sounded from as high as can be reached to the groundline in order to locate suspicious areas that should be cored or drilled. If ultrasonic equipment is available, it should be used to locate areas of suspected decay.

3.4.6.3 Drilling and Coring

After sounding or other evaluation, the pole should be drilled or cored downward at an angle of about 45 degrees at the groundline or slightly upward above ground so that water cannot collect in the holes. Determine shell thickness and depth of preservative treatment. Poles that sound "good" should be drilled or cored at the groundline or, preferably, about a foot below the groundline near or below the widest check. Then, if the wood is solid, the poles may be rated as "good" until cultures confirm or reject the estimate of condition. If rot is found, drill holes at third or fourth points around the circumference and measure shell thickness, depth of preservative treatment, and pole circumference. This information may be used to determine whether the pole should be replaced, retained in place, reinforced, or treated in place.

Poles that sound suspicious should be drilled or cored in suspicious areas and near the widest check at or below groundline. If the shell is inadequate, schedule the pole for reinforcement or replacement. If the shell is adequate, remove cores (fig. 3.4) at third or fourth points; depending on shell thickness, schedule the pole for replacement, stubbing, supplemental treatment, or reinspection.

3.4.6.4 Digging Inspection

Dig the pole out to a depth of 18 inches in wet climates and deeper in dry climates. Clean the surface of dirt and examine for rot, probing suspicious areas for soft wood. Scrape the surface with a dull tool or shovel to remove all rotten wood.

Drill or core the pole below the largest check to inspect for internal rot. If present, determine shell thickness and preservative penetration. Determine from this information what action should be taken.

3.5   References for Chapter 3

1. Agi, J. J.
    1976. Detecting damage. AE Concepts in Wood Design, Nov.-Dec. American Wood Preservers Institute.

2. American Institute of Timber Constructon.
    1978. Roof slope and drainage for flat or nearly flat roofs. AITC Tech. Note No. 5.

3. Eslyn, Wallace E.
    1968. Utility pole decay--Part I: Appraisal of a device for nondestructive detection of decay. Wood Science and Technology, Vol. 2, p. 128-137.

4. Eslyn, Wallace E., and Joe W. Clark.
    1979. Wood bridges--decay inspection and control. U.S. Dep. Agric., Agric. Handb. No. 557.

5. Forest Products Research Society.
    1979. Proceedings, metal plate wood truss conference. FPRS Proceedings No. P-79-28.

6. Gerhards, Charles C.
    1978. Comparison of two nondestructive instruments for measuring pulse transit time in wood. Wood Science 11(1):13-16.

7. Graham, R. D., M. M. Wilson, and A. Oteng-Amoaka.
    1976. Wood-metal corrosion--an annotated survey. Res. Bull. 21, Forest Research Laboratory, School of Forestry, Oregon State Univ.

8. Graham, R. D., and Guy G. Helsing.
   1979. Wood pole maintenance manual: inspection and supplemental treatment of Douglas-fir and western redcedar poles. Research Bull. 24, Forest Research Laboratory, School of Forestry, Oregon State Univ.

9. Helsing, Guy G.
   1979. Recognizing and controlling marine borers. Bull. SG49, Extension Marine Advisory Program. Oregon State Univ.

10. Helsing, Guy G. and Robert D. Graham.
    1981. Control of wood rot in waterfront structures. Bull. SG 55, Extension Marine Advisory Program, Oregon State Univ.

11. Kuenzi, Edward W. and Billy Bohannan.
    1964. Increases in deflection and stress caused by ponding of water on roofs. For. Prod. J. 14(9):421-424.

12. Lerchen, Frank H., James H. Pielert, and Thomas K. Faison.
    1980. Selected methods for condition assessment of structural, HVAC, plumbing, and electrical systems in existing buildings. NBSIR 80-2171, Center for Building Technology, National Engineering Laboratory, National Bureau of Standards.

13. Scheffer, T. C. and A. F. Verrall.
    1973. Principles for protecting wood buildings from decay. USDA For. Serv. Res. Pap. FPL 190, revised 1979.

14. Selbo, M. L.
    1962. A new method for testing glue joints of laminated timbers in service. For. Prod. J. 12(2):65-67.

15. Wilcox, W. W.
    1978. Review of literature on the effects of early stages of decay on wood strength. Wood and Fiber 9(4):252-257.

CHAPTER 4   EVALUATION*

4.1   General**

As indicated earlier, there are several different events which can bring about the need for an evaluation of the load-carrying capability of a timber structure; these include:

    (a) partial or complete collapse of the structure,
    (b) development of a state of unserviceability of the structure due to excessive deflection, vibration, or cracking,
    (c) changes in the use of the structure,
    (d) changes in the applicable building code, particularly with respect to prescribed loads,
    (e) fire damage,
    (f) reduction in the strength of wood structural members due to damage, modifications, partial failure, deformation, decay, insect attack or settlement of the structure.

The selection of approaches and methods to be used in structural evaluation is somewhat influenced by the particular reason for carrying out the evaluation; however, in general, there are four distinct phases:

    1. inspection of the condition of members and connections, and of the structural environment,
    2. determination of the loads on the structure,
    3. structural analysis to determine the effects of loads on individual members and connections,
    4. assessment of the ability of members and connections to resist the applied loads.

These evaluation steps may be followed by design of reinforcement or by specification of use or occupancy restrictions if the structure proves to be unsafe or unserviceable.

4.2   Inspection**

Although the inspection of timber structures is described in Chapter 3, a few points are worth emphasizing here. Regardless of whether the

---

* Chapter coordinator: K. F. Faherty, Professor of Civil Engineering, Marquette University, Milwaukee, Wisconsin.

**Authors: F. J. Keenan, Manager and A. T. Quaile, Associate, Timber Engineering and Wood Technology Division, Morrison, Hershfield, Burgess and Huggins, Limited, Toronto, Canada.

structure is a building (residential, commercial, agricultural or public), a bridge, a retaining wall, a cooling tower, or any other particular type, there are several specific items which must form a part of the inspection. These items are as follows:

(a) The condition of the wood and connections must be ascertained in order to judge whether they are as sound or as good as new materials.

(b) The quality of the wood and other elements must be determined in a manner that permits assignment of safe or reasonable allowable stresses. For timber, this usually means that the species must be determined and the individual members must be graded in accordance with one of the current visual stress grading rules. The quality of steel, concrete or masonry components may not be so easily determined and physical testing may be required.

(c) The dimensions of the individual elements and of the structure in general are required in order to permit a structural analysis. Where structural drawings are available, the existing layout and dimensions may be checked against these drawings. Where drawings are not available, the inspection must include a survey of all relevant dimensions.

(d) The service environment of the structure must be noted in order that allowable stresses may be properly assigned. Where environments are unusual or severe in terms of temperature, moisture, acidity or proximity to soil, the conditions should be carefully noted along with the type and degree of protection which has been or can be provided for the wood.

(e) Any information related to the magnitude and distribution of loads on the structure should be obtained at the time of inspection. This information must include a listing of the building materials in order to calculate dead loads and it must also include any features which might affect the magnitude or distribution of superimposed loads. For example, either an adjacent structure or forest shelter can influence the snow accumulation on a structure. Similarly, the roughness of terrain in the vicinity of the structure can influence the magnitude of applied wind loads.

A valuable supplement to the inspection work is information which can be obtained from people who know the history and maintenance record of the structure. A description of previous structural problems, repairs and modifications, and a load history of the structure will assist the inspector in locating specific problem areas.

4.3   Determination of Loads*

In preparation for the structural analysis phase of an evaluation, it is necessary to have a complete description of the loads which act or

---

*Authors: F. J. Keenan and A. T. Quaile.

EVALUATION

might reasonably be expected to act upon a structure during its service life. The load types and their sources are described below:

(a) Dead loads include the weights of all permanent materials in or on the structure including such items as framing members, connections, deck, flooring, roofing, walls, mechanical equipment, and electrical systems. Many structures support permanent fixtures which do not fall into any of the above classifications, and these too, must be taken into account. The groundwork for calculation of dead loads is done in the field at the time of inspection and care should be taken to itemize all dead loads.

(b) Loads due to occupancy and use are determined in relation to the intended purpose of the structure. Floor loads may vary greatly depending upon whether the structure is intended to support bedroom furniture, for example, or highway transports. Recommended loads for specific uses can usually be obtained from the applicable local or model building codes (12,17,19,20) or from published standards (1). In some cases, a floor may be required to support specific machinery or equipment in addition to other occupant loads and the weights of these items must be obtained from the user.

(c) Snow loads govern the design of many roof structures in areas other than the southern states. Recommended snow loads for various localities are provided by building officials or by the model codes. These recommended loads are based upon observations of snow depth on the ground. Roof snow loads are usually less than published ground snow loads for any given area because a considerable portion of the accumulation is blown off. This is not the case in some mountain valleys where snow accumulation on a roof can be equal to that on the ground. Roof snow loads can also be much greater than ground snow accumulations when variations in the roof level lead to drifting or sliding. This must be taken into consideration when a roof is multi-leveled or curved, has parapet walls or obstructions, or is adjacent to a higher structure. This non-uniform accumulation is critical in the design and safety of many roof elements. A notable example is the bowstring truss; roofs utilizing this type of construction are highly susceptible to accumulations of snow on one side only, and major collapses have occurred because of this, even though the roofs were safe under the recommended uniformly distributed snow for the area. In heated structures, melt water from warmer roof areas (due to heat loss through the roof) can refreeze on colder parts of the roof or on the eaves; this can cause high ice loads or ice damming leading to water backup.

(d) Wind loads govern the design of many structural elements and much of the annual damage to structures is caused by wind. The prescribed reference wind pressures for any given area are published in the model building codes. For application to a structure, these reference pressures must be adjusted for gust effects and for the shape of the structure under investigation and then used to check the stability of the structure as a whole and to verify that all elements can safely resist the horizontal or uplift pressures. A structure with openings, or one that is not totally enclosed must also be investigated for the effect of internal pressures where they might compound the effects of external pressure. For tall slender structures such as towers, the dynamic effects of vortex shedding give rise to a fluctuating force, acting at right angles to the wind direction, which must be taken into account. For rounded structures such as silos, the pressures can depend upon the roughness of the surface.

(e) Earthquake loads seldom govern the design of timber structures if all other potential loads have been properly taken into account. Nonetheless, all structures should be designed to resist moderate earthquakes without significant damage and to resist major earthquakes without collapse. Seismic zoning maps are published with the model codes showing the regions in which horizontal ground acceleration has a reasonable chance of exceeding a certain value. For simplicity, these zones are numbered from zero to four with the higher numbers representing the more severe conditions.

(f) Rain loads occur on roofs where insufficient allowance has been made for drainage. On flat roofs, care must be taken to provide a cambered or shaped surface which will not form depressions under the dead load. If this is neglected, particularly when unseasoned wood is used, serious ponding loads can occur and these can result in overstress and perhaps failure when added to the other design loads.

(g) Other loads which may act upon structural elements include those due to thermal or hygroscopic expansion, differential settlement of foundations, and pre- or post-tensioning systems.

While thermal movement is relatively small in wood, it can create problems if totally neglected, particularly in very long structures or those in which other more responsive materials form an integral part of the system. The tendency of wood to expand or contract across the grain with changes in moisture content is well known and this must be considered in structural details. Some examples are given in section 4.6.4.

The loads applied to structural elements by foundation movement should not be overlooked in an investigation. Differential settlement can result in direct damage to column base connections, diaphragms, or bracing and can cause secondary damage to other components.

Occasionally, structural components are either installed or reinforced with materials that apply direct loads to the elements. For example, roof trusses may be strengthened by the addition of post-tensioning wires or cables. The member loads imposed by these systems must be taken into account in the structural evaluation.

4.4 Structural Analysis*

In the evaluation of a structure, the structural analysis is used to determine:

(1) what types and magnitudes of forces are applied to the individual elements by the loads on the structures, and

(2) the stresses and deformations produced in the elements by the applied forces.

For example, the snow on a trussed roof will induce bending, compression, and shear forces on a top chord of a truss and tension forces on a bottom chord. The integrity of the structure depends upon the ability of the various members and connections to resist all of these induced forces.

To perform the structural analysis, it is necessary to model the structure in a way that realistically reflects the response of the entire assembly to the applied loads. The model must include an accurate geometric analog and must be able to deflect in proportion to the stiffnesses of the actual components. The choice of a model is not always clear because joints are seldom either perfectly fixed or pinned. Many times the joints in a timber structure are "semi-rigid" and the analyst must consider whether a pin-connected or rigid joint analog is more appropriate in a particular case.

The current availability of high speed computers at relatively low cost has simplified the task of structural analysis and thus enabled a more detailed examination of structures than was possible with hand calculations. In the past, the difficulty of performing detailed analyses of statically indeterminate systems encouraged designers and investigators to make simplifying assumptions. In the case of large timber trusses, the assumption of pin-connected joints has proven to be reasonable, but for smaller plate-connected lumber trusses this assumption is overly conservative. A simplifying assumption which errs on the unsafe side is one in which knee braces (even though present in a structure) are not considered to be part of the system supporting snow and dead loads. A detailed analysis of roof trusses with knee braces indicates that some relatively large forces can be transferred by the knee braces into the columns during deflection of the main roof trusses under gravity loads. This, combined with their intended function of resisting lateral wind or earthquake forces, can cause serious overstress in the braces, columns, or truss connections.

Analysis of wind effects on structures is generally more time consuming and difficult than that for gravity loads and, again, the computer is an invaluable tool.

Regardless of whether the structure is complex or relatively simple, the analysis is a critical step in relating external or applied loads to member stresses and it requires careful consideration by the investigator.

---

*Authors: F. J. Keenan and A. T. Quaile.

The following design parameters are usually calculated in the analysis of timber structures. This list is not exhaustive, but should form a reasonable starting point for any particular structure.

(a) Lumber decking--deflection and bending stress using appropriate assumptions of continuity based on deck-laying patterns, bearing stress, shear stress (necessary only if the decking is laid on edge rather than flat), stress reversal due to wind uplift, localized high wind suctions at edges of a structure, ability of fasteners to resist uplift forces on decking due to wind suction.

(b) Plywood sheathing--everything as for lumber decking plus rolling shear stress in the plane of the plies.

(c) Joists and purlins--bending stress, shear stress, bearing stress, deflection, lateral stability. On a sloping or curved roof, oblique joists or purlins should be analyzed for biaxial bending except where there exists some other means of resisting the component of load in the plane of the roof, as by diaphragm action, by blocking to the eaves, or by tie rods to the ridge beam.

(d) Diaphragms and shear walls--nail forces due to shear and flexure, axial stress in eave and ridge members, racking loads in "web" materials.

(e) Beams and girders--bending stress, shear stress, bearing stress at supports and at points of concentrated applied loads, deflection, additional normal stress due to axial forces applied to beams, lateral stability, stress reversals.

(f) Compression members--slenderness in two planes, axial compression stress, bending due to end moments, eccentric compression forces or transverse bending, possibility of axial stress reversal due to uplift, capacity of lateral supports to provide a minimum of 4 to 5 percent of the axial capacity of the member. The presence of bow or misalignment in existing columns should be taken into account in calculating bending moments.

(g) Connections--check various possible directions of load application. Note that friction should never be considered as a contributor to the capacity of a joint because its contribution cannot be counted on at all times; if drying shrinkage occurs, friction can be greatly diminished. However, in analyzing a collapse, the possible role of friction should be taken into account.

(h) Axial stresses in bracing members.

(i) Foundations--horizontal and vertical reaction forces including possible uplift forces.

Some constructions such as nailed-laminated beams or columns, T-shaped bracing members, and segmented truss chords are not adequately covered by the current editions of model codes and standards. Nonetheless, the codes do permit rational analysis based on well-established principles of mechanics or on generally established theory. Alternate methods include studies on model analogues and evaluation of full-scale structures or prototypes by load test. For some unusual structures, analytical guidance is provided by the "Timber Design and Construction Handbook" (24).

4.5 Estimating Load-Carrying Capacity of Structural Elements

4.5.1 General*

The final step in the evaluation of a structural member is to check its ability to resist the applied loads while maintaining a reasonable factor of safety against failure and to check that deflections are within recommended values. The successful completion of this step for any member assumes that the forces and stresses have been accurately calculated and that allowable stresses and the modulus of elasticity for the member are known. The latter values are usually published in national standards or model codes (11,12,17-20). However, these allowable stresses must be tempered by the inspection observations relating to wood condition and quality, actual dimensions, and the service environment. When the loading history on a given structural member is not known with any accuracy and if the member is used under a new loading condition for which it was not originally intended, 90 percent of the normal design stress values otherwise permitted is recommended. It is also important to know the date the structure was designed and built when trying to assess the allowable design stresses and principles used in the design as both have changed materially over the past 25 years. For example, for a West Coast Douglas Fir Dense Select Structural Beam and Stringer, the allowable stresses given by the West Coast Lumber Inspection Bureau were:

In 1956--

Bending = 2,050 psi
Tension = 2,050 psi
Horizontal shear = 120 psi
Compression parallel to grain = 1,500 psi

In 1977--

Bending = 1,900 psi
Tension = 1,100 psi
Horizontal shear = 85 psi
Compression parallel to grain = 1,300 psi

It must also be noted that lumber and common glulam sizes are generally smaller than they were 10 or 15 years ago. For a 2 x 8 in 1981, the approximate change in cross-sectional properties from those of a 2 x 8 prior to 1971 are: cross-sectional area, 11 percent less; section modulus, 14 percent less; and moment of inertia, 17 percent less.

Surface appearance of wood can be deceptive. Decay may have begun through the end grain and may not be readily apparent. It may be necessary to take small wood cores with an increment borer for visual examination. It is highly recommended that all test holes in members

---

*Authors: K. F. Faherty, Professor of Civil Engineering, Marquette University, Milwaukee, Wisconsin and B. F. Hurlbut, Chairman of Board, HKM Associates, Billings, Montana.

exposed to adverse conditions be plugged with treated wood plugs or dowels so as not to introduce decay at these locations.

A site inspection must be made to determine to what extent the wood remains viable and to determine if only a portion of the member needs to be cut out and replaced or if it will be necessary to replace the entire member. Wooden members showing decay in any form, whether at initial or advanced stages, should be assumed incapable of resisting tension or bending stresses of any magnitude. Wood in its initial stage of decay may be capable of resisting compressive stresses if the decay is made inactive and the moisture content is maintained below 20 percent. It is important that all affected portions of the wood be replaced or that decay be made inactive.

The changes in dimensions of wood members due to changes in the equilibrium moisture content (EMC) may be an important consideration in the analysis of an existing structure and must be considered in structural details. For example, the intended effect of applying steel clamps to timber members can be negated over a few periods of moisture fluctuation. When the wood tries to expand, it is restrained by the clamps, thereby causing a compressive "set" in the wood; when it later contracts, the clamps become loose and ineffective. Similarly, bolted sideplates in which bolts are set far apart across the grain can result in splitting of the wood along the grain as it attempts to contract (see examples in section 4.6.4). Figure 4.5.1 is helpful in determining the probable change in dimensions with changes in the EMC. Figure 4.5.2 shows the relationship between EMC, the ambient temperature and the relative humidity. A moisture meter is a valuable inspection tool.

Some treatments of wood reduce the strength of the wood and its connections (see 7.4.4.3). For solid sawn lumber pressure-impregnated with fire-retardant chemicals, all design values should be reduced by 10 percent. For fastenings installed in this wood, the design values should also be reduced by 10 percent provided the wood has been dried after treatment in accordance with the American Wood-Preservers' Association Standard C20, Structural Lumber: Fire-Retardant Treatment by Pressure Processes, or Standard C27, Plywood: Fire-Retardant Treatment by Pressure Processes, as applicable (10). However, for structural glued laminated timber, the effect on strength is dependent on the species and treatment combinations involved and the effect on strength must be determined for each treatment. The manufacturer of the treatment should be contacted for specific information on adjustments to all recommended design values to account for the effect of fire-retardant treatment.

For wood treated by an approved process and with an approved preservative, no reduction need be made in the design values for solid sawn lumber, glued laminated structural timbers, and their fastenings. However, if the member is pressure-impregnated with a preservative salt to the heavy retentions required for "marine" exposure, the wood tends to become hydrolyzed and the impact load-duration factor should not be applied.

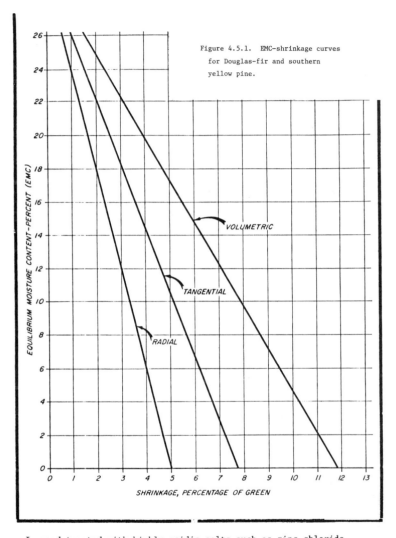

Figure 4.5.1. EMC-shrinkage curves for Douglas-fir and southern yellow pine.

In wood treated with highly acidic salts such as zinc chloride, moisture is the controlling factor in the corrosion of fastenings. Therefore, wood treated with highly acidic salts is not recommended for use under conditions of high humidity.

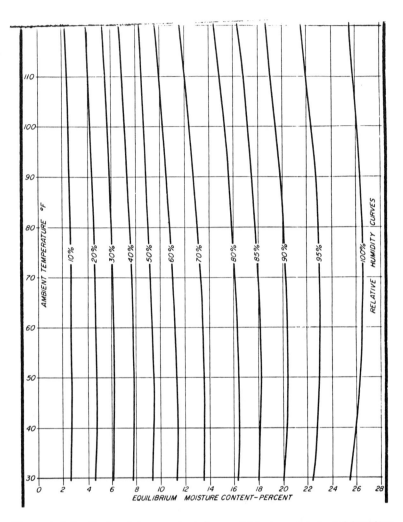

Figure 4.5.2. Wood moisture content for variable temperature and humidity.

4.5.2. Solid Wood Members*

The assignment of allowable stresses is relatively straightforward for nondeteriorated elements of solid sawn wood if the species of wood and grade of lumber are known. The allowable stresses and moduli of elasticity are published in national design standards and model codes (1,11,12,17-20). If grade stamps (2) are not visible and the wood quality cannot be determined, it may be desirable to contact a wood technologist or the regional grading agency for assistance in determining the species and grade of wood.

Checking and splitting of sawn wood members is a problem that has given rise to large expenditures for repair and reinforcement even though the problem may be much less serious than for other forms of deterioration in the member. To evaluate seasoning checks, reference should be made to the depth and location of checks permitted by the appropriate grading rule. Often checks that appear serious to the owner of a structure are well within the permitted limits and no reduction of allowable stresses or repair is necessary.

While seasoning checks penetrate only part of the member thickness, splits extend through the full thickness and penetrate from the member ends toward mid-length. For the effect of splits (or checks) on the horizontal shear strength of members, reference may be made to ASTM Standard D245 (2). Often splits and checks are found to be within acceptable limits for member strength but not for connection strength. The effect on connection strength is more fully discussed in Section 4.6.4.

Splits and checks in wood can exhibit a change in wood color indicating the age of the check. A light color indicates a more recent split or check.

Beams and rafters notched on the bottom (tension side) may have ultimate strengths much less than uncut members of equal size. If the notch is cut from the bottom upward in the member to its mid-depth (fig. 4.5.3) the ultimate strength of the notched beam will be approximately 25 percent of that of an uncut beam. The notch reduces the

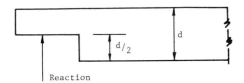

Figure 4.5.3. Notched Beam

---

*Authors: K. F. Faherty and B. F. Hurlbut.

effective shear strength of the member and also produces stress concentrations conducive to splitting.

4.5.3 Glulam Members*

The assignment of allowable stress for glulam members can generally be made from available records that indicate the stress combination of the glulam members and, thus, the original design stresses can readily be established by reference to the applicable laminating standards such as AITC 117-79[1,2] (27). If records are not available it may be necessary to obtain the services of a member of the AITC inspection bureau staff.

Industry practice with respect to the fabrication of axially loaded glulam members is to utilize a constant grade of lamination throughout the entire cross-section. In the fabrication of bending members, in order to achieve maximum utilization of available resources, the industry has used a graded lamination approach. In the zones of high stress (i.e., the outer tension and compression zones) the higher grades of lumber are utilized to resist the high bending stresses. In other areas subject to lesser compression and tension stresses (i.e., the inner compression and tension zones) lower grade laminations are utilized. This results in a graded member with the quality of the lamination used in the various zones being matched with the bending stresses induced by the design loading.

Allowable glulam stresses through the end of the 1960's were based upon the United States Forest Products Laboratory (USFPL) clear wood testing of small clear straight-grained specimens. Full-size glulam member testing by AITC, USFPL, and other cooperators beginning in about 1968 showed that then-allowable tension stresses were too high to be justified. Present allowable tension stresses parallel to the grain are about 40 percent of what was allowable 20 years ago.

Extensive full-scale testing by the laminating industry also showed that most failures in glulam beams are initiated in the tension laminations and that, therefore, these laminations are critical to the strength of a flexural glulam member. Thus, the strength of a bending member is controlled by the quality of the laminations in the outermost tensile zone of the beam. Therefore, special tension lamination grades

---

*Authors: K. F. Faherty and D. W. Neal, Owner, Neal Engineering Associates, Portland, Oregon.

[1]For purposes of this paper all references with respect to glulam stress combinations and stress values will be referenced to the AITC specification. It is noted that the AITC is the national nonprofit technical trade association representing the structural glued laminated timber industry and that AITC is responsible for establishing basic design stresses for all glulam members. These design stresses published by AITC are further referenced in all the major model building codes

[2]For Canadian design standards in glued-laminated timber, the reader is referred to reference (13).

have been developed for use in these highly stressed tension zones. New tension lamination requirements were established in the early 1970's. Any knots (particularly edge knots), notches, daps, holes, etc. which disrupt the stresses in the tension laminations of a flexural member will substantially reduce its load-carrying ability.

If the top and/or bottom laminations of a glulam beam are considered to be destroyed or ineffective in carrying load, then the effect on the load-carrying capacity of the beam is twofold. First, the cross-sectional properties (area, section modulus) are reduced by the change in effective beam depth and, second, the maximum allowable bending stress must be reduced to that of the lower-quality inner laminations.

Little glulam was produced in North America prior to World War II. Some of the early glulam production used backyard techniques with little or no quality control. Nailing the laminations to provide clamping pressure was common and waterproof adhesives were not commonly used.

Prior to the establishment of the AITC in the early 1950's, little record of adhesives used was kept. Most of the glulam made during that period was with adhesives that were not suitable for conditions where the moisture content was high. These adhesives are no longer in use. Prior to 1972, most AITC laminators used a water-resistant casein adhesive and some supplied a phenol-resorcinol waterproof adhesive as an alternate. Since 1972 only waterproof adhesives have been used by AITC laminators. For beams manufactured in Canada by the hot-press method (16), it should be assumed that regardless of the condition at the time of inspection, the beams will eventually become totally delaminated if they are exposed to moisture. Delamination of the adhesive bond was a perennial problem prior to the use of waterproof adhesives.

Wood separation (in tension perpendicular to grain) may be confused with delamination if it occurs near the adhesive line. If an adhesive line separates without fracture of wood fibers, then it may be assumed that the problem is a failure in the glue (delamination). An examination may be difficult if the opening is small; therefore, it may be necessary to remove plugs cut perpendicular to the glue line at the separation. The glue lines can then be more readily examined for cause of failure.

Any wood is weak in tension perpendicular to the grain (cross-grain tension), allowable stresses being as low as 15 psi. If any fiber separations occur from shakes or checks, the cross-grain tensile strength is virtually zero. Cross-grain tension may occur in several types of structures due to particular types of loadings. Three such cases are:

(a) A curved structural member in which the loading tends to increase the radius of the member (straighten it out) causing stress across the grain which is known for this type member as radial tension (see fig. 4.5.4). A load

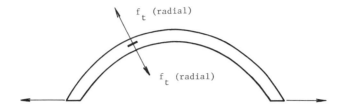

Figure 4.5.4. Curved Member

    which decreases the radius of the curved member causes compression perpendicular to the grain which has a much higher allowable stress.

  (b) A truss chord in which the connection of web members to the chord loads the chord perpendicular to the grain of the chord member. The cross-grain tension results from normal truss action.

  (c) A member that is attached to a second member by a connector loaded perpendicular to the direction of its grain.

Repair of delamination in shallow members, whether the result of failure in the glue line or due to excessive cross-grain tensile stress, is generally accomplished by using mechanical reinforcement. This reinforcement might be continuously threaded lag screws or reinforcing bars embedded in an epoxy. They are placed perpendicular to the wood fibers. For deep members, it may be necessary to allow for cross-grain expansion and contraction. This can be accomplished with through-bolts equipped with spring-loaded washers.

4.5.4 Rehabilitation of Fire-Damaged Members*

The information provided in Chapter 2 on the effect of fire damage to heavy timber members and the information provided in the preceding section on the make-up of glulam members permits the structural designer to apply this knowledge to the rehabilitation of fire-damaged heavy timber members. Attention will first be given to the application of these concepts in assessing residual strength of glulam members. Subsequent discussion will address the problem of relating fire damage to residual strength of solid sawn heavy timber members.

The first decision confronting the evaluator is to estimate the amount of wood required to be removed (by sandblasting or other means).

---

*Author: T. G. Williamson, Vice President, Engineering, Laminated Fabricators, Inc., Indianapolis, Indiana.

EVALUATION 173

Specifically, one must determine how much wood below the char layer should be removed. Realistically, one desires that a minimal amount of wood be removed so as to retain as much of the original section properties as possible. The removal of the char layer plus approximately 1/4 inch or less of wood below the char-wood interface is recommended. The resultant exposed surface will appear to be normal wood.

If the design capacity of the structural member is controlled by compressive strength or stiffness, there is no reduction required to these basic design parameters if at least 1/4 inch of material below the char is removed. That is, wood at this depth has not been exposed to temperatures sufficiently high to degrade its compressive strength or modulus of elasticity.

However, if the strength of the member is governed by its tensile strength, then the determination of amount of wood to be removed below the char does have an influence on the strength of the member. For a depth of approximately 0.1 to 0.5 inch below the char, the residual strength is reduced by approximately 10 percent, while at a depth of approximately 5/8-inch there has been no reduction in tensile strength. Thus, if the char layer and an additional 0.625 inch of wood is removed, the evaluator can utilize 100 percent of the original basic allowable tensile stress in calculations. However, removal of this amount of wood may significantly reduce the cross sectional properties of the member. Therefore, it is generally recommended that approximately 1/4 inch of wood or less below the char-wood interface be removed and that the evaluator use 90 percent of the basic allowable design stress in calculating residual tensile strength of the member. This should adequately compensate for the effect of temperature. For the effect of temperature on wood strength, the reader is referred to Chapter 2.

4.5.4.1 Axially Loaded Members

A straightforward analysis can be accomplished for structural glulam members that are axially loaded, such as a column or a compression or tension chord of a timber truss. Because axially loaded members have primary stresses uniformly distributed across the member, the use of laminations of only a single grade throughout the cross section results in all laminations theoretically being stressed equally. Solid sawn timbers of a specified grade are also assumed to have the same strength properties throughout the cross section of the member.

An illustrative example is given in figure 4.5.5. It is assumed that the column has been damaged by fire and that 1/2 inch of char has resulted. The column is to be rehabilitated by the removal of the char plus an additional 1/4 inch of wood. Investigation of construction documents indicate that the column was fabricated using No. 5 Douglas-fir stress combination in accordance with AITC standard 117-76. As prescribed by these specifications, the column consists of all L-1 grade material with allowable compressive stress of 2,200 psi and a modulus of elasticity of 2,100,000 psi.

174   WOOD STRUCTURES

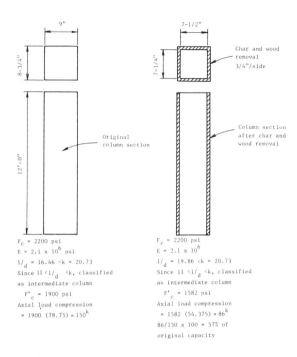

Figure 4.5.5. Evaluation of a Glulam Column Damaged by Fire

The evaluator is cautioned at this point that a reduction in lamination width such as would be caused by the removal of the char could result in a change in grade of the laminations. In grading the original L-1 lamination used in this illustrative example, the allowable knot size would have been based on its ratio to the width of the lamination. Therefore, if the lamination width is subsequently reduced with no corresponding reduction in knot size, the resulting ratio of knot size to lamination width could be such as to require the lamination to be downgraded to a lower-strength category. This would be a difficult interpretation to make in the field as only one face of the member can be seen, whereas, in actual grading practices, it is possible to view all faces of all laminations so that the volume of the cross section displaced by knots can be determined.

As is shown in the illustrative problem, the least width is reduced to 7-1/4 inches and the l/d ratio for this column is 19.86. Reference to the 1977 edition of the NDS (18) shows that this column would be classified as an intermediate column and its allowable compressive stress would be 1,582 psi. Its total allowable axial load is then calculated to be 86 kips. Its original design load would be 150 kips, so that the column capacity would be reduced by 43 percent.

The capacity of the rehabilitated column (86 kips) can then be compared to the actual design loads to determine the feasibility of reusing the column after removal of the fire damage by sandblasting.

A similar analysis can also be applied to a solid sawn timber column of a known species assuming the resulting grade of the member can be determined after the char has been removed.

The residual strength of fire-exposed compression or tension chords of a glulam truss fabricated using single-grade laminated members or a solid timber truss chord is analyzed in the same manner. However, full analysis is more complex because the effect fire has on the connections used in truss fabrication may become the controlling design consideration. The effect fire has on the strength of any connection is most difficult to determine without a thorough investigation of the affected connection, since the amount of damage is dependent on the quantity of metal and the surface contact of metal with fire along with other factors. All that can be suggested at the present time is that a thorough examination be made of all connections with metal fasteners. Damage, both externally and internally, should be determined and engineering judgment be used in evaluating the residual strength of the connection.

4.5.4.2 Glulam Bending Members

Determining the residual strength of a glulam bending member damaged by fire is complicated by several factors. First, as previously discussed, the effect of high temperature induced in the wood below the char depth during the fire exposure results in a loss in residual tensile strength as compared to virtually no loss in compressive strength. Thus the evaluator must consider a reduction in basic design flexural stress based on the effect of heat on the wood below the char zone and thus upon the depth of damaged wood removed (see Chapter 2). Secondly, glulam bending members are fabricated so that the highest quality laminations are placed in the zones of highest bending stress. In other areas of the member, subjected to lesser flexural stresses, lower quality laminations are used. Thirdly, restrictively graded tension laminations are used in bending members (see fig. 4.5.6) and damage to these laminations in a fire severely reduces the allowable design stresses that can be assigned to the members.

As an example illustrating the determination of the residual bending strength of a glulam beam damaged by fire, a beam is assumed having an original cross section of 8-3/4 by 42 inches (fig. 4.5.6) that has charred to approximate depths of 1/2 inch on the bottom face and 1/4 inch on the sides. It is further assumed that the construction records show the beam to be a 24F-3 southern pine combination in accordance with AITC 117-76. Furthermore, assume that it has been determined by the evaluator that he will remove, by sandblasting, 1/2 inch of wood and char from the sides of the beam and 3/4 inch from its soffit face. This results in an effective cross section of 7-3/4 by 41-1/4 inches (fig. 4.5.6). The original section modulus of 2,572 inches cubed is reduced to 2,198 inches cubed, which is about 85 percent of the original section modulus.

176   WOOD STRUCTURES

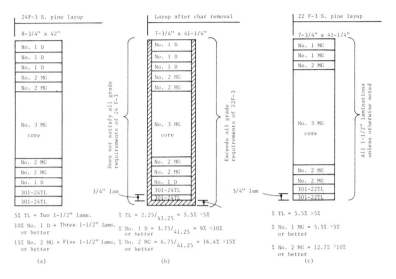

Figure 4.5.6. Evaluation of Glulam Beam Damaged by Fire

The evaluator next needs to determine the allowable design bending stress for the reduced section. Figure 4.5.6(a) shows the original beam layup which is in accordance with the requirements of AITC 117-76. These requirements for the outer tension side are that:

(a) Ten percent or more of material be graded as No. 1 dense southern pine.
(b) In case of a member 42 inches deep, an additional requirement is that 5 percent of the depth of the member be graded as 301-24 tension laminations.
(c) For this combination, it is also required that 15 percent of the depth be No. 2 medium grain (MG) material or better.

The above three requirements are satisfied as shown in figure 4.5.6(a). Also shown is the cross section which results after removal of the char and wood, figure 4.5.6(b).

An evaluation of the cross section which has been reduced to 41-1/4 inches shows that 2-1/4 inches of the tension laminations remain which comprises 5.4 percent of the depth and satisfies the 5 percent requirement for 301-24 tension laminations. Also, 6-3/4 inches of No. 2 MG or better material remains which is 16 percent of the depth and meets the requirement of 15 percent for 24F-3 combination. However, only 9 percent of the depth of the member (3-3/4 in.) is graded as a No. 1 dense or better and thus does not satisfy the 10 percent requirement of the 24F-3 combination. Therefore, the inner tension zone may be slightly overstressed at 2,400 psi and a lesser design stress would need to be used. It is possible, however, that the original 10 percent No. 1 dense or better requirement as given by AITC 117-76 may not have been the controlling criterion for establishing the allowable bending stress of 2,400 psi in the original combination. To further assess the effect of the grade of lamination on strength of glulam member, the designer is referred to ASTM D 3737-79 (5) which provides more detailed information for establishing the allowable stresses.

Figure 4.5.6(c) shows that the grading requirement for a 22F-3 bending stress is justified for the reduced cross section. Thus an allowable stress of 2,200 psi is suitable for the beam after removal of char and wood.

Combining this new allowable stress of 2,200 psi with a further reduction of 10 percent for bending stress (due to the removal of material only 1/4 in. below char) gives a final allowable bending stress for the section of 1,980 psi. And the rehabilitated beam has a residual allowable flexural stress of 82.5 percent and a residual allowable flexural strength of 70.5 percent of that of the original beam.

In addition to analyzing the member for bending stress it is also necessary to check the beam for shear stress and for lateral and vertical deflection. The effective shear area of the reduced cross section is 319.7 square inches which is 13 percent less than the shear area of the original cross section. For vertical deflection, the only significant change in the parameters for deflection calculation is the smaller moment of inertia of the reduced cross section. For the illustrative example, the moment of inertia is reduced from 54,022 to 45,330 inches to the fourth power. Thus the reduced cross section provides approximately 84 percent of the moment of inertia of the original beam to resist deflection. Lateral deflection needs to be considered only if the compression edge of the beam is not supported laterally throughout its length.

Based on the above considerations of bending and shear strengths, and vertical and lateral deflections, the evaluator can determine the acceptability of reusing the member.

Other types of glulam members, such as tudor arches, also involve the use of graded lamination combinations, and these become very complex to analyze with respect to residual strength after fire damage. It is the recommendation of the authors that, if a building evaluator is involved in a rehabilitation project utilizing arch-type members, he contact the American Institute of Timber Construction for guidance in assessing residual strength characteristics in these arch members. While the AITC does not offer consulting services, they would be able to provide guidance to the evaluator in these unique situations.

In addition to the concepts for assessing allowable glulam design stresses as presented here, the evaluator is referred to reference (5). This standard provides the evaluator with more detailed criteria for establishing glulam design stresses assuming the grades of lamination throughout the member are known.

4.5.4.3 Solid Sawn Timbers

In the case of structures utilizing glulam timber framing, records are generally available to indicate the stress combination of the glulam members and, thus, the evaluator can readily establish the original design stresses by reference to the applicable standards. However, in the case of solid sawn timbers, this information is often not available. Thus, determining the original grade and corresponding design stresses of solid sawn members which have been damaged by fire is more difficult, since the char layer masks the solid sawn timber to such a degree that it is not possible to visually determine the grade of timber or species involved.

The evaluator must first determine the species of wood used and, second, must determine an applicable design stress for the original material. The only feasible way at present to determine the grade of solid sawn timber is to remove the char and then visually regrade the member. Thus, a decision would have to be made to salvage the material prior to actually analyzing the load-carrying capacity of the member. After the char has been removed, the evaluator can attempt to determine the species, regrade the material, and estimate the applicable design stresses based on current lumber grading rules.

If it is assumed that the records defining the original grade of solid sawn timbers utilized are available, the designer still is faced with the problem of regrading the members after the removal of fire-damaged material to determine whether the resultant cross section will still satisfy the applicable grading rules. For example, assume that the original grade of solid sawn timber permitted the occurrence of a one-third cross sectional area knot at the center of the cross section. However, upon removing a specific amount of wood by sandblasting, this original maximum size knot could now conceivably occupy more than one-third of the cross section. The member with the reduced cross section would then fall into a lower grade classification and, thus, have lower applicable design stresses. Obviously, the evaluator would also be required to use the reduced section properties after removal of material in assessing residual load-carrying capacities.

It is noted that the primary emphasis of the preceding concept has been on reassessing the structural capabilities of the wood which has been damaged by fire. However, other approaches to the rehabilitation are also available if design calculations indicate that these members with reduced cross section are not capable of carrying design loads. Some other possible approaches are: reinforcing the damaged timbers, modifying the basic plan by introducing additional load-carrying elements, or field testing the distressed timbers to assess the remaining structural capacity.

The evaluator should also remember that a thorough inspection of the connections is necessary if a complete assessment of the residual strength of the structure is to be correctly made. As stated earlier in the section, this can be most difficult.

4.6 Estimating Load-Carrying Capacity of Connections

4.6.1 General*

Many types of fasteners have been used to join wood members, to fasten the wood elements to the foundation, or to support load. The simplest type of fastener over the years has been the common wire nail and spike. This type of fastener has been augmented in recent years by staples, box nails, threaded hardened-steel nails and spikes, and other types of coated or deformed-shank nails. Information on nail sizes, types, and definitions can be obtained in references (4,7,8,22).

The types of fasteners for wood members that have typically been used are: machine bolts, lag bolts, shear plates and split rings, connectors, dowels, toothed and pronged plates, and specialty items such as the old TECO toothed rings, clamping plates, and spike grids. These fasteners typically transfer the load from one member to the other by single or multiple shear which parallels the faces of the joined pieces. Further details on the fasteners can be obtained from the manufacturers' publications or references (18 and 26).

4.6.2 Use Conditions*

The allowable load for any fastener or connector is dependent on the specific gravity of the wood being joined, the moisture content of the wood and its physical dimensions and the arrangement of the fasteners or connectors. Allowable loads are usually based on criteria that establish both a deformation standard (amount of movement between joined pieces) and the ultimate strength of the joint. Values of allowable design loads for fasteners and connectors typically used are published in the NDS (18) and model building codes. Some connectors have allowable loads established by the manufacturers which have been approved by the model codes. Some of these connectors are joist hangers, column caps, mud-sill anchors, metal-plate connectors, etc.

The allowable design value for a fastener is dependent on the moisture content of the wood throughout the history of the joint. Therefore, as a minimum, it is important to try to establish the probable moisture content of the wooden members at the times of construction and inspection. Undesirable moisture contents can reduce allowable loads by 50 percent or more.

Changes in dimension (shrinkage) of timber members because of a change in the equilibrium moisture content (EMC) may result in less effective connections. Bolts and split rings may lose their effectiveness because of wood shrinkage, resulting in a loose fit. Usually it is intended that trusses built up of several layers of wood are to have bolts tightened after being in service for a year or longer. Unfortunately, this is seldom done. Deep members, such as glulam girders, may be connected to unyielding vertical walls or columns. If the top and bottom of the members are fastened by through bolts, wood shrinkage is restrained, which may result in horizontal splits. An uninformed

---

*Authors: K. F. Faherty and B. F. Hurlbut.

inspector may consider this to be a horizonal shear failure which may not be true. Decay or other deterioration in the wood surrounding the fastener or connector will make it incapable of transferring force through the joint.

Deterioration of the fastener or connector may not be easily determined; however, it will be necessary to estimate its condition in order to assign an appropriate strength. Nails may have severely rusted or corroded if the wood has not remained dry.

Other conditions to be aware of are deformation at a connection, such as a sag, twist or rotation, and changes in the alignment of members. Usually some evidence of change in alignment is associated with light and dark wood where the lighter color would indicate a recently exposed wood surface. Splits or checks in the wood can also exhibit a change in wood color indicating either a recent or an old condition. Changes in alignment generally occur in a direction which tends to relieve stress in the joint, although this change may contribute to a greater stress in related structural components. A beam hanging in a saddle on the side of a girder may, by rotation, increase any torsional stress in the girder. A bracket on a column may cause increased eccentricity of the load on the column as a result of rotation. Damage to a connection should be examined for example, for possible impact from a lift truck instead of for excess stress from the type of connection. There has been recent evidence of connection failures, primarily in trusses, which have happened as a result of the method of design which was not in agreement with the method of fabrication. Bowstring trusses have usually been designed by assuming pin connections at all panel points, although the bottom chords may be continuous through two or more panels. Secondary stresses may be induced in the members and connections because of this continuity and vertical movement of the truss joints.

Load sharing in some structures may permit an inadequate connection to apparently function properly, when in reality other parts of the structure have taken over its function and are probably overloaded because of the failure of the connection. For example, a purlin-to-beam connection may have failed, but the deck takes over and actually serves as a hanger to help hold the purlin in place. This, of course, may overstress the deck and cause adjacent members to be overloaded.

4.6.3 Allowable Loads for Fasteners and Connectors*

Allowable loads for nails, spikes, screws, lag bolts, bolts, split rings, and shear plates are published in "National Design Specification for Wood Construction" (NDS) (18). Some of these fasteners have their allowable loads published also in model codes. Basic performance information is given in the "Wood Handbook" (26).

Design values given by NDS are based on seasoned wood which remains dry and a normal duration of load. All design values must be adjusted to account for, as a minimum, the moisture content of the wood and the

---

*Author: K. F. Faherty.

duration of load. Additional adjustments may be necessary depending on the type of fastener used and the adjustments recommended are given in the NDS.

For existing wood structures, it will be necessary to determine the density of the wood and to assess the condition of the joint (wood and fastener) as described in Section 4.6.2. In addition, it is important to determine the actual size of the sawn lumber and glulam members, since current (1981) sizes are generally smaller than those prior to 1970. This may affect the allowable load for the fastener.

4.6.3.1 Nails, Spikes, Screws, and Staples*

Design values are published in the NDS for common wire nails and spikes, threaded hardened-steel nails and spikes, box nails and screws. Design values for most of these fasteners are published also in the model codes. Values given are for both lateral and withdrawal loads. However, it is recommended that these fasteners not be loaded in withdrawal if possible.

In addition to the adjustments mentioned in Section 4.6.3, it may be necessary to adjust allowable values for the type of material used for the side plates and for the amount of fastener penetration into the member holding the point of the fastener.

Design values as published by NDS for common wire nails and spikes and screws have remained essentially unchanged during the past 20 years. In recent years, values have been added for threaded hardened-steel nails and spikes and box nails. NDS makes no distinction between the performance of annularly or helically threaded nails and spikes.

If allowable loads are not available for wire staples they can be closely approximated by treating the staples as pairs of slender nails with shank diameters equal to those of the legs of the staples.

Basic performance information on these fasteners is given in the "Wood Handbook" (26). A critical study of the many theories previously advanced on the behavior of nailed, spiked, and stapled joints was made in "Nailed Joints in Wood Structures" (15). It is pointed out that complete joint evaluation based on a single all-embracing formula is not feasible. Thus, it is not justified to evaluate joint performance singularly on the basic ultimate load-carrying capacity. Furthermore, it is not appropriate to design joints only on the basis of joint-slip limitations. Therefore, both the characteristic ultimate load and the corresponding deformation characteristics need to be given consideration for given conditions.

---

*Authors: K. F. Faherty and E. G. Stern, Earle B. Norris Research Professor Emeritus of Wood Construction, Virginia Polytechnic Institute and State University, Blacksburg, Virginia.

In order to evaluate the performance of nails, spikes, and staples and their joints in existing structures, it may be desirable or even necessary to perform full-size tests in accordance with existing or proposed consensus standards of the American Society for Testing and Materials. The proposed methods of testing fasteners (9) include dimensional tests, hardness tests, tension tests, static and impact bending tests, and coating weight and adherence tests. The performance of fasteners in wood and wood-base materials is determined according to ASTM standard methods of testing mechanical fasteners in wood (3). These methods include tests on the nail, spike, and staple withdrawal and lateral resistance. The performance of nails designed to fasten wood and wood-base materials to concrete and masonry elements can be determined in accordance with an existing standard which is being revised at this time (6).

4.6.3.2 Lag Bolts and Bolts*

A few years ago, the effects of grouping bolts on the joint performance were again given consideration. Specific procedures were prescribed for a reduction in the strength of a joint with more than two fasteners in a row. Test results reported in the late 1960's by Cramer (14) showed that the load a row of fasteners is capable of resisting does not increase in direct proportion with an increase in the number of fasteners in a row. For example, five bolts in a row may have a strength of only 76 percent of that of a single bolt multiplied by five.

Since 1973, a group reduction factor has been included in the NDS code and it applies equally to lag bolts, bolts, and connectors. Previously this effect had been given consideration during the establishment of design procedures (21,23); however, it was not included as a separate parameter in the design procedure. A reduction factor to account for the effect of group action should be included when rating existing groups of lag bolts, bolts, and connectors.

Most of the change in design values for bolts and lag bolts which has occurred over the past 20 years is due to a reduction in the thickness of wood (2-, 3-, and 4-in. nominal sizes) by 1/8 inch. However, there has also been a change in the procedure for calculating the allowable design value for two-member bolted joints (single shear). For example, in a two-member joint in which the members are of equal thickness, the allowable load prior to 1977 was equal to one-half the tabulated design value for a piece twice the thickness of one of the members. In the 1977 NDS (18), the value is to be one-half the tabulated design value for a piece the thickness of one of the members. For a joint of two 1-1/2-inch-thick southern pine (medium grain) members and a single 5/8-inch bolt, the parallel-to-grain design value was 1,000 pounds. The current design value (1977 NDS) is 590 pounds.

Placement of lag bolts and bolts is an important factor when determining the allowable load of an existing structural joint. If minimum and maximum end and edge distances, spacing in row and between rows,

---

*Author: K. F. Faherty.

and net section requirements are not met, then the allowable load should be reduced in accordance with the designer's judgment.

4.6.3.3 Connectors*

The group reduction factor discussed in the section on bolts applies also to split rings and shear plates (connectors). And, as was the case with bolts, a group reduction factor was not incorporated in the NDS code for connectors until 1973; however, it had been given consideration during the establishment of design procedure prior to 1973 (21,23).

Changes which have occurred in the last 20 years in the design values for connectors have been essentially the result of a decrease in thickness of wood members by 1/8 inch. This change in thickness of dimension lumber first appeared in the 1971 NDS code.

In evaluating existing split-ring and shear-plate joints, it is important that the placement of these connectors be determined. Measurements for end distance, edge distances (loaded and unloaded) and spacings in rows and between rows should be made. These measurements are necessary since the allowable connector load can be dependent on them in addition to net section requirements and lumber grade and thickness. A most helpful publication on design values and placement recommendations for connectors can be obtained from the Timber Engineering Company (25).

4.6.4   Examples of Problem-Type Connections**

The following series of connection details, figures 4.6.1 through 4.6.4, illustrate connections which have caused problems and should be looked for in an inspection and avoided in design.

Detail Number 1--An abrupt notch in the end of a wood member creates two problems; one is that the effective shear strength of the member is reduced because of the end notch. The second is that the exposure of end grain in the notch will permit a more rapid migration of moisture in the lower portion of the member and may cause the indicated split.

Detail Number 2--The condition shown is similar to Number 1, but perhaps is not as evident.

Detail Number 3--In this case, clip angles or concealed internal anchors are connected to the upper portion of the beam. Similarly to Detail Number 1, this condition reduces the effective shear strength of the member and enhances the formation of a split.

---

*Author: K. F. Faherty.

**Authors: K. F. Faherty and B. F. Hurlbut.

184   WOOD STRUCTURES

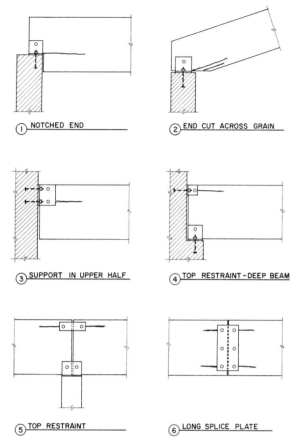

Figure 4.6.1. Connection details 1 through 6.

<u>Detail Number 4</u>--The beam is bearing on a masonry base and is laterally supported at the upper portion by clip angles or similar hardware. With a deep beam, the shrinkage due to drying reduces the depth of the beam and can create a split at the upper connection since the bottom is restrained.

<u>Detail Number 5</u>--A top splice plate is attached to the ends of the beams as shown. If the bolt holes are not slotted to allow movement of the ends of beam at the top, a negative moment can develop. If the connection has not been designed to resist such a moment, a split can occur.

EVALUATION

Detail Number 6--This type of connection has an effect similar to that of Detail Number 4. This may be a splice over a column or the end of one beam being supported by a second beam. As the wood shrinks, the steel side plates resist shrinkage and can, thus, cause the beam to split. This condition can be particularly hazardous if one beam is supported by a second beam, since a split at the bolt holes reduces their effective strengths.

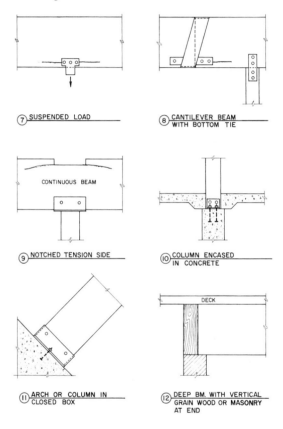

Figure 4.6.2. Connection details 7 through 12.

Detail Number 7--Supporting moderate or heavy loads with through bolts near the loaded edge of a member can result in a split through the wood at the bolt level, even though the bolts have been properly designed for allowable loads perpendicular to grain.

Detail Number 8--Some designers have developed the tension connection on a cantilever saddle system by using bolts in a single line in both members being joined. As the supporting beam shrinks, the saddle wants to move downward. However, the tension bolts in the supporting member restrain it; therefore a split may develop.

Detail Number 9--A cut has been made in the top of a beam over a support to provide space for a hardware connection or for the passage of a conduit. This is serious in any continuous member and particularly so for a glulam member, since the tension laminations are critical in the performance of the structure.

Detail Number 10--Many designers try to conceal the base of a column or an arch by placing concrete around the connection. If the concrete is in contact with the ground, moisture migrates into the lower portion of the wood and causes decay.

Detail Number 11--This detail has an effect similar to that of Detail Number 10. The base of an arch is placed in a closed steel box in which moisture can accumulate and decay will result. This can be avoided by providing weep holes or any other method which permits the drainage of water.

Detail Number 12--In this detail, vertical wood framing or masonry is built in at the end of a relatively deep beam with a deck or other structural roof system over the top. As the beam shrinks it causes a tilting up of the roof at the eave, since the vertical wood or masonry does not shrink. This produces a bowl effect that can seriously impair proper drainage of the roof. In recent years, considerable emphasis has been placed on the ponding effect on roofs. The construction shown in this detail will add to the loading caused by ponding.

Detail Number 13--It was common practice in the past to place, in the end of a beam, a drift pin which extended downward into the top of a column. This provided a clean, concealed connection. However, a drift pin in the end grain of the column does not provide adequate uplift resistance for roofs which are subjected to negative pressures as a result of wind. Furthermore, water tends to follow down the length of the drift pin with the resultant possibility of a pocket of high moisture content and as a result a high probability of decay.

Detail Number 14--A support bracket is attached to the side of a column. As shown, the bolts have been placed close to the loaded edge of both column and bracket. Splitting can result due to excessive tension across the grain, which can cause an abrupt failure.

Detail Number 15--Shrinkage is again an important problem in the detail shown. The girder can shrink, permitting the top surface of the girder to move downward. Shrinkage can occur also in the beam supported by the hanger, thus causing an additional downward movement of the roof surface at this point. A bowl-type effect will result similar to that caused by Detail Number 12. In addition if the beam is placed close to the wall and is of appreciable length, a vertical deflection of the

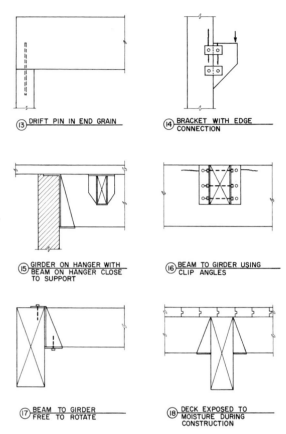

Figure 4.6.3. Connection details 13 through 18.

beam will make it difficult to maintain a level surface for a roof or a floor since the wall will not deflect vertically.

Detail Number 16--A common and simple way to connect a beam to a girder is by the use of clip angles and through-bolts. If the clip angles are long and the beam and/or girder shrinks, the rigid clip angles can cause the wood to split. A better solution is to use a saddle-type hanger since bolts can be placed so that when shrinkage of beam and/or girder takes place, no splits will develop as a result of the hanger and bolts. The saddle-type hanger is also a better solution since it can be connected to the girder prior to erection. Then when the beam is placed, it can be set in the saddle and whatever bolts are required can be installed at a later time.

Detail Number 17--A girder supporting beams on only one of its sides should be carefully analyzed. If no restraint against rotation is provided to the girder, as shown in the drawing, the girder may rotate and it could possibly fail. A horizontal tie between the beam and girder at the bottom of the hanger is recommended.

Detail Number 18--If, during construction, the roof deck above the girder is permitted to become wet, there is a strong possibility that it will swell and thereby force the beams away from the girder. This is an erection problem that the specification writer should be aware of and he/she should write a specification to inform the builder of this possible problem.

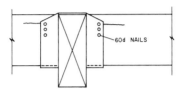

(19) TOP RESTRAINT - NAILS

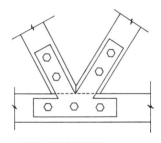

(20) FIXED ANGLE GUSSET

Figure 4.6.4. Connection details 19 through 20.

Detail Number 19--This detail is similar to Detail Number 4 in that a deep beam is restrained at the top and rests on the hanger at the bottom. Since large spikes or nails are used, designers feel that the spikes and nails cannot restrain the top of the beam sufficiently to cause it to split as the beam seasons in place. However, in fact, large spikes can develop forces of sufficient magnitude to cause the beam to split.

Detail Number 20--A connection for a truss joint is fabricated from steel plates. The steel gussets have a fixed angle between truss diagonals as a result of welding. As the truss is loaded and deflects, some angular change occurs between the diagonals. Since the gusset is rigid, it restrains the diagonals from rotating, which may cause them to split along the bolt rows.

4.7   Estimating Load-Carrying Potential of Overall Structure*

It is as important to assess the overall strength and stiffness of a wood structure as it is to insure that the individual structural elements have adequate strength and stiffness. When a structure is loaded by a combination of loads such as dead, live, and wind or seismic, it is necessary that structural components such as diaphragms and shear walls be adequate to carry the loads applied to them. It is also equally important that the structural components are satisfactorily tied together and that the structure as a whole is properly anchored to its foundation. Sufficient strength, rotational resistance, and stiffness must be developed if the structure is to perform satisfactorily as a unit in carrying loads, to which it is subject, to its foundation.

Some details to look for in assessing the ability of the structure as a whole to resist, in particular, the lateral loads applied to it by wind or seismic actions are:

(a) anchorage of foundation walls to footings,
(b) anchorage of sole plates and header joists to foundation wall or flat slab,
(c) anchorage of corners of shear walls to foundation by providing a tie between studs and foundation,
(d) fastening of studs to sole plates or floor framing,
(e) fastening of rafters, ceiling joists, and trussed rafters to wall studs,
(f) tying the two sides of a gable roof together at the ridge beam,
(g) that adequate strength is provided in roof, floors, and walls so that they can act as diaphragms and shear walls,
(h) that continuous chords are provided for flanges of diaphragms and shear walls,
(i) that transverse and longitudinal bracing is provided for trusses and trussed-rafter roof systems,
(j) that adequate x-bracing is provided if diaphragms and shear walls are not provided.

In the assessment of the load-carrying behavior of the overall structure, some attention should be given to the possibility of progressive collapse. In this mode, a local failure in a single element or joint can precipitate a much more extensive (if not complete) failure of the structure. The collapse progression can be stopped using one of two general methods: (a) Inclusion of strong zones in the structure, or (b) inclusion of weak zones. An example of the first method is the doubling up of structural components at intervals along the length of a structure which consists of a highly repetitive system of structural elements at close spacings. The second method is more difficult to use but includes devices such as discontinuous bracing or construction

---

*Authors: K. F. Faherty, F. J. Keenan, and A. T. Quaile.

joints in roof decking. In both cases, the failure does not progress beyond the special zone.

4.8   Reporting on a Structural Evaluation*

One of the final steps in a structural evaluation is the preparation of a report. A complete report should include each of the following items, along with any other information specifically requested:

   (a) an introduction outlining the purpose and scope of the report,
   (b) a description of the structure including its history, if applicable,
   (c) a report of the site investigation,
   (d) a report of the structural analysis including design loads, design methods, and allowable stresses used,
   (e) a summary of results showing specific problem areas and excessive stress conditions, and
   (f) recommendations for any further action which should be taken.

The conclusions drawn from a completed structural evaluation are either an affirmation or a denial of the structure's adequacy for its intended purpose. If the structure is not adequate, the options available include restricting the occupancy or use pending reinforcement, provision of temporary shoring or support, or provision of immediate permanent reinforcement. Except in extraordinary cases, the decision to condemn a structure should not be made by the structural evaluator; instead he/she should provide sufficient information about the relative costs of reinforcement and replacement in order for the owner or user to make a rational decision to repair or to replace the structure.

Restrictions as to the use and occupancy can be a desirable option in many instances. For example, the safe load on a floor may be determined from the structural evaluation and, if this is less than that desired, restriction to the safe level may be imposed. A structure with a roof that is unsafe under heavy snow loads may be used under monitored occupancy, provided that the safe load is clearly established and that provisions are made for maintaining a check on the snow build-up and for evacuating the building when it reaches a predetermined snow depth. Similarly, temporary monitoring of wind velocities can be a reasonable option if moderate winds can be tolerated by the structure and the stability is adequate for the seismic zone in which it is located.

For public buildings, it is not recommended that snow removal be an acceptable long-term condition for occupancy because the work of snow removal is itself hazardous and because severe snowstorms can prevent the movement of workmen to the site. For structures in which snow removal is considered an acceptable option, the effects of removal over parts of the roof area should be studied; continuous beams with splices in one or more spans can be seriously overstressed by loading that is unevenly distributed.

---

*Authors: F. J. Keenan and A. T. Quaile

Electrical snow melting coils cannot be depended upon to reduce snow loads because of the danger of power failures during a storm and also because, with possible future energy shortages, there is no assurance that adequate energy will be available for melting the snow.

4.9 References for Chapter 4

1. American National Standards Institute.
    1972. Building code requirements for minimum design loads in buildings and other structures. Am. Natl. Stand. ANSI A58.1. New York, N.Y.

2. American Society for Testing and Materials.
    n.d. Standard methods for establishing structural grades and related allowable properties for visually graded lumber. ASTM D245. Philadelphia, Pa.

3. American Society for Testing and Materials.
    n.d. Standard methods of testing mechanical fasteners in wood. ASTM D1761.

4. American Society for Testing and Materials.
    n.d. Standard definitions of terms relating to nails for use with wood and wood-base materials. ASTM D2478.

5. American Society for Testing and Materials.
    n.d. Standard method for establishing stresses for structural glued laminated timber (glulam) manufactured from visually graded lumber. ASTM D3737.

6. American Society for Testing and Materials.
    n.d. Draft revision of standard method of test for strength of anchors in concrete and masonry elements (ASTM E488). Comm. E06 on Performance of Building Constructions.

7. American Society for Testing and Materials.
    n.d. Standard definitions of terms relating to collated and cohered fasteners and their application tools. ASTM F592.

8. American Society for Testing and Materials.
    1980. Draft product standard for wire nails, spikes, and staples. ASTM Comm. F16 on Driven and Other Fasteners.

9. American Society for Testing and Materials.
    n.d. Draft standard method for testing nails. ASTM Comm. F16 on Fasteners.

10. American Wood-Preservers' Association.
    n.d. Manual of recommended practice (see current edition). Bethesda, Md.

11. Building Officials and Code Administrators International, Inc.
    1978. Basic building code. Chicago, Ill.

12. Canadian Standards Association.
    1980. Code for engineering design in wood. CSA Stand. 086-M. Ottawa, Ontario.

13. Canadian Standards Association.
    1980. Structural glued-laminated timber. CSA Stand. 0122-M1980.

14. Cramer, C. O.
    1968. Load distribution in multiple-bolt tension joints. J. of the Struct. Div., Am. Soc. of Civ. Eng., Vol. 94, No. ST5, Proc. Paper 5939, May.

15. Ehlbeck, J.
    1979. Nailed joints in wood structures. Bull. No. 166, Wood Research and Construction Laboratory, Virginia Polytechnic Institute and State University.

16. Fox, S. P., E. N. Aplin, and F. J. Keenan.
    1976. Hot press glulam--a chronic problem. Information Rep. No. VP-X-157, Can. For. Serv. Western For. Prod. Lab.

17. International Conference of Building Officials.
    1979. Uniform building code. Whittier, Calif.

18. National Forest Products Association.
    1977. National design specification for wood construction. Washington, D.C.

19. National Research Council of Canada.
    1977. National building code of Canada. Ottawa, Ontario.

20. Southern Building Code Congress International, Inc.
    1976. Southern standard building code. Birmingham, Ala.

21. Stern, E. G.
    1940. A study of lumber and plywood joints with metal split-ring connectors. Bull. No. 53, Eng. Exp. Sta., Pennsylvania State College.

22. Stern, E. G.
    1967. Nails--definitions and sizes, a handbook for nail users. Bull. No. 61, Wood Research and Wood Construction Laboratory, Virginia Polytechnic Institute and State University.

23. Stern, E. G.
    1969. Discussion of "Load distribution in multiple-bolt tension joints," by Calvin O. Cramer. J. of Struct. Div., ASCE, Vol. 95, No. ST2, February.

24. Timber Engineering Co.
    1956. Timber design and construction handbook, 1st ed. McGraw-Hill Book Co.

25. Timber Engineering Company.
    1973. Design manual for Teco timber connector construction. Publ. No. 109. Washington, D.C.

26. U.S. Department of Agriculture.
    1974. Wood handbook: wood as an engineering material. Agric. Handb. No. 72, rev. U.S. Gov. Print. Off.

27. U.S. Department of Commerce.
    n.d. Structural glued laminated timber. Product Standard PS56 (see current edition).

CHAPTER 5   METHODS OF REPAIR*

5.1   <u>Considerations in Selecting the Right Repair Procedure</u>**

Previous chapters have provided both qualitative and quantitative information for evaluating the condition of a structure. The purpose of this chapter is to describe methods of repair that have proven successful and to delineate the range of practical application for each method.

A number of factors may influence the selection of the right repair method. However, the first and foremost is invariably economics. Economic considerations often play a more important role in repair than in most other areas of structural engineering because repair is often neither an anticipated nor a budgeted item. The pressure for completing repairs as inexpensively as possible can therefore be quite intense. In cases where the damaged portion is affecting operations, the problem can even be multiplied because of production losses. It is difficult, if not impossible, to rate general repair procedures according to cost. The variables associated with specific applications can differ so greatly from case to case that an inexpensive method in one instance can be expensive in another. As a consequence, no attempt will be made to rate repair methods by cost. Rather, successful repair methods will be described and examples given in which the costs have been found to be competitive. It is anticipated that the descriptions provided here will give the engineer the tools for determining the most cost-effective repair procedure for his specific application.

While economic considerations tend to be the overriding concern in selecting the right repair method, other factors may also have considerable influence in certain situations. Perhaps the most important of these is structural integrity. The engineer needs to know the strength levels that can be attained in a given repair situation. Ideally, the repair should restore the timber to its original strength. However, in some cases such restoration may not be possible. There are several conditions related to structural integrity which must be evaluated in this regard. Overload or overstress factors may be critical, whether they result in excessive flexural, compressive, tensile, shear or torsional stresses, or in excessive deflection. Such overloads often result from prior modification of a structure. Typical examples include roof overlays and repairs, usage changes such as converting office space to storage, and the addition of machinery or permanent equipment. In other cases, errors in design, detail, or execution may

---

*Chapter Coordinator: R. Richard Avent, Professor of Civil
   Engineering, Mississippi State University, Mississippi State, MS.

**Author: R. Richard Avent.

METHODS OF REPAIR

influence the repair decision. Examples are: faulty assumptions in the original design; design errors; neglecting deformation considerations; improper connection design; poor detailing such as notches, daps or holes which reduce the section or provide crack-initiation sites; laterally unstayed members; inaccurate cutting, fitting or joining; and joint eccentricities.

Problems related to structural integrity may be caused by short-term occurrences such as fires, earthquakes, wind, or mechanical damage. On the other hand, long-term causes such as chemical attacks, weathering, high in-service temperature, high humidity, or checking, splitting, and shrinkage may cause the problems. Finally, degeneration from decay or insect attack must be considered. Each may require substantially different repair methods.

Associated with structural integrity is serviceability. Some repairs may be very effective for a few years but then deteriorate. Others may be effective for an indefinite period. Fortunately, most repair methods can be evaluated as to their structural integrity and such methods will be documented here, where possible.

Another factor to be considered is the physical configuration and associated environment. For example, small checks may be effectively repaired with methods entirely different from those used to repair large cracks. Or, an obviously simple repair technique for timber protected from the outside environment may be entirely unsuitable for underwater timber structures. The influence of such factors can only be evaluated in a specific application.

Consideration must also be given to the functional activities surrounding the structure. It may be necessary to continue certain activities within a structure during repairs. In such cases the repair procedure must keep disruptions to a minimum. Another example would be a structure which must continue to carry significant loads during the repair phase. Repair methods must reflect this situation.

Future utilization of the repaired structure may greatly influence the repair method chosen. In particular, the expected lifetime of the unit must be evaluated along with potential future uses of the structure. All too often in the past, structures have been repaired without serious thought to the future. Instances of roof repairs which increased the dead load on the structure have been approved with the understanding that the building would be used for only a few years. Years later the building would still be in use with severe safety problems because of the increased loads. Functional uses of buildings also change over a period of years. Repairs need to reflect these possibilities whenever possible.

The availability of craftsmen, material and equipment may influence the selection of repair procedures. For example, some methods are quite specialized (even patented). Obtaining the needed items may thus add significant cost to an otherwise desirable repair method. Other factors including equipment accessibility, experience of local craftsmen, or even the degree of inspection and supervison required could be the deciding factor in selecting a repair method.

Safety must also be considered when selecting the right repair method. The use of volatile materials may not be appropriate. Repairs requiring welding of metal plates near wood or flammables should be avoided. Proper bracing, temporary supports, and protection to those below the repair activity must also be provided.

A final consideration may be aesthetics. Often the structural components are hidden and require no such consideration. However, many designs in wood emphasize its natural beauty with exposed structural elements. Completing the repair without destroying the aesthetic appeal is sometimes the most challenging part of the repair process. The repair of historic structures offers notable examples where aesthetics may be of paramount importance. It should be noted that historic structures may have numerous specialized requirements in addition to aesthetics; these are discussed in section 1.7.

With the above list of considerations, the engineer is ready to evaluate repair methods for his specific application. The following sections describe successful repair techniques that have been proven through both research and practical applications. Following these general descriptions, specific examples of successful repairs are presented.

5.2 Description of Alternative Repair Methods

5.2.1 Mechanical Repair of Structural Damage*

Methods of timber repair are influenced to a considerable extent by the type of member or assembly and the manner in which members or components are stressed. In addition, the relationships, both geometric and stress-related, between individual members and between components at their interface with other dissimilar parts of a structure can significantly influence the repair methodology. Suitability of connections is of paramount importance, as they are frequent sites of failure or distress.

Principal types of members or assemblies considered herein include the following: sawn timber members (surfaced and unsurfaced), glued laminated wood members, poles and pole structures, trusses and similar framed members and assemblies, conventional light framing (rafters, joists, bracing, wall framing), and structural plywood.

The following general types of repair methods and materials are described as suitable methods of mechanical repair of structural damage, with comments as to conditions of suitability.

---

*Authors: James Warner, Consulting Engineer, Mariposa, California and Henry Sanders, Vice President, Wheeler and Gray, Consulting Engineers, Los Angeles, California.

5.2.1.1 Conventional Fasteners

These fasteners are the traditional fasteners of wood and have applicability to all types of repair for all types of members, provided they are used correctly with due regard for end and edge distance, spacing, splitting, section reduction and similar factors all related to good structural design practice. Such factors as well as safe loads for the various conditions and direction of application are competently given in ref. (13). Typical connectors are:

1. Bolts, nuts, and washers (machine or carriage).
2. Lag bolts or screws.
3. Wood screws.
4. Nails and spikes.

5.2.1.2 Manufactured Connection Hardware and Devices

These fasteners and devices encompass a broad and diverse range of hardware items of standard manufacture. Some have been in conventional use in timber design for many years while others are of relatively recent origin, based principally on specialized design and fabrication of sheet metal. All of these have application in the area of repair and rehabilitation of wood structures, depending on the corrections desired and the configurations and joint relationships which are to be improved. They may be roughly categorized into three general families or general types.

"TECO" Shear-Connector-Type Hardware

These devices, developed many years ago by the Timber Engineering Company, derive their principal use as devices to increase the shear or bearing transfer characteristics of bolts through bolted joints. Most of them require special cutting tools to cut grooves or circular recesses into the wood members at the shear interfaces of joints. The fasteners are then driven into the recesses cut for them and the joints are drawn together by tightening the bolt through the center of the fastener system. They are useful in increasing the capacity of bolted joints provided that proper end distances, edge distances and spacing between fasteners are used. Complete descriptions of these devices, together with engineering data for design and capacity tables are given in Part V of ref. (13). The principal items in this group are:

1. Split-ring connectors: Round rings, discontinuous at one point to permit some adjustability. They are used primarily for close fitting joints of sawn timbers; they are inserted in circular grooves, one-half the width of the ring in each half of the joint, to provide increased bearing.

2. Shear Plates: Round plate devices with flanges at the periphery of the plates to engage grooves cut into the wood. They provide greater bearing in the wood as well as providing the added benefit of metal-to-metal bearing at the shank of

the bolt. They are usually used for close-fitting joints in sawn timbers.

3. Toothed Rings and Spike Grids: These are malleable or cast iron devices with teeth cast integrally on the grid of the metal on either one or both sides, with a bolt-size round hole in the center. They are suitable for use where space exists between adjoining members, and in loose-fitting types of joints. The pointed teeth are seated into the wood members by drawing the members together with the bolt in the center. These devices are available for sawn timbers, irregular members, and pole structures.

Manufactured Hardware Devices of Sheet Metal or Light Steel Plate

This classification includes a wide variety of fasteners and devices for a great range of applications. They are factory-produced in production sheet metal fabrication facilities. All of them are based on using nails, spikes, screws, or bolts depending on the capacity of the device, its specific application and other considerations involved in its design. These devices are of great assistance in the repair or rehabilitation of wood structures due to the wide variety of configurations available. Most manufacturers distribute illustrated catalogues with tables giving working values for the various units. The most reliable engineering data are given in the Research Reports published by the Research Committee of the International Conference of Building Officials (ICBO), whose evaluation requires the manufacturers to provide competent engineering design and physical testing of the devices to obtain working loads.

A partial listing of types of devices readily available under this grouping includes, but is not limited to the following:

1. Framing anchors.
2. Joist hangers.
3. Purlin and beam hangers.
4. Hinge connectors.
5. Tie straps and framing ties.
6. Wall braces.
7. Metal bridging.
8. Tie-down or overturning anchors.
9. Post caps and post bases.
10. Beam and girder seats.
11. Angle clips of various sizes and configurations.

Punched Truss Plates

These devices are flat sheet metal plates which have been punched to form raised pointed teeth. They are principally used as gusset plate connectors for trusses and other framing joints where the framing members are in the same plane and the connectors are applied to the side grain of the members. They are of limited use in repair or rehabilitation due to the fact that such devices are best applied in presses which apply uniform massive pressures forcing the teeth into the wood.

METHODS OF REPAIR

While the use of such fasteners may be limited in field repairs, there are brands of punched truss plates made especially for field use. They can be fastened by either a hammer or hand operated gripper. In addition, certain types of punched truss plates are made for use with special short nails.

5.2.1.3 Specially Designed and Fabricated Hardware or Devices

Such hardware is usually used for heavy or special structures where standard manufactured hardware cannot answer the requirement. These must be developed by careful structural analysis and design. Examples include:

1. Straps, rods, and similar tension devices.
2. Clamps.
3. Dowels and pins.
4. Gussets.
5. Wedges and compression devices.
6. Special connection assemblies and member splices.

5.2.1.4 Member Augmentation Methods

The devices listed in previous sections are frequently employed to strengthen, reinforce, or replace damaged members. Specific examples are illustrated in section 5.4 with a more general discussion included here.

Scabbing or splicing is often used for damaged members requiring an increase in effective cross section. Additional members are added to the deficient element and connected with bolts or other hardware. Such repairs should be individually designed to insure adequate load transfer.

Another effective repair technique is to augment the structural system with steel. Typical of such reinforcements is the splicing of steel sections to existing timber or adding tension rods or straps. When tension rods are added, they are sometimes post-tensioned in order to relieve the stress in the damaged member. Care must be taken in such cases to insure that redistribution of stresses does not cause damage in other locations.

A common problem with timber structures is longitudinal splits parallel to the grain. Some such splits can be disregarded while others may have a serious weakening effect. Reference (12) provides the following guidelines on longitudinal splits.

"At the ends of the members stressed in compression parallel to the grain, checks and splits may be disregarded, provided there is no evidence of slip from wedging action of connectors and bolts."

"For members stressed in tension parallel to the grain, splits outside the connector area that are approximately parallel to the grain may be disregarded. By approximately parallel to the grain is meant that the angle of slope of split does not exceed the permissible slope of grain for the grade of structural lumber."

Examples are shown in Figure 5.2.1.1.

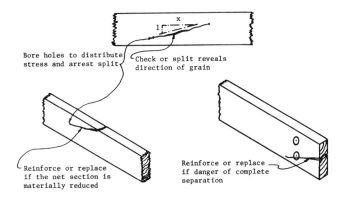

Figure 5.2.1.1   Unless net section is reduced, disregard split when the slope of grain (1 in x) is within the limitations of the structural grade. A split probably is not harmful unless part of the joint is in danger of separation from the main piece. In the latter case it is necessary to reinforce or replace defective member.

"For tension members with connectors loaded parallel to the grain, a single split within the connector area up to 3/16 inch opening of the splits for a connector of 2-5/8 inches or less and a split of 1/4 inch for connectors with diameters over 2-5/8 inches up to 4 inches may be disregarded. Splits opened up in excess of these widths should be carefully investigated."

"All measurements of opening of splits should be made at the end of the piece. The dimensions given are limitations for end distances that are standard or less. For greater than standard end distances, the split may increase in the proportion of actual length of end distance to standard end distance."

"If there is more than one split in the connector area, the total opening of the splits should not exceed one and one-half times that for one split. If splits or checks are located in the connector area within 1/4 inch from tangents to a connector 2-5/8 inches or less in diameter or 1/2 inch from tangents to a connector over 2-5/8 inches and up to 4 inches diameter they should be given further consideration. Other splits or checks should be treated as single defects."

"For end splits in either tension or compression members with connector loads acting in a direction other than parallel to the grain, the suggestion above for treating tension members should be applied at the ends of the pieces" (fig. 5.2.1.2).

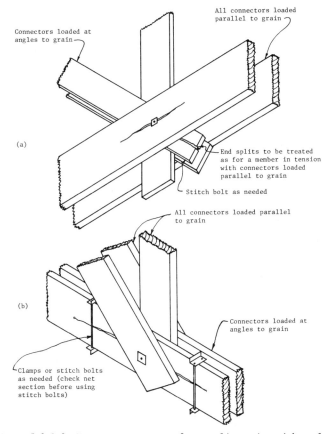

Figure 5.2.1.2 Treatment necessary for a split varies with angle at which the working load is applied

"When an extensive split occurs within a connector area at a joint other than at the end of a member, installation of stitch bolts or clamps on each side of the joint containing the split should be considered. When stitch bolts are used, insertion at large knots should be avoided and the member should be investigated to determine that the required net section is maintained."

"If a split in a member shows the slope of grain to be greater than that permitted in the structural grade required, critical members stressed in tension should be replaced or reinforced. If a split develops in lumber having a slope of grain permitted in the structural grade but is of such a nature that it may tend to split the member into two pieces, the member should be replaced or reinforced."

"Normally, checks are of relatively little importance but, if they are of an extensive and deep nature, their seriousness may be evaluated from the suggestions on splits on a basis of their relative importance compared with splits."

"In connector construction, shakes in the connector area that are approximately parallel to the contact faces and within a zone having a depth 1/4 inch greater than the depth of penetration of the connector into the wood, or which eventually may result in separation of a portion from the rest of piece, should be carefully investigated (fig. 5.2.1.3). Other shakes may be disregarded as such and be evaluated as a check or split."

"In general the data under connectored construction will apply to bolted construction except that splits within the bolt area should be further investigated. Other splits may be disregarded."

"When used at the end of a piece, stitch bolts should be placed between 2 inches and 3 inches from the end. Small 3/8- or 1/2-inch-diameter bolts are suggested. Ordinarily, when bored at a critical stress section of a member, the area of cross-section removed by the hole for the stitch bolt should not exceed the cross-sectional area occupied by the maximum knot permitted in the structural grade. In drawing up the stitch bolts they should be tightened only to the point where the bolts begin to take tension. No attempt should be made to close a split or check as this may extend the split on the other side of the joint. In servicing structures, stitch bolts should be tightened as well as other bolts."

"In wide splice pads and other axially loaded members, it is desirable to place saw kerfs between rows of connectors or use two narrow pieces to control any possible splitting. Where the kerf runs into the end but not the entire length of the piece, a hole (not exceeding 3/8 in. to 1-3/8 in. diameter, depending on the width of the piece) should be bored at the end of the kerf to avoid its extension. When making kerfs, insofar as possible, the edge distances with respect to connectors should be maintained as required at the edge of the piece (fig. 5.2.1.4). The use of saw kerfs is more important in grades that permit steep slope of grain than in those requiring relatively straight grain because the split, tending to follow the slope of the grain, is more likely to be dangerous."

"When repairs or reinforcements are made, the distribution of stresses between members should be rechecked to avoid improper design. When repairing one leaf of a multiple-leaf member, preferably all leaves

METHODS OF REPAIR

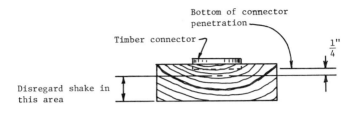

(a) Flat grain piece

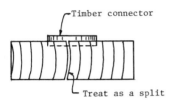

(b) Vertical grain piece

Figure 5.2.1.3 In flat grain pieces a shake below the connector may be disregarded unless there is danger of separation from the main body. In vertical grain a shake across the thickness of the piece has the appearance of a split and can be treated as such.

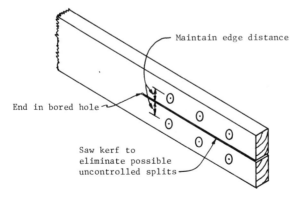

Figure 5.2.1.4 Saw kerf entirely through a piece controls splits and checks. A bored hole at the end of the kerf distributes the stress concentration and stops extension through splitting

should be spliced between the same panel points to avoid eccentric distribution of loading caused by differential deformations of the various leaves" (fig. 5.2.1.5).

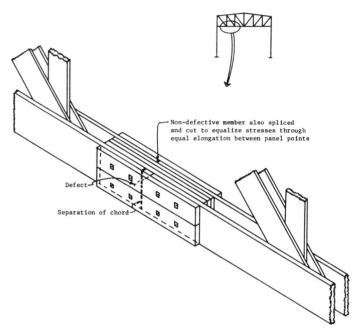

Figure 5.2.1.5 Where necessary to splice one chord at a defect, splice both and cut original lumber entirely through to make splice take the load and distribute the stress equally. Chords can be cut after splicing by boring adjacent vertical holes across the narrow face of the piece. If space between the chords is not adequate for split-ring insertion, use toothed ring connectors or bolts only.

Clamping and stitch-bolting (11) are methods often used to arrest cracking and splitting. Examples of clamping are shown in figures 5.2.1.6 and 5.2.1.7. The clamps should have bolts parallel to and in close contact with the wide faces and steel straps across the narrow face. Side plates should be long enough to throw cross bolts well back from the check providing good anchorage.

METHODS OF REPAIR

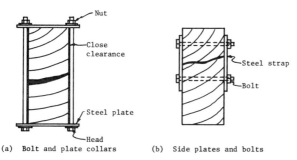

Figure 5.2.1.6 Bolt and plate collars lessen further checking in serious cases. Side plates and bolts are useful but less effective since the net section is reduced

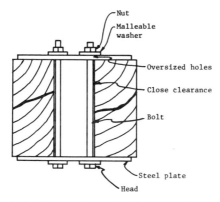

Figure 5.2.1.7 Two-leaf members may be repaired by the use of two plates and two bolts. Collar type repairs do not reduce net section and may be installed at a moderate cost

Stitch bolting as shown in figure 5.2.1.8 can be used instead of clamping. However, stitch bolting may be undesirable in members with large width-thickness ratios since the holes may unduly reduce the net section.

Another method used to control splitting is the use of metal bands. Jessome and Kennedy (10) reported that banding was successful in restoring artificially induced splits. However Avent, et. al. (5) found that load-induced splits could not be effectively repaired by

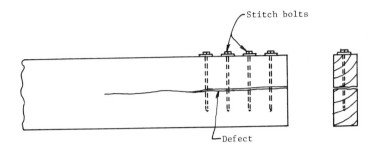

Figure 5.2.1.8 Stitch Bolts on a longitudinal split

banding. The basic disadvantages are (1) difficulties in determining which damaged areas are amenable to banding repairs; (2) the possibility of ineffective repairs is great with the current state of knowledge; (3) there is a lack of standard procedures to insure the proper number, placement, and prestressing of bands; and (4) moisture variations may cause significant shrinkage and swelling of wood, leading to loose and ineffective bands. It is therefore recommended that banding not be used as a repair technique for timber structures.

A final alternative is a full member replacement. Replacement offers the advantage of completely removing the defect. However, the cost may be high since member replacement requires supporting the structure while the member is replaced. As such, member replacement is usually considered after other repair alternatives have been exhausted.

5.2.2 Repair of Structural Damage with Epoxy Adhesives*

5.2.2.1 Epoxy Compounds for Structural Repairs

Epoxy compounds, consisting of basic resins and resin-hardening agents properly blended or modified with flexibilizers, extenders, diluents, and/or fillers, depending upon the end-use requirements, have revolutionized many aspects of industry, including all phases of construction. The basic resins are related to bakelite or formica and may be solidified by many reactants, such as the amines and polyamides which are related to nylon ([1]). The first use of epoxies in this country was by the paint industry prior to 1950, but was soon followed by the aircraft industry. These two fields pioneered in developing the potential of epoxy compounds.

---

*Author: R. Richard Avent, Professor of Civil Engineering, Mississippi State University, Mississippi State, Mississippi.

Some of the unique properties of epoxy compounds which have been responsible for their rapid development are as follows:

1. Rapid hardening at normal temperatures from a liquid to a solid.
2. High degree of adhesion to most clean surfaces.
3. Toughness which combines durability with resistance to cracking, and superior chemical resistance to most acids, alkalies, and solvents.

There are numerous epoxy compound formulators who generally purchase the ingredients from a relatively small group of chemical manufacturers. The versatility of the epoxy resins, especially with regard to compounding techniques, allows the formulators to produce a myriad of products with an extremely wide range of properties.

In highway and other construction, the formulations are broken down into three types (2). These types are as follows:

Type A - A polysulfide-modified system blended with mineral filler

Type B - A clear or light colored amine- or polyamide-cured system

Type C - A coal-tar-modified system

There are a number of Type B formulations which have been specifically developed to fill hairline and larger cracks in concrete structures by use of fully automated equipment. This equipment mixes the two epoxy compound components immediately before the resin leaves the injection nozzle. The same type of epoxy resin has been recently applied in strengthening both laminated wood and sawn lumber having various cracks and recesses. While other adhesives are commercially available, epoxies are primarily used in repair because of their strength and quick cure rate.

5.2.2.2 Classification of Defects Suitable for Epoxy Repair

It is convenient to group defect types into general categories based on similarities in damage and repair capabilities. The following list illustrates the variety of defects for which epoxy repair may be suitable.

Type 1: End Splits. Common defects found in wood structures include end splits parallel to the grain. They range from small checks only partially penetrating through the thickness, figure 5.2.2.1, to cracks large enough to insert your hand, figure 5.2.2.2. This type of defect occurs frequently in trusses at bolt and split ring connections. However, it can also be found in beam or column framing members, figure 5.2.2.3, or in glulam beams and arches. In some cases the crack will extend only a short distance longitudinally, but it can also extend over a major length of the structural element.

Figure 5.2.2.1. Partially Penetrating Small End Splits

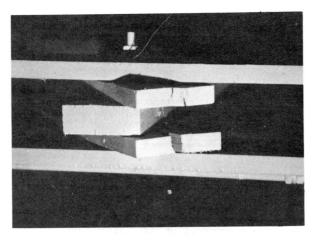

Figure 5.2.2.2. Large End Splits

Structurally, the defect can greatly weaken joints in trusses as the crack widens, or cause reduced shear strength in beams and glulam members.

Figure 5.2.2.3. End Splits in Columns

Figure 5.2.2.4. Typical Interior Longitudinal Splits

Type 2: Interior Longitudinal Splits. This defect is similar to Type 1 except that it always extends through a major portion of the member. Typical examples are shown in figure 5.2.2.4. It has little effect on tensile and compressive strength but can significantly reduce shear strength associated with bending. Such a defect may be found in truss members or beams, and between laminations of glulam members.

Type 3: Broken Members. Occasionally a member will break completely apart as illustrated in figure 5.2.2.5. While not a common type of defect, it can result in catastrophic consequences to the structure. This defect is usually associated with tension or bending stresses in sawn lumber, particularly around knots and bolt holes.

Type 4: Decayed Members. Decay is usually associated with exposure to moisture. A typical example is shown in figure 5.2.2.6. Significant decay destroys the load capacity of the member in every respect. It is probably the most difficult defect to repair by any method other than replacement. Decay may be found in any type of timber structure and occurs often in waterfront construction.

Figure 5.2.2.5 Broken Member

Figure 5.2.2.6. Decayed Joint

METHODS OF REPAIR 211

5.2.2.3 Typical Steps in the Epoxy Repair Procedure

The epoxy repair method can be described as a four-step process. The procedure is outlined as follows.

1. <u>Special Member Preparations</u>--In cases where excessive damage or deterioration has occurred it may be necessary to jack the member back to a more normal position. Based on field observations of buildings in need of repair, however, jacking has generally not been needed except in isolated instances. Jacking should not be done unless the sag exceeds tolerable limits. In addition to being expensive, jacking may cause stress reversals, and adjoining building elements such as roof decking, etc., may be damaged.

For some repairs it may also be necessary to add splice plates. The primary example is the case of a broken member. The epoxy repair technique is most effective with lapped joints in which the epoxy is injected between the lapped surfaces to provide shear resistance. For broken members, splices are tack nailed on both sides of the broken member which is to be later injected with epoxy. Since most failures occur at joints that already consist of lapped members, it is generally not necessary to provide additional splice plates.

For decayed sections, the moisture source must be eliminated or the element protected from moisture to prevent repeated deterioration. It may also be necessary to remove the decayed wood and add reinforcement such as steel, fiberglass rods, or replacement wood. In other cases, such as underwater repair, an encasing mold might be needed around the component. The epoxy can then be pumped into the damaged area, forcing the water out.

Finally the area to be repaired must be cleaned. All dust and debris should be cleared. In addition, the area should be checked for moisture content. While epoxy formulations exist which adhere in moist environments, the presence of moisture may mean that decay will continue in encapsulated or adjacent areas and result in continued deterioration. If the moisture content exceeds 20%, the member should be dried out before repair and the source of the high moisture eliminated.

2. <u>Joint Sealing</u>--The area to be repaired must be completely sealed on the exposed surfaces except for injection ports. Figure 5.2.2.7 shows injection ports in place after the sealant has been applied to a truss joint. Large cracks, edges of lapped joints, areas around bolts, and areas around injection ports are usually sealed in this fashion. The sealant used should be a high viscosity epoxy with a putty-like consistency. After sealing these visible openings, it is recommended that, for porous wood surfaces, a thick epoxy paint be applied to the entire surface area to be repaired. This application fills hair-line cracks and small holes not generally visible. The permeability of some wood makes this application necessary to insure a leak-proof joint.

Figure 5.2.2.7. Sealed Joint Ready for Injection

Figure 5.2.2.8. Epoxy Injection in Progress

3. <u>Epoxy Injection</u>--The final step is to pressure-inject a low-viscosity two-component epoxy into the sealed joint through the injection ports (fig. 5.2.2.8). This can usually be done by attaching the nozzle of the injection gun to a single port and letting the other ports serve as vents. As epoxy fills the joint, the venting ports leak epoxy and are sealed off by clamping. After all ports are sealed, epoxy injection should be continued to ensure penetration into the fibers. Care must be taken not to inject with too much pressure or the seal might break. A maximum nozzle pressure of 40 psi (276 kPa) is recommended. Should a leak develop during the injection, a quick-drying patching cement can be applied. Since it is essential to finish the injection before the epoxy initially sets, a patching cement with a 2- to 3-minute pot life is needed. Injection is usually accomplished with automatic equipment. Typically, two positive-displacement pumps geared to the specified mix ratio feed the separate components into a nozzle. Mixing is accomplished by forcing the epoxy through static mixing brushes in the nozzle. As long as the flow is not interrupted for more than a few minutes, injection can progress for hours without damage to the nozzle.

4. <u>Finishing</u>--The initial cure time for many epoxies is less than 1 hour. Final cure is usually accomplished in a few days. After curing, any temporary supports may be removed. If aesthetics are important, the injection ports are removed and the sealing gel is sanded smooth. Then paint can be applied to finish the repair.

5.2.2.4 Strength and Serviceability of Epoxy-Repaired Structures

In recent years a considerable research effort has been conducted to evaluate the effectiveness of epoxy repair on timber structures. The factors studied include: crack width, timber age, joint shear strength, grain orientation, weather resistance, load duration, effect of decay, temperature effects, epoxy penetration, reinforcing techniques and full-scale truss and beam repair evaluation. The purpose of this section is to outline current knowledge on the effectiveness of epoxy repairs in various situations.

Epoxy repair has been found to be most effective when utilized to provide shear resistance. In one of the first published studies on epoxy-to-wood strength, Avent, et al. (3) experimentally evaluated both bond characteristics and the strength of epoxy-repaired bolted connections. Both tension and shear-bond tests were performed, and various factors affecting strength were evaluated. For dense southern pine with relatively smooth surfaces, the ultimate bond stress in tension was on the order of the allowable tensile stress for the wood. In shear, the ultimate stress of the bond corresponded to the ultimate shear stress of the wood. It was found that crack widths of up to 3/4 inch (19 mm) could be treated without a decrease in the failure stress. For larger cracks, the mass of the epoxy generated such high heat as to "honeycomb" the epoxy and thus reduce its ultimate strength. Of greater significance was the result of bolted and spliced joint tests. A procedure of first loading an undamaged joint to failure, repairing the failed joint, and then reloading the joint to failure was utilized.

For both bending and tension specimens after repair, the joints consistently reached an ultimate strength of 30%-60% greater than the original strength of the undamaged specimen. This was due to the fact that the joint was effectively changed to a rigid joint, with a resulting even distribution of stresses over the cross section. It was thus shown that epoxy repair of lapped joints restored the joint to better than its original condition.

In a later study, Avent et al. (4) investigated the epoxy repair method on full-scale trusses. For the trusses constructed with new timber, the epoxy repair method restored strength to a level approximately equal to the original. The test results showed the epoxy repair method to be an effective repair procedure for the most common types of truss failures: broken members, end splits at bolt and split ring connections, and longitudinal splits in members. Not a single epoxy-repaired joint in sound timber failed during the load tests. The repair method was somewhat less successful for badly rotted members. The epoxy repair increased the load capacity. However, the repairs were marginal at best. The primary problems associated with these repairs were the difficulty of detecting all rot, of detecting partially weakened joints, of adequately sealing joints, and of poor bonding achieved in highly rotted material. It is recommended that repairs should be made at all joints of a truss in such condition, even if visible damage is minor. Because of the extensive deterioration of most of the old timber trusses, a firm conclusion was not drawn as to the effect of timber age. However, based on other joint tests (3) along with the results from one old truss that was not rotted, it can be tentatively concluded that old trusses can be effectively repaired with epoxy unless damaged by significant decay. The redistribution of forces associated with epoxy repair was found to be small at the most critical points in the truss--the connections. However, some members experienced stress increases on the order of 20% due to the bending induced by fixing certain joints. It was concluded that these stress changes were not significant because the bending stresses themselves are relatively small and joints rather than members tend to be critical. In addition, many structures with rigid joints (such as welded steel trusses) have been traditionally analyzed and designed as pin-connected members without ill effect.

Avent et al. (6) have investigated weathering effects on epoxy-repaired timber structures. Two types of effect were considered: (1) The effectiveness of epoxy repair procedures for joints in a severely deteriorated state; and (2) the effectiveness of epoxy-repaired joints exposed to weathering after repair. To measure the effectiveness of repair for weathered joints, both full-scale truss tests and joint tests were conducted. Severely weathered members can be effectively repaired. However, special care must be taken to ensure that the area to be repaired is completely sealed. It is essential to completely pressure-inject the entire area so that all voids are filled. An incomplete seal allows leaks which lead to less than full penetration. Another problem is to determine which joints need repair. In some cases, joints have rotted on the inside (around split rings) where detection is difficult. For severely weathered structures, it is recommended that all joints be repaired even if visible decay is not present. Special care should be taken to dry the decayed area before repair. If not, the excessive moisture may allow the decaying process to continue after repair. The source of moisture must be eliminated, otherwise decay will continue.

To determine the effect of weathering on joints already repaired with epoxy, two series of tests were initiated. The first was an accelerated aging test which showed the epoxy-repaired specimens to have a relatively constant strength through 120 cycles of wetting and drying. Lack of comparative data makes it difficult to extrapolate these results into an effective number of years of exterior exposure. However, a second test series has been initiated in which epoxy-repaired joints are being exposed to natural weather conditions over a period of several years. These joints will be tested at regular intervals to determine weathering effects. These tests have not been conducted long enough as yet to draw final conclusions, but after one year the joints have shown no decrease in strength.

An extensive study of the strength of epoxy-injected joints by Avent and Baylot (2) has led to the development of design formulas. Theoretical formulations, including finite element analysis, and experimental studies were utilized to develop design criteria. The three most important parameters were found to be: lap length, thickness of middle member, and glueline thickness. However, experimental evaluation indicated the glueline thickness to be less important than various theories had indicated. Formulas developed are valid for parallel-grain bond. Additional experimental work is needed to develop formulas for various grain angle orientations.

Another promising aspect associated with epoxy repair is its use combined with fiberglass or metal reinforcing. This technique is particularly useful in historic structures or those in which aesthetic concerns are vital. Pioneering work in this area has been done in Europe (9). The epoxy repair procedure in general follows that previously outlined except that, prior to injection, reinforcing elements are placed in the timber. Typical examples are shown in figure 5.2.2.9. Holes are drilled and reinforcing bars inserted to form: (1) either a strong connector pattern between sound and unsound wood, or (2) a stiffening internal truss or frame for increasing load-carrying capacity. Epoxy injection (or in some cases only pouring into the zone is required) bonds the reinforcing members to the surrounding wood. Since all reinforcing is internal, only the holes used to insert the rods need to be refinished. This method has been successfully used in several cases. However, little load testing and evaluation data are available to document the degree of structural integrity gained in such procedures.

Epoxies have also been used for restoration and conservation of elements of historic structures. This may take two forms. One involves impregnation of decayed or insect-damaged sections with epoxies to consolidate the deteriorated areas. The second involves patching cracks or larger voids. Both have the primary purpose of conserving the architectural fabric rather than replacing it. The procedures may,

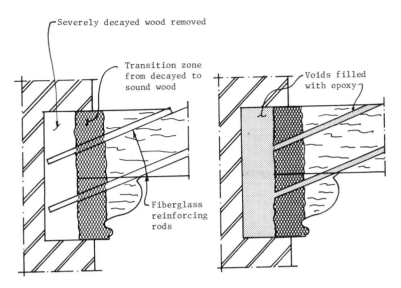

(a) Joint with decayed wood removed and reinforcing in place

(b) Epoxy grout injected into all voids

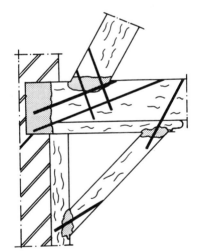

(c) Multi-joint section after repair

Figure 5.2.2.9 Reinforced epoxy repair

however, fulfill additional functions such as arresting decay, or forming a water barrier.

Phillips and Selwyn (15) have studied a number of epoxies to establish their suitability for such functions and to illustrate their use. They point out, however, that such restorations can serve only limited structural functions. That is, they may be adequate to carry load when the member is stressed only in compression, but that resistance to bending and shear is quite low.

While the use of epoxy repair in timber is a recent innovation, a significant amount of verification exists to justify its use. Examples of successful repairs can be found in references (7,8,16) and are illustrated in a later section.

5.2.3 Repair of Piles*

5.2.3.1 Types of Damage

Damage to pile structures may be caused by a number of environmental effects.

(1) Corrosive and pollutant attacks such as oil spills, particularly oil of aromatic nature, tend to wash creosote from the exposed surfaces of creosoted piles. This makes it easier for marine borer larvae to settle and enter the piling.

(2) Sunlight exposure effects. Continuous exposure will cause shrinkage, cracking, and checking of raw wood. Sunlight tends to degrade the residual creosote from the surface of creosoted wood piles.

(3) Water penetration effects. If the wood is kept continuously wet there is no effect, except in the case of exposure to sea water, where a rarely found marine fungus may soften outer wood fibers. This poses only a very slight problem of durability.

(4) Wave and current effects. Severe storms could destroy a wooden coastal structure.

(5) Effects of exposure to fire. Wood will shrink, crack and eventually char upon reaching its ignition temperature.

(6) Ice effects. A sufficient buildup of ice on wood batter piles could crack them.

(7) Marine organisms. Untreated wood of all kinds is susceptible to attack and destruction by many kinds of marine borers, such as the Teredines, _Limnoria_, _Chelura_ (limited action) and in some areas Pholads. Creosoted wood is susceptible to attack by _Limnoria tripunctata_ (Menzies). They are found in all ocean waters, on both coasts, from Southern Alaska to the tip of South America. One species

---

*Author: Carrol M. Wakeman, Consulting Engineer, Rancho Palos Verdes, California.

of Teredo is found in the fresh water of Lake Maracaibo, on the
northern coast of South America.

(8) Periodic wetting and drying. Under these conditions portions of
the wood will be subject to attack by decay fungi.

Untreated pine and fir piles usually last no longer than 2 years in
the ocean, often less than 1 year where marine borers are present in
great numbers. Treated piles (creosoted) have a life expectancy
averaging 8 to 10 years where Limnoria tripunctata are present.
Limnoria tripunctata (Menzies) are found all along the Atlantic seaboard from the southern tip of Africa and South America to the
Arctic Circle. In the Pacific Ocean, Limnoria tripunctata are active
from the southern end of the South Island of New Zealand to several
hundred miles north of Vancouver, Canada. Limnoria tripunctata were
selected as an example because they are the only known species of
Limnoria which will attack and destroy heavily creosoted piling.

Creosote penetrations in the sapwood vary from 3/4 inch to a little
over 1 inch (20 pcf, full cell treatment, retention determined by
assay) on properly selected piles of Douglas-fir. On the other hand,
southern pine is easy to treat and creosote penetrations up to 3 inches
are obtainable with the same treatment as for Douglas-fir.

No untreated wood, commercially available for piling, either domestic
or imported, will resist borer attack for more than a few years. One
species, greenheart (Ocotea rodiaei) may, even though untreated, last
2 or 3 years longer than treated Douglas-fir, in the same water. But
in the end, the borers will win.

All bearing piles for use in a marine environment should be pressure
treated with creosote to a 20-pcf retention, with the retention determined by assay or with chromated copper arsenate (CCA) to a retention
of 2.5 pcf in the outer 1 inch of Douglas-fir piles and, in southern
pine piles, to 2.5 pcf in the outer half inch and to 1.5 pcf in the
next 1-1/2 inches, with retentions determined by assay. Some chemicals, such as pentachlorophenol, are not considered suitable for use
in seawater.

It has been found that, for piles to be used in waters where Limnoria
tripunctata are active, a dual treatment involving both an approved
waterborne preservative and creosote gives better protection than does
either preservative alone. Requirements for the dual treatment are
given in AWPA standard C3.

5.2.3.2 Methods of Repair and Protection

Barrier Wraps

Synthetic wrappers (PVC) for the protection of wood piles in the ocean
have been found to be effective for protecting piles from further
damage by borers.

METHODS OF REPAIR

Wood piles can be protected from marine borer attack, in place, by wrapping with flexible PVC (polyvinyl chloride) sheeting (cigarette fashion) from 12 inches below the mud line to 12 inches above the highest tide line. In preparing a wood pile for a jacket it is important that all sharp protrusions be removed. In the case of barnacles, sharp edges can be smoothed by various simple hand or mechanical devices.

Modular kits are available which permit fast and positive application from above or below water. This system effectively isolates wood piles from their environment. Marine borers attacking the piles, encapsulated under the wraps, die from lack of oxygen within 48 hours while the PVC sheath prevents further intrusion. This system has been successfully used on both coasts of the United States, in Germany, Australia, the Bahamas and elsewhere for more than 20 years. The United States Navy has also used the system on numerous projects.

Metal wrapping to protect wood piles from marine borer attack has had a measure of success (cupro nickel and aluminum), but is too expensive when compared to the 30-year-plus life expectancy of PVC jackets.

Concrete encasement

Wood bearing piles which have been damaged by marine borers (either partially or totally destroyed) can easily be restored to their design capacity, in place, by encasing them in nylon jackets (adding steel reinforcement if necessary) and filling the jacket with a tremie concrete. A similar successful method is to enclose the damaged portion of the pile with a fiberglass form and fill the space between the form and the pile with a hydrophyllic epoxy.

Wood piles which have been severed by marine borers have been successfully restored, in place, by literally dozens of methods. Almost all of these are based on external reinforcement, such as heavy-wall steel pipe, overlapping the upper and lower pile sections, followed by corrosion protection of the steel by a plastic wrap. Small-scale tests in the laboratory have proved the efficiency of these methods with respect to restoring their full design capability in both the bearing and bending capacity of the repaired piles.

In-place treatment of timber cracks

If such cracks make the pile structurally unsound and also expose it to internal marine borer action, the crack can be bolted together with form-fitting steel washers on each side of the pile by one or more bolts. The entire area should then be jacketed with a PVC (polyvinyl chloride) jacket as described in the foregoing section.

Replacement

The wood pile is cut several inches below the drift pin and the lower section is bent away from the pile cap which permits this section to be removed by a crane. The upper part of the pile is pried off the drift pin and the latter cut off or hammered up flush with the bottom of the pile cap. A replacement pile is set in the same hole as the one from which the damaged pile was removed and driven to refusal. The cut-off is made several inches below the pile cap and the pile is pulled into place. Shims are inserted to fill the space between the top of the pile and the pile cap and a hole is drilled into the pile top to receive the drift pin which is hammered into place.

5.2.4   References for Section 5.2

1. AASHO.
    n.d.   Standard guide specification for epoxy resin systems. MZW-65, AASHO Specifications for Highways, Washington, D.C.

2. Avent, R. R., and J. T. Baylot.
    1978. Theoretical and experimental investigation of the strength of epoxy-injected joints. Eng. and Ind. Res. Stn. Rep. No. MSSU-EIRS-CE-79-4, Mississippi State Univ., Mississippi State, Miss.

3. Avent, R. R., L. Z. Emkin, R. H. Howard, and C. L. Chapman.
    1976. Epoxy-repaired bolted timber connections. Journal of the Structural Division, ASCE, vol. 102, No. ST4, April. pp. 821-838.

4. Avent, R. R., L. Z. Emkin, and P. H. Sanders.
    1978. Behavior of epoxy repaired full-scale timber trusses. Journal of the Structural Division, ASCE, vol. 104, No. ST6, June. pp. 933-951.

5. Avent, R. R., L. Z. Emkin, P. H. Sanders.
    1980. Bonding for structural repairs at timber connections. Journal of the Structural Division, ASCE, vol. 106, No. ST1, January. pp. 87-98.

6. Avent, R. R., R.R.A. Issa, and J. Baylot.
    1978. Weathering effects on epoxy repaired timber structures. Proceedings, Structural Use of Wood in Adverse Environments, Vancouver, B.C., May.

7. Avent, R. R., P. H. Sanders, and L. Z. Emkin.
    1978. Epoxy repair of timber structures comes of age. Construction, Adhesive Engineering Company, vol. 13, No. 2, September. p. 6

8. Avent, R. R., P. H. Sanders, and L. Z. Emkin.
    1979. Structural repair of heavy timber with epoxy. Forest Products Journal, vol. 29, No. 3, March. pp. 15-18.

9. Building Research and Practice.
    1978. Restoring timber with plastics. Building Research and Practice, May/June.

10. Jessome, A. P., and D. E. Kennedy.
    1960. Strengthening ring-connected timbers with steel strapping. Canadian Consulting Engineer, November.

11. Ketchum, V.
    1945. Timber maintenance methods. Western Construction News, January.

12. Ketchum, V., T. K. May, and F. J. Hanrahan.
    1944. Are timber checks and cracks serious? Engineering News-Record, July 27, pp. 90-93.

13. National Forest Products Association.
    1977. National Design Specification for Wood Construction, NFPA, Washington, D.C.

14. Nordby, G. M.
    1962. Guide for use of epoxy compounds with concrete. J. of ACI, vol. 59, No. 9, September.

15. Phillips, M. W., and J. E. Selwyn.
    1978. Epoxies for Wood Repairs in Historic Buildings. Heritage Conservation and Recreation Service Publication No. 1, U.S. Dep. of Inter., Washington, D.C.

16. Sanders, P. H., L. Z. Emkin, and R. R. Avent.
    1978. Epoxy repair of timber roof trusses. Journal of the Construction Division, ASCE, vol. 104, No. CO3, September. pp. 302-321.

5.3 Quality Control to Insure Adequate Repair*

One of the major difficulties associated with timber repair procedures is quality control. Engineering inspections are necessary but may not be enough. While inspections may verify that the correct number of bolts or bracing components have been added, it may prove difficult to determine if all split rings have been placed properly, or if the epoxy has completely penetrated the damaged area. This concern is often emphasized by the fact that workmen may have relatively little experience with repairs as opposed to their experience in new construction. The purpose here is to describe appropriate quality control measures to ensure adequate repair.

Engineering inspection is a primary measure. In unusual repair situations, timely inspections, particularly at key points in the repair process, can preclude later difficulties. Inspectors should not only be checking the repair itself but also be looking for any damage previously undetected. In some cases the removal of coverings during repair may reveal wider damage than originally anticipated. For example, decay around split rings may not be visible until members

---

*Author: R. Richard Avent, Professor of Civil Engineering, Mississippi State University, Mississippi State, Mississippi.

are pulled apart during repair. In other cases cracks and fiber damage may not be exposed until repair is well along.

In many cases laboratory testing is not possible for wood repair in contrast to concrete repair where test cylinders can be taken. For example, lack of quality control can result in serious problems for epoxy-repaired members. Many epoxies are very sensitive to mix proportions. The standard injection equipment consists of two positive-displacement pumps driven by a single motor geared to obtain the proper mix. The two epoxy components are mixed at the nozzle; thus a fairly continuous flow prevents hardening of the epoxy in the nozzle. However, crimped lines, malfunctioning pumps, or line blockages can sometimes occur. In severe cases the epoxy will not harden at all, but in other cases the problem may result in soft spots within joints. Frequent collecting of small samples in containers will verify if the epoxy is hardening as expected, and this is routinely done by contractors on an hourly basis. The detection of weak but hardened materials is much more difficult. One method is to inject shear block specimens at the beginning of operations and after the repair of every fifth member. A shear specimen (fig. 5.3.1) is cut into four shear blocks after curing and each is tested in single shear. The failure stress level should be approximately equal to the ultimate shear strength of the wood. This level of shear strength indicates a high-quality bond.

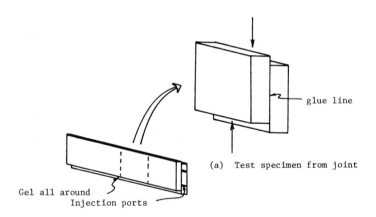

(a) Test specimen from joint

(b) Joint for quality control tests

Figure 5.3.1 Shear block specimen

Another quality control problem is that of determining epoxy penetration into voids. Special sampling techniques are currently in the development process, but none have proved completely satisfactory as yet. This problem is often heightened because there are two types of epoxy repair: structural and non-structural. Non-structural repairs are associated with sealing in applications such as water-proofing, crack sealing to prevent contamination, and cosmetic repairs. Many contractors are familiar only with this type. The approach to non-structural repairs is to inject from port to port without undue concern for complete penetration. Often air voids are trapped by such an approach. The key to a successful structural repair is to fill all voids. To ensure complete penetration, it is best to inject from only one port while letting all others serve as vents. The successive bleeding and capping of these ports gives a high degree of confidence in the amount of penetration. An average repair often involves at least 12 ports and many have considerably more. However, without close supervision of the injection operation, a contractor may revert to his usual approach for non-structural repairs, especially since the different goals of these types of repair are usually not appreciated. Close supervision thus becomes the primary method of quality control.

For some cases, it may be necessary to proof-load the repaired structure to ensure adequate repair. Special caution is urged because overloads can cause fiber damage in wood which is not readily observable. A lower load on a second application could cause failure.

In special cases radiography or radiation-transmitting devices may prove useful in examining internal repairs otherwise not visible. However, such equipment may not be readily available.

Whatever the repair application, quality control is an important part of the process. Consideration to quality control should be given at an early stage to ensure a successful repair.

5.4 Examples of Successful Repairs

5.4.1 Beams*

5.4.1.1 Epoxy Repair of 80-Foot-Long Delaminated Beams in a Gymnasium

Construction was nearing completion on a community college gymnasium in western Oregon when a series of open joints showed up in one beam, (fig. 5.4.1.1(a)). Later it was determined that the open joints had been noticed when the beam was being wrapped, but the yard foreman shipped it anyhow without saying anything to the management. After several conferences it was agreed that the joints should be injected with epoxy and, at the request of the architect, all of the beams would be load tested to 1-1/2 times the live load. The only dead loads not on the beams at that time were the backboards. The local structural concrete bonding firm had the only readily available injecting equipment so that an epoxy which is commonly used for concrete repair was injected.

Author: Dick W. Ebeling, Dick W. Ebeling, Inc., Portland, Oregon.

224   WOOD STRUCTURES

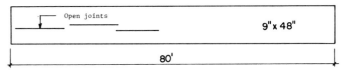

(a) Original delaminated beam

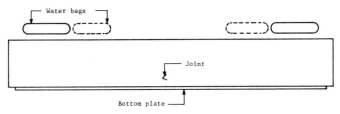

(b) Load testing and reinforcing used

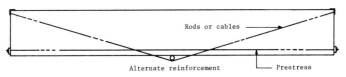

(c) Alternative reinforcement suggestions

Figure 5.4.1.1. Repair of delaminated roof beam

For test loading, two of the large rubber-impregnated nylon bags used by the trucking firms for glue transport were rented. These were about 4 feet in diameter and 30 feet long when full and made a simple and accurate loading mechanism when used with a metered water pump. Shoring was placed below the beams with a 2 to 3 inch clearance.

The first test was made with the bags in the dotted position on one of the undamaged beams (fig. 5.4.1.1(b)). At approximately the full test load, the beam failed in bending and dropped to the scaffolding. The beams had been erected in early October but, due to excessive rain, it was not possible to put any roofing on until late February. The beams were for interior use so casein glue had been used throughout including the finger joints. Inspection after the failure revealed that a horizontal finger joint in the second soffit lamination had most of its glue dissolved and this apparently triggered the bending failure in the tension laminations.

The beam was jacked back into position and all breaks injected with epoxy. A steel cover plate sized to carry all the tension (with the top portion of the beam carrying the compression due to bending) was attached to the soffit with full-thread lags and shear plates. The test was successfully repeated and the beam has been in service for over 10 years now, with no signs of any problem. Some of the initial abrupt tension breaks near the center have separated by about a quarter of an inch. However, this is reasonable as they are not expected to carry any load although the stress level is high until they fail.

All the other beams were then reinforced with a thinner cover plate before testing. The retesting and final testing were done with the bags in the position shown by the solid line in figure 5.4.1.1(b). This loading stressed the beam to 1-1/2 times the design load in shear but only to the design load in bending. During testing, some of the joints in the beam which had been open originally, and had been epoxy-injected, reopened. Upon inspection it was found that in the thinner joints the epoxy had penetrated only a short distance. By holding the joints open about one-sixteenth of an inch while reinjecting the epoxy, full penetration was achieved. Penetration was checked by core borings. The beam then passed its load test and the roof was accepted.

Several things can be noted from this repair: first, if a standard epoxy for use with concrete is used, the joints must be held slightly open for the feathered-out wood fibers are more resistant to flow than are concrete cracks. If epoxies of a lower viscosity are available, the opening of the joints will not be needed. Second, extreme caution must be used in load testing beyond the design load because of the variability of the strength of wood.

Although not used in this case, an alternative reinforcing scheme as shown in figure 5.4.1.1(c) could have been utilized. Rods or cables can be used to provide a modified king post type of load-carrying member having a higher capacity than the original beam. Alternately, prestress cables anchored into bearing blocks at the beam ends can be used to reinforce the beam.

5.4.1.2 Replacement and Reinforcement of Decayed Beam Ends

A warehouse at a papermill on the Oregon coast was constructed with projecting beam ends as shown in figure 5.4.1.2(a). After a few years, loose paint on the beam ends was noted. Closer inspection revealed extensive rot and some insect damage, probably from termites. To repair, two procedures were used. When the rot visible to the naked eye stopped within 5 feet of the wall, a steel moment splice with fill was used (fig. 5.4.1.2(a)). If the rot extended further, a 30-foot section of the beam was replaced as shown in figure 5.4.1.2(b). All of this repair work was done without disturbing the roof other than detaching the joists and reanchoring them. Shoring was used to support the roof at the working points. The location of rot was determined by inspection of the new cut ends, with cores taken in case of doubt and usually examined without a lens.

226 WOOD STRUCTURES

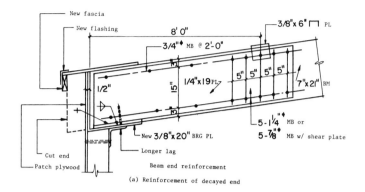

(a) Reinforcement of decayed end

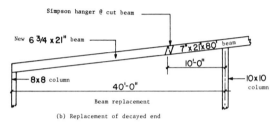

(b) Replacement of decayed end

Figure 5.4.1.2. Replacement and reinforcement of decayed beam ends

5.4.2 Roof and Bridge Decking

5.4.2.1 Transverse Post-Tensioning of Laminated Decks*

The structural deterioration and subsequent reduction of load-carrying capacity in vertically laminated decking is commonly caused by the loss of load transfer capacity of the nails. Transverse post-tensioning of the decking, in effect, replaces the need for nailing by introducing sufficient interface friction between laminations to transfer the required loads.

The system of post-tensioning was developed primarily for use in bridges, but it can be applied to any deteriorated laminated decking. Its effectiveness lies in distributing loads, primarily moving concentrated loads, to a number of laminations. Therefore in situations

---

*Author: R. J. Taylor, Research Engineer, Ministry of Transportation and Communications, Ontario, Canada.

where the loads are predominantly uniformly distributed, its effectiveness may be reduced to one of only introducing better structural integrity.

The details of a rehabilitation by post-tensioning is described in Chapter 8 on case histories, as applied to the Hebert Creek Bridge in Ontario, Canada. The effectiveness of post-tensioning almost doubled the width of load distribution obtained on the Hebert Creek Bridge.

5.4.2.2 Repair of Inadequate Roof Diaphragm*

The warehouse shown in figure 5.4.2.1 was built for a public agency in the Portland, Oregon area a number of years ago on a design-build contract. Due to a combination of errors, five trusses collapsed under dead load only. Checking by the city, the insurance company, and the owner's new consultant after the disaster revealed that replacement of the trusses was necessary. The successful contractor replaced the trusses without disturbing the roof joists, sheathing, roofing, columns, or beams. In working out the design for the replacement trusses, it was found that the system of columns and knee braces was not quite adequate for lateral loading and neither was the plywood diaphragm system in the roof. However, by combining the two systems, the bracing would be adequate if the end portions of the diaphragm were blocked rather than unblocked. (The original diaphragm had been properly nailed but no blocking had been used.) One way of obtaining these higher values is with cleats rather than the usual nailed blocking. This is rarely done because normally it is not cost competitive due to the necessity of working from below. For this case, however, the method was desirable because it did not disturb the roofing. Plywood cleats (fig. 5.4.2.1) were attached to the bottom of the roof sheathing in the six bays at each end of the building at joints between plywood panels in the roof diaphragm. A new joist was spliced to the eave joists in the central portion of the roof to create chords for the diaphragm action.

5.4.3 Columns*

5.4.3.1 Erection Damage to Laminated Column

During construction a contractor failed to adequately brace a 30-foot-long 10-3/4- x 10-1/2-inch glued laminated column. When it fell, several splits opened up over most of the length. Upon checking, it was found that a nailed or bolted laminated column would carry the design load. The column was clamped at 2-foot intervals as shown in figure 5.4.3.1 to simulate bolted behavior. As the building is a cold storage plant with a normal temperature of 10 degrees below zero, the wood would rapidly drop to its final moisture content. Therefore, one or two bolt tightenings after the building was in service were all that would be required.

---

*Author: Dick W. Ebeling, Dick W. Ebeling, Inc., Portland, Oregon.

228   WOOD STRUCTURES

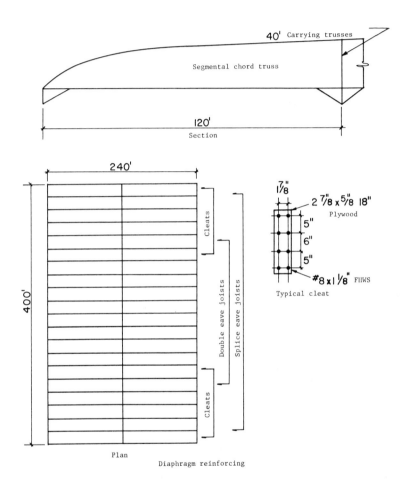

Figure 5.4.2.1. Repair of inadequate roof diaphragm

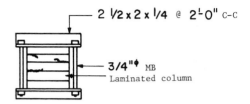

Figure 5.4.3.1 Repair of damage to laminated column

5.4.3.2 Impact Damage to Glulam Column

A car traveling at a high rate of speed failed to negotiate a dog-leg street intersection and crashed through the wall of a church. (The car was totaled but the driver was released after a few hours in the hospital.) A laminated column supported the roof and the balcony at the point where the car entered. The portion of the column below the balcony and the anchorage was shattered. The insurance company felt that the damage was only below the balcony, but the owners of the church followed the engineer's advice to replace the entire column. When the upper portion of the column, which was concealed by the balcony and seating was uncovered, it too, was found to be thoroughly shattered. Damage due to impact such as this or by dropping members may extend for a considerable distance beyond the immediate area of impact.

5.4.4 Trusses*

5.4.4.1 Mechanical Repair of Joints

A large northwest plywood mill had been built on a design-build contract. The roof system was panelized plywood on 2- by 6-inch subpurlins, supported on 50-foot purlin trusses at 8-foot centers (fig. 5.4.4.1(a)) with 90-foot carrying trusses supported on columns. Following a heavy snowfall, a number of the purlin trusses collapsed. In the ensuing investigation the end joint (fig. 5.4.4.1(b)) of the purlin truss was found to be overstressed. Eccentricity had not been considered in the design. Calculations indicated that the safe load for the original joint was only 13 psf.

Repairs were in two phases: the immediate work needed to get the plant back into operation, and the reinforcement of the structure to reduce the probability of future trouble. Fifty purlin trusses were replaced with heavy truss-joists and 11 others, where the existing trusses could not be removed or the truss joists could not be installed, had beams put alongside the original truss to carry the load in the first phase.

---

*Author except where noted: Dick W. Ebeling, Dick W. Ebeling, Inc. Portland, Oregon.

230   WOOD STRUCTURES

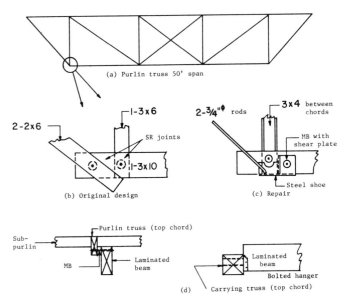

Figure 5.4.4.1. Mechanical repair of truss joists

The second phase was to reinforce 447 purlin trusses by rebuilding the end joint, figure 5.4.4.1(c). A few had other damaged diagonals and were repaired by splicing in new sections between the good ends. The main trusses also were reinforced as shown in figure 5.4.4.1(d). Heel plates were extended, extra splices were added, splice plates were added at joints where the rod and dap reduced the chord section too much, extra rods were added, angles were added to reinforce some bending members, and corbels were added for extra bearing. At the end of the work the roof had a safe capacity of 80% of the live load it should have had originally.

Suspended scaffolds along with night and weekend work were used as much as possible in the second phase to allow uninterrupted operation of the plant.

5.4.4.2 Beam Spliced to Upper Chord

One truss with a configuration similar to that shown in figure 5.4.4.2 had the support for a 48-inch dust and chips blow pipe added near one end some years after the building had been in operation. This blow pipe had a 62-foot span on one side of the truss with 150-foot span on the other. When the pipe plugged up, an overload occurred, and the

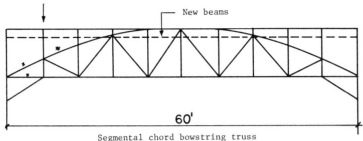

Figure 5.4.4.2. Beam spliced to upper chord of bowstring truss

pieces marked "X" split into kindling wood. The wall leaned out about 3 inches and the remaining upper bowstring chord started to buckle. Luckily there were stacks of plywood under the truss so it dropped only about 6 inches. The loading of the blocked pipe was much more than the original truss capacity. Two beams of the required new capacity were installed, one on each side of the original truss and just below the roof joist with new column struts to the floor. The column struts were stabilized by being bolted to the original columns.

5.4.4.3 Mechanical Repair of Heels and Splices

Several plywood mills constructed 20 to 30 years ago developed brash tension failures in the lower chords or in the splice plates of roof trusses. Usually these failures were in areas near the dryers where the temperatures had been in the 120° to 150° range for years. The failures were nearly straight breaks across the member. Some typical replacement splices are shown in figure 5.4.4.3 where the rod length is varied to get the connections into sound wood. One mill also had several trusses developing problems at the heel because of change in bearings. The original bearing was on a 12- by 12-inch column. Due to changes, a steel beam was used to span to a new column location. As the new bearing was much narrower, the upper chord end was subjected to considerable splitting forces. The addition of the full 12-inch bearing plate eliminated this problem.

5.4.4.4 Reinforcing of Lower Truss Chord with Rods or Cables

The same mill described in section 5.4.4.2 had another truss develop several full-width splits extending completely through the lower chord. As the wood had a one-to-sixteen slope of grain, the lower chord was split into several long pieces. As there was no other damage, it was decided to add longitudinal rods for the full length. This technique works well with bowstring trusses which are essentially tied arches. As long as enough wood remains for the web connection, most of the load can be taken on full-length rods. In this case the lower chord was

232 WOOD STRUCTURES

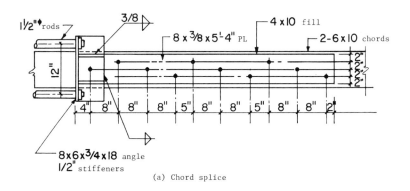

(a) Chord splice

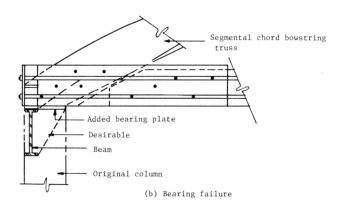

(b) Bearing failure

Figure 5.4.4.3. Mechanical repair of heels and splices

tied together with stitch bolts and angles to carry the shear loads, figure 5.4.4.4.

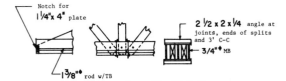

Truss with split chord
Figure 5.4.4.4. Repair of truss with split chord

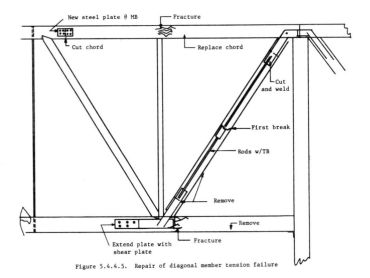

Figure 5.4.4.5. Repair of diagonal member tension failure

5.4.4.5 Reinforcement of Failed Diagonal Tension Member

A large southern plywood mill had a sudden tension failure in the main diagonal at one end. This caused bending failures in the chords as shown in figure 5.4.4.5. The diagonal was replaced by rods welded to the end plate gussets. The lower chord connection was extended to assure sound wood with the damaged portion being removed. The upper chord was replaced for the two end panels with a bolted splice joint at the interior joint. As there were piles of plywood stacked under the truss it dropped only 3 feet which prevented other damage to the truss and kept the roof repairs to a minimum. The joists were not removed when the truss was jacked back into place. The plywood was renailed as necessary and the roofing patched.

5.4.4.6 Reinforcement of Inadequate Joints

Occasionally it will be necessary to correct errors in fabrication. Some possible repairs are illustrated. A split ring connection was too near one edge so that edge and end distances are inadequate; one web member is replaced and the new connection is placed properly as shown in figure 5.4.4.6. The old connection is left with one web cut off as it is not possible to add rings for the web left in place, but the addition of a bolt through the original web will probably compensate for the deficient ring. An alternate solution is to add a bolt to each web member as shown in figure 5.4.4.7.

If splice plates are badly checked or the rings improperly placed, new splice plates can be used. In the example shown in figure 5.4.4.8, the new splice plates are counterbored to accommodate the existing bolt heads, and the new rings are located to clear the existing rings with adequate distances.

A repair by adding a steel shoe for inadequate end distance in the lower chord at the heel is shown in figure 5.4.4.9. An example of reinforcement to the column-to-truss anchorage is shown in figure 5.4.4.10. This was needed because of inadequate welds on the original. Shown in figure 5.4.4.11 is a reinforced heel where the original column bolts had pulled out due to truss failure during erection.

Considerable bracing can be added by using light strapping, such as 0.35- x 1-1/4-inch perforated steel strapping which is available in 600-foot rolls. It may be applied either inside or outside, nailed to girts or purlins. This strapping can also be used for hold-downs at the posts by running the strapping over a truss in a "U" configuration and nailing both ends to the post. With the usual metal siding and roofing, some lateral strength can be added by increasing the nailing to the purlins and installing sheet metal screws in the lap joints of the siding between purlins. Rods the full length of these trusses have been used to reinforce the lower chords; the load in the rods should not exceed the tension in the lower chord at the center.

5.4.4.7 Epoxy Repair of End Split at Truss Joints*

A typical problem with old timber trusses involves longitudinal splits, particularly through the connectors at joints. Several major repair projects consisting primarily of this defect type have been repaired with epoxy. Avent, et al. (1,2) reported the details of one such project at Robins Air Force Base, Georgia. The purpose of this section is to describe a repair project in Denver where heavy timber tension chord joints and splices were successfully repaired with epoxy.

---

*Author: R. Richard Avent, Professor of Civil Engineering, Mississippi State University, Mississippi State, Mississippi.

METHODS OF REPAIR

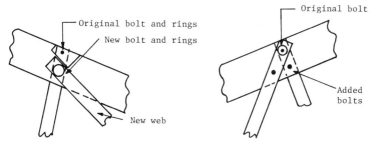

Weak joints

Figure 5.4.4.6. Repair of deficient joints by adding split rings

Figure 5.4.4.7. Repair of deficient joints by adding bolts

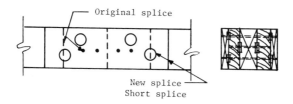

Figure 5.4.4.8. Repair of splice by adding new plates

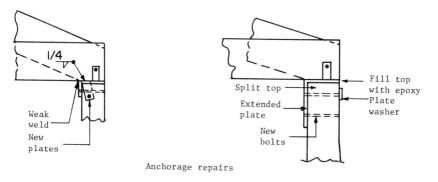

Anchorage repairs

Figure 5.4.4.9. Repair by adding steel shoe to heel of truss

Figure 5.4.4.10. Reinforcement of column to truss anchorage

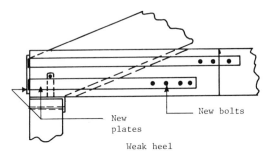

Figure 5.4.4.11. Reinforcement of heel at column to truss anchorage

The building was a hangar at the Denver Airport. The roof trusses were bowstring type with a laminated compression chord and heavy timber tension chords and bracing. Typical members ranged from 3 by 6 inches to 3 by 12 inches. The trusses spanned 180 feet and were spaced at 30 feet. A total of 120 deficient joints were repaired. A typical damaged section is shown in figure 5.4.4.12. Occasional breaks such as shown in figure 5.4.4.13 were also found. The repair procedure consisted of sealing the damaged joints with gel and pressure-injecting a low-viscosity epoxy into the opening. For members with transverse cracks (fig. 5.4.4.13), splice plates were added before sealing and injecting. A typical joint ready for injection is shown in figure 5.4.4.14. The trusses were 30 feet above the floor, requiring all sealing and injecting to be done on rolling scaffolding as shown in figure 5.4.4.15. Since no detectable sagging was found, the trusses were not damaged beyond repair. The regions that were epoxy-repaired will now resist live load stresses created by any snowstorm.

Liberal use of injection ports insured satisfactory penetration throughout a joint. In addition, shear block specimens similar to those shown in figure 5.3.1 were taken after every fifth joint was repaired to verify epoxy strength. The average shear stress at failure for all specimens was approximately 1,200 psi.

5.4.4.8 References for Section 5.4.4

1. Avent, R. R., P. H. Sanders, and L. Z. Emkin.
    1979. Structural repair of heavy timber with epoxy. <u>Forest Products Journal</u>, vol. 29, No. 3, March. pp. 15-18.

2. Avent, R. R., P. H. Sanders, and L. Z. Emkin.
    1978. Epoxy repair of timber structures comes of age. <u>Construction</u>, vol. 13, No. 2, September. p. 6.

METHODS OF REPAIR 237

Figure 5.4.4.12. Typical Damaged Chord in Hangar

Figure 5.4.4.13. Broken Lower Chord Member

Figure 5.4.4.14. Broken Member with Coverplates Ready for Injection

Figure 5.4.4.15. View from Floor Level of Sealing and Injection Process

5.4.5 Arches

5.4.5.1 Wood Arch Relamination by Epoxy Grouting*

The services of consulting engineers were acquired by the U.S. Coast Guard to perform an engineering evaluation of, and to design the repair of the glued laminated wooden arches in hangars at the U.S. Coast Guard Air Base, Elizabeth City, N.C. An evaluation of the arches revealed that there was considerable decay and that repairs were thus necessary. The engineers were asked to design an arch repair procedure that could be expected to effectively accomplish the needed repairs in the least amount of direct "work-on-arch" time.

The hangars were built during World War II and are presently utilized as the Coast Guard's only major aircraft overhaul and repair facility. Each hangar is 220 by 400 feet in overall plan and is framed with laminated wood arches spaced at 20 feet and spanning 175 feet (fig. 5.4.5.1). Arch rise at mid-span is 45 feet clear above the hangar floor. Each arch actually consists of 2 adjacent sister arches. When combined, they have a width of 15 inches and a depth of 32-1/2 inches. Original construction material for the arch laminations was 1- by 8-inch dense southern yellow pine. The length on curve of the arches is approximately 200 feet. Purlins resting on ledgers fastened to the arches support the diagonal roof sheathing which is covered with built-up roofing.

The investigation of the condition of arches revealed that the most significant and widespread deficiency was wood decay caused by leaks in the roofing and flashing. However, there also existed some minor insect damage, fire damage, and delamination. An evaluation of the loadings on the arches indicated that a moment connection was required at the arch crown joint along with shimming and tightening of the arch base connectors in order to prevent movement at the crown joint and tearing of the roof. Other minor repairs included reinforcement, replacement of purlins and ledgers, and tightening of bolts.

Three types of wood arch repairs by epoxy grouting were specified, depending upon the location and extent of decay. The simplest repair was the Type 1 repair, which was used when the decay occurred at the top surface of the arch and was less than 12 laminations deep. The decayed wood was replaced with new 1- by 8-inch laminations with a minimum 16-inch lap on each side of the preceding new lamination. The wood used for the lamination replacement was No. 1 dense southern pine.

The Type 2 repair was used when the decay was located at the top surface of the arch and extended through 12 or more laminations. In this repair, through-bolts were used in the vicinity of the decayed area to tie the sister arches together so that there was a sharing of the loading during the repair. The decayed wood was replaced with new 1- by 8-inch laminations with a minimum 1-inch lap on each side of the preceding lamination. It was determined to be less expensive on the

---

*Author: LCDR Ronald F. Silva, U.S. Coast Guard Headquarters, Washington, D.C.

Figure 5.4.5.1. Repair of Laminated Wood Arches in Progress

Figure 5.4.5.2. Exposed Arch With Decayed Wood Cutout

deeper Type 2 repairs to provide a 5/16-inch thick steel top plate and use 1-inch lamination laps rather than use 16-inch lamination laps and replace considerably more wood. The steel plate was attached to the top of the arch with lag bolts and shear plates, and extended approximately 7 feet on either side of the top new wood lamination.

The Type 3 repair was used when the decay existed in the center of the arch. As in the Type 2 repair, through-bolts were installed, the decayed wood cut out, new 1- by 8-inch laminations with 1-inch laps were installed, and a steel plate was attached to the top of the arch with lag bolts and shear plates. The only difference between repair Types 2 and 3 was that the new laminations increased in length from the bottom of the decayed area to the middle and then decreased in length to the top of the decayed area for repair Type 3.

The first step in the repair procedure was to expose the decayed portion of the arch as shown in figure 5.4.5.2. This included opening the roof, removal of fire draft stops and/or removal of exterior siding on end arches. Once the arch was exposed, through-bolts were installed as required by the repair type.

Next, the damaged laminations were cut out full width to sound wood. Further cuts were made for the required laps, again based on the repair type. The existing wood contact surfaces were thoroughly cleaned in preparation for the new laminations.

The required lengths of the new laminations were carefully measured and they were cut to size. Metal shims 1/6 inch in thickness were secured to the laminations to maintain grouting space between laminations. The new laminations were set up on a sawhorse on the hangar deck and vertical holes were drilled through the laminations to allow the epoxy grout to flow from one lamination joint to another during injection. Also, injection ports were drilled 7 inches on center in each lamination joint and all surfaces were cleaned of wood shavings.

The new laminations were then transported up to the arch. Before the new laminations were set in place, a thick coat of non-sag epoxy joint sealer was applied to the inner face of the sister arch. The new laminations were then set in place and nailed securely to the existing wood. Plastic injection tubes 2 to 3 inches long were inserted in previously drilled injection ports as shown in figure 5.4.5.3. All surface joints were then sealed with non-sag epoxy joint sealer and left overnight (12 hours) to set up.

The next phase of the repair was the mixing of the epoxy grout (DURALITH LV as manufactured by the Dural International Corp.). The epoxy grout was mixed in batches which were sized based on the application rate of the injection crew. The epoxy grout was poured into reusable plastic cartridges which were inserted into hand operated or pneumatically operated caulking guns.

Once the caulking gun was loaded, the epoxy grout injection procedure was ready to begin. The grouting was accomplished by injecting the grout into the lowest port, maintaining a slow steady pressure. When

METHODS OF REPAIR

Figure 5.4.5.3. Arch Section Ready for Injection

the grouting material reached the next highest port, the lower port was closed with a wood plug. The same procedure was followed on the next higher port until the repair was completely grouted and all voids were filled.

The next day the joint sealer was sanded off and the surface prepared and painted. The top steel plate was installed if required by the repair type. Finally, the roof was patched or the siding replaced as necessary to complete the job.

Control testing by a testing laboratory was required for the repair grouting of the arches. Two test blocks were prepared, grouted, and cured using the same procedures as were actually performed on the arches. The test blocks were 4 feet long and six 1- by 8-inch laminations deep, with lamination scarfs located at the center of the test block, where the tension test specimens would be taken. The test blocks were then delivered to an independent testing laboratory. Specimens were prepared and tested in accordance with ASTM D 143 for tension parallel to grain and shear parallel to grain. The test results were satisfactory.

A total of 46 separate arch repairs were made on the 42 pairs of arches in the hangars. The repairs ranged in size from 2 linear feet of laminations replaced to 1,715 linear feet. One particular arch had 1,970 linear feet or approximately 25% of its laminations replaced. In total there were 7,196 linear feet of new laminations grouted into place.

Although the repairs varied considerably in size and complexity, a systematic approach to their accomplishment facilitated the scheduling of Coast Guard activities on the hangar deck. Once the arch was exposed, it was possible to perform most of the repairs in a 4-day operation as follows: first day, cut out decayed wood; second day, set up new laminations and seal joints; third day, inject epoxy grout; fourth day, sand and paint.

A great deal of the success of this project must be attributed to Mr. Ralph Silverbrand of Brandway Constructors, Inc., Hawthorne, N.J., who supervised the project and was responsible for developing the field techniques used in epoxy grouting of the wood arches on this project.

Once the initial learning curve was passed, the repair method described herein was considered an effective method for repairing in-use glued laminated wood arches. The requirement of minimal disruption of hangar deck activities was met. The repair method required only minimal direct "work-on-arch" time and could be performed systematically. The repaired arches have been in service since 1977 and no visual signs of distress have been noted.

5.4.5.2 Decayed Arch Base*

A church discovered that they had a problem with their arches when a boy idly stuck his knife into the arch to the full length of the blade. The plans had placed the arches on the footings with the slab poured around the base of the arch. The arch which had the decay had an open gap between the slab and the footing. This arch was cut above the floor and a moment splice applied to a new base section anchored on top of the slab. All of the other arch bases were cut at the floor line and a new base shoe installed on top of the slab as shown in figure 5.4.5.4.

5.4.5.3 Radial Stress Cracks in Arches*

Several tudor arches with exposed rear faces on the leg developed a feathering out of the surface wood fibers on the rear of the arch. This was not a structural defect but it gave an unsightly appearance. To correct this defect, cover boards were nail-glued to the back face with the edge surfaced down to match the arch width as shown in figure 5.4.5.5. This stopped further feathering and improved the appearance.

5.4.5.4 Repair of Improperly Placed Arches*

After completion of roofing and removal of the wrapping, the arches in figure 5.4.5.6 were found to have been laid up with the higher-grade

---

*Author: Dick W. Ebeling, Dick W. Ebeling, Inc., Portland, Oregon.

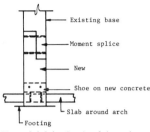
Figure 5.4.5.4. Repair of decayed arch base

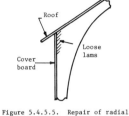
Figure 5.4.5.5. Repair of radial stress cracks

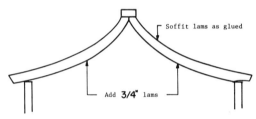
Figure 5.4.5.6. Repair of improperly placed arches

soffit lamination on the top rather than the bottom surface. To correct the appearance and the under-strength, two 3/4-inch-thick laminations were field glued to the soffit with casein glue. Twelve-penny casing nails, one for each 2 square inches, were used for gluing pressure. The arches were then load tested to design load before acceptance.

5.4.6 Pole Buildings*

The pole buildings considered here are those structures whose foundations consist of wood poles or posts, usually pressure treated, set into holes in the ground which are then back-filled. Similar methods may apply to construction with shallow pile footings or to some sidehill stilt-type construction with concrete footings under the posts. These buildings are usually of an intermediate style of construction between frame buildings similar to houses and the light industrial-type roof supports at 20 feet or so. Some of the problems to be discussed do not occur with the original concept of pole supports at 10-foot

---

*Author: Dick W. Ebeling, Dick W. Ebeling, Inc., Portland, Oregon.

centers both ways or with the pole foundation for a house platform where there are no trusses and the poles are short.

Many of the problems encountered with pole buildings have been due to improper construction practices or failure of the builder to obtain an engineered design for the building. Some are due to the need to bring a building into compliance with the code due to change in occupancy from an agricultural to a commercial use, or arrival of a building inspector to stop work until the design is properly performed and a building permit obtained. Adding posts in the walls to share in the lateral loads, adding intermediate trusses to reduce the load on trusses or purlins, and adding bracing in the roof or walls are all solutions that have been used to increase the capacity of an inadequate design to that required for code-specified loadings.

As some codes do not recognize earth backfill around posts for lateral support although they may allow more movement at the groundline in some types of structures, one solution is to excavate the earth from around the post and backfill with concrete. If part of the problem is inadequate bearing area, steel pins through the post projecting into the new concrete can be used. If the concrete footing is too small, the top can be chipped out and a collar poured. These techniques are shown in figure 5.4.6.1.

On a recent job several large trusses that were not properly guyed blew down during construction and broke the posts off near the ground. As a solution, the pole was replaced in the footing. To remove the pole stub from the footing the contractor used a household vacuum cleaner and a section of downspout to supply needed oxygen to the fire he started in the top of the stub after drilling some large holes for a starting point. No damage from the fire was apparent in the concrete.

A frequent problem which is easier to solve during construction is the failure to get the proper pole size for lateral or vertical loading due to the unsupported height. Several solutions have been used. One is to bolt two poles together with sufficient bolts to ensure beam action (fig. 5.4.6.2(a)). The bolts are uniformly placed for the full length using twice the spacing that would be required at the ground. Second, the use of short stubs tied to the main columns to share the bending may be adequate (fig. 5.4.6.2(b)). Third, the addition of a knee brace to partially fix the top of the column will reduce the bending in the column and its unsupported length, although its effect on the trusses must be checked (fig. 5.4.6.2(c)). Fourth, on one job where there was a floor, braces were added underneath to eliminate the movement at the ground surface as shown in figure 5.4.6.3. This allowed the existing posts and undersized footings to carry the only horizontal load at the ground which they could do because they no longer had a moment at the groundline.

On one job it was found that the stub provided was too short. It was necessary to splice the stub to provide adequate length as shown in figure 5.4.6.4. Also, the addition of a concrete floor slab with the post anchored into it will allow a reduced depth or width of footing which will upgrade existing footings.

METHODS OF REPAIR

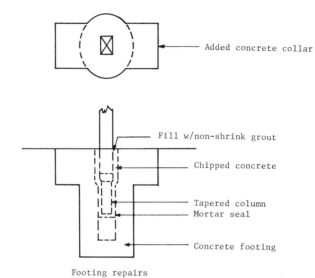

Footing repairs

Figure 5.4.6.1. Enlarged collar around pole to improve bearing.

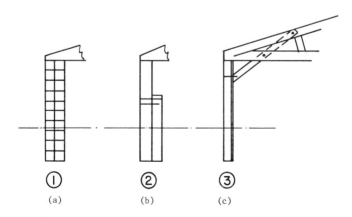

Figure 5.4.6.2. Repairs for improper pole size

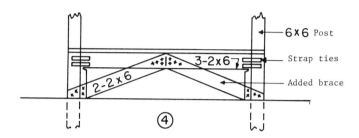

Figure 5.4.6.3. Bracing of undersized poles below floor level

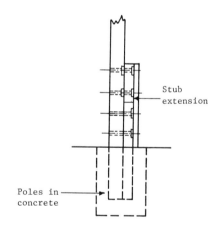

Figure 5.4.6.4. Stub extension to pole

METHODS OF REPAIR 249

5.4.7   Repair and Maintenance of Piles*

The Port of Los Angeles, with its inventory of over 40,000 creosoted wood piles, continually faces the problem of economical maintenance of its wharf structures. Approximately 200 to 300 piles are destroyed in this harbor annually by <u>Limnoria</u>. In order to minimize costs, the Port of Los Angeles Testing Laboratory started a program of applied research about 20 years ago to devise ways and means to protect the piles from premature destruction. Many of these studies involved the application of various types of barriers in order to increase the life of in-place piles for an indeterminate period. Many obstacles in fastening the wrappers securely were finally overcome in the year 1957 when an outside investigator devised a unique method of applying prefabricated barrier units under water.[1]

After several years of investigation, flexible polyvinyl chloride sheeting has been found to be the most economical material to encapsulate the piling. The purpose of this barrier is to create a lethal environment, in which the borers cannot live, by eliminating dissolved oxygen in the water under the wraps. This is often referred to as the "stagnation principle." To achieve this effect, it is necessary to have the barrier fit the general contour of the pile rather snugly. This may be easily accomplished by attaching the vertical edges of the sheeting to a dowel and rotating the dowel to tighten the barrier sheet against the pile (fig. 5.4.7.1).

To prove the efficacy of the method, untreated soft pine test sticks were inserted under the wraps in such a manner that the sticks could be withdrawn and examined, after the pile had been in service for a year or so, in order to confirm the presence or absence of marine borers under the wrapping. There has been no reported incident of any attack, even after 15 years exposure of these sticks under the wrapper in borer-infested waters. The "stagnation principle" of using wrappers has since been successfully used in the Gulf of Mexico, along the California Coast, and in Pearl Harbor. Nearly all of the applications have been made by contract; the largest one was recently completed for 60,000 lineal feet of protection at the Port of Los Angeles.

Application of these in-place barriers is not only economical and extremely practical, but has the advantage of preserving wood piles under deck structures, such as warehouses, loading and unloading facilities, etc., where the replacement of piles destroyed by marine borers would be highly expensive and very difficult unless all or part of the superstructure was removed. In addition, none of the installing operations interfere with the use of the wharfs for regular port functions, because all work is performed below the deck surface.

---

*Condensed from an article "The Case of the Proliferating Punctata," by Carrol M. Wakeman and Frank J. Steiger, <u>Wood Preserving News</u>, September, 1966.

[1]<u>Extending Service Life of Wood Piles in Sea Water</u>, C. M. Wakeman and L. L. Whiteneck, ASTM Special Technical Publication No. 275, 1959.

Figure 5.4.7.1--Installing barrier wrap on pile.

Figure 5.4.7.2 illustrates a section of piling which had been partially wrapped in the year 1962. This pile was withdrawn from service 3 years later. At the time the lower portion of the pile received its barrier coat, there was insufficient attack on the upper section to warrant wrapping, and hence, the upper half of this specimen remained without a protective wrapper up to the year 1965. The advantage of the barrier is obvious from the diminished diameter of the unwrapped upper portion of the sample which suffered attack during that period.

The average harbor engineer is apt to be a little startled by a proposal to protect wooden piles with 30-mil sheets of polyvinyl chloride. The question immediately arises in his mind as to how this material will survive the impact of floating debris. Parenthetically, it should be remembered that the only vulnerable area, insofar as debris damage is concerned, is the intertidal zone. We can answer this question by citing the 18-year experience at the Port of Los Angeles where not a single case of damage to the wrappers has been found in more than

METHODS OF REPAIR   251

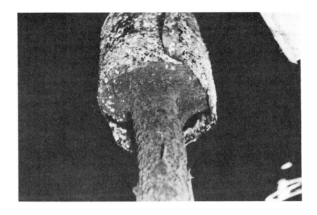

Figure 5.4.7.2--Section of wood pile after 3 years in sea water.
Lower portion had been protected with barrier wrap.

100,000 lineal feet of barrier-coated piles, despite the fact that an
average of more than 14 tons of floating debris is removed daily from
the harbor. This accumulation varies from derelict skiffs, log rafts,
and free floating logs and driftwood, to empty oil drums, crates, and
other flotsam.

In a recent situation, dramatically confirming the strength of polyvinyl chloride jackets, a violent storm occurred in the early part
of 1966 in Avalon Bay, Santa Catalina Island, where piers are supported by pressure-creosoted Douglas-fir piles protected from marine
borers with PVC barriers. Within a half hour after the storm struck,
Avalon Bay was one heaving maelstrom of wreckage. Enormous waves tore
entire piles out from under the piers. Fourteen boats were sunk
immediately and all of this debris acted as battering rams against
the piles where barrier coatings had been placed. A later inspection
indicated that not a single wrapper on the piles remaining in the pier
had been torn or otherwise damaged by storm action.

How long will polyvinyl films last in a marine exposure? Frankly,
no one knows for sure, but there has been a successful history of this
material at the Port of Los Angeles for more than 18 years--PILE-GARD
is one example. It is now generally conceded that wrapping can be
considered as a permanent barrier rather than mere "first aid."

In general, barrier wrapping should be installed after a reliable
inspection has shown that marine borer attack has penetrated the outer

creosote layers of the piling. Much depends upon the practical economics of the particular situation.

It is not economical to encase piles prior to driving because, particularly in rock formations, the jackets are invariably damaged. While wrappers can easily be repaired by divers after the driving, the cost is usually greater than in-place installations after first attack has been noted.

Engineers often apply the reasoning that, if jacketing a partly destroyed pile is feasible and economical, why not apply barriers for the entire length of all newly driven piles. First, for some mysterious reason apparently instinctive to borers, not every pile in a given wharf system will be attacked at the same time, if ever.

To illustrate, a certain wharf was constructed in 1957, using 467 heavily creosoted Douglas-fir piles. By the year 1965, only 5 of that number had deteriorated sufficiently to warrant wrapping. It is quite apparent here that it would have been uneconomical to have installed barriers on 467 piles where only 5 were required after 8 years of exposure. Secondly, even after the attack has begun, the destroyed area may be localized for a number of years. Sometimes a relatively short jacket applied to each pile will protect it indefinitely and no further barriers will be needed.

Another reason why polyvinyl chloride barriers should not be applied to new piles is borne out by the fact that, when first driven, all recently pressure-treated creosoted piles have a tendency to "bleed"; that is, there is a slow exudation of excess creosote down the outside of the piles from the outer wood cells. Polyvinyl chloride, while completely resistant to a marine environment, is susceptible to blistering in the presence of free creosote. But, after piles have been exposed to sea water for several years, there is very little danger of free creosote being present, particularly when the outer portion of the pile has been removed by action of the _Limnoria_. While a thin layer of polyethylene (completely resistant to creosote) can be interposed between the polyvinyl and the wood pile, this is an added expense. It is far less costly to let nature remove the creosote before applying the barrier. At the present time, the Port of Los Angeles finds it economical to delay wrapping borer-infested piles until the attack has reached about 10%.[2]

Prior to the adoption of barriers for extending the life of creosoted wood piles at the Port of Los Angeles, an economic study was made of an alternative choice of pile replacement. The average cost of pile replacement in our facilities is $450 per pile. The conclusions of this study were: "the justifiable expenditure (for barrier application) per lineal foot of exposed pile . . . is $9.86." In an environment where the piles will last only 15 years, as described elsewhere in this paper, timber pile wrapping may reduce annual pile maintenance

---

[2]_Evaluation of Wood Piles in Sea Water by Standard Rating System_, E. V. Dockweiler and H. E. Stover, ASTM Special Technical Publication No. 275, 1959.

cost by as much as 50%. The justification for pile wrapping must be evaluated in terms of local environment, pile service life, as well as wrapping cost; however, a good pile inspection program will readily establish the facts for use by the engineer.

CHAPTER 6 MAINTENANCE*

6.1   Introduction**

6.1.1   Maintenance Philosophy

When this commentary was in the planning stage, it was assumed that it would be rather a simple matter to locate authorities on maintenance of timber structures. Calls to several major companies, owners of acres of buildings and other structures worth millions of dollars, failed to locate one with a comprehensive and systematic maintenance program. This was not the case for machinery and production equipment where downtime reflected immediate economic losses. When productivity is the issue, there are as many as four levels of maintenance ranging from routine to emergency maintenance.

But when it comes to a structure, or the building that houses the plant, maintenance suffers from a low priority image. Maintenance is the first to feel the impact of any economic belt tightening. The philosophy seems to be, "Let it fail and then replace," but in fact it is more economical to maintain than to replace.

6.1.2   Literature Review

To determine what information is currently available on maintenance of wood structures, a computer literature search was conducted. Five categories, or files, were searched, including Compendex (Engineering Index) and National Technical Information Service. Twenty-five set item descriptions (key words) were used to identify the scope of the search.

Out of more than 300 citations, less than 50 addressed wood in any depth. Of those that did, many were directed more toward factors which influence serviceability, inspection, and repair (chapters 2, 3, and 5) than toward maintenance. Many of the authors listed in the citations are contributors to this commentary.

The subject of wood finishing listed the greatest number of citations, most of which appear in section 6.6.6 of this chapter.

There are some excellent publications that deal with maintenance, but do not focus on specific materials such as wood. One publication (1) on port maintenance covers the entire facility, and does have a chapter

---

*Chapter coordinator: Roger L. Tuomi, Principal Research Engineer, USDA, Forest Service, Washington, D.C.

**Author: Roger L. Tuomi

on wood for waterfront structures. Another is specific to historic preservation (2). Both provide valuable background on the benefits of a sound maintenance program, and the steps necessary to implement and administer it.

Some very good articles on maintenance of wood structures appear periodically in digests, technical notes, and newsletters published in other countries. These include: Canadian Building Digest, Division of Building Research, National Research Council of Canada; Rebuild (a newsletter), Division of Building Research, Commonwealth Scientific and Industrial Research Organization (CSIRO), Australia; information papers and digests, Building Research Establishment and Princes Risborough Laboratory, England; and technical notes, Timber Research and Development Association (TRADA), England.

The conclusions of this literature search are that there is not a great wealth of information published specifically on the maintenance of wood structures, and those that are cannot readily be located and retrieved.

6.1.3   What is Maintenance?

Maintenance includes all the activities necessary to prolong the life of property, and is often called by different names. Routine maintenance is the work of keeping things in a good state of repair, such as the simple tightening of bolts in a joint to correct for shrinkage after the wood has reached equilibrium moisture content.

Preventive maintenance is a term often used to define activities that reduce the probability of a problem developing, such as adding lock washers to the above joint to minimize the need for retightening.

Preservation is the application of measures to sustain the structure essentially as existing, without changing appearance or materials of construction. Maintenance may have slightly different meanings depending upon purpose, value, and function of the structure, but the goals are common--insure the continued use of a structure at a desired level of efficiency and capacity.

6.1.4   Why Have a Maintenance Program?

When maintenance is ignored or discontinued, the life span of structures is dramatically reduced. Premature failure of buildings and other structures results in high costs for repair or replacement plus potential danger to life safety. To avoid this, an effective maintenance program should be established and attentively followed.

The benefits to both owners and society include (1) longer service life, (2) reduced frequency and extent of repair, (3) improved safety, and (4) extension of the wood resource (about $1 billion worth of wood products are lost each year due to poor construction practice and lack of maintenance).

6.1.5 The Maintenance Program

Since all structures vary in age, type, environment, exposure, etc., no single maintenance program can be applied to all. Maintenance is a common sense approach to avoiding problems. Once established, it should become as routine as changing the motor oil in an automobile according to the odometer reading or season. Some key elements in a sound maintenance program include record keeping, inventory of facility or structure, evaluation of condition and potential hazards, systematic inspection at prescribed frequencies, and corrective actions.

6.1.5.1 Records

Records on buildings and other structures have an uncanny knack for getting lost, particularly during change of ownership. They can be extremely valuable documents that should be maintained in a safe place. There are several types of records that serve different purposes, but all are important. Some important ones are discussed.

<u>Age of the Structure</u>

As mentioned elsewhere, age in itself is not a factor in the structural integrity of buildings or other structures. But knowing when the structure was built will help to identify design methodology, design stresses, material sizes, fastener types, and craftmanship to name a few. The discussion of old wood structures, along with special concessions for World War II, presented in Chapter 1, and the case histories in Chapter 8 will be helpful in understanding the design philosophies and methodologies at the time older structures were built. Maintenance requirements and conditions to be aware of will differ with age and construction details.

<u>Original Construction Information</u>

Original construction plans, specifications, and as-built drawings along with engineering calculations are useful to determine if the structure is being used and loaded as intended. Increased load demands or alterations (often nonengineered) to the structure will identify critical areas or elements that require special attention and observation.

<u>Modifications and Alterations</u>

Too often modifications and alterations are made without the benefit of an engineering analysis. Mechanical equipment, crane rails, hoisting equipment, and suspended ceilings are often installed whenever convenient without regard for their impact on structural integrity. This should be avoided at all costs. Modifications and alterations should require a complete engineering evaluation. Plans, specifications, and calculations should be retained along with the original construction records.

Change in Use or Environment

Many structures are converted from one use to another. For example, an unheated warehouse might be converted to office space with a conditioned environment. A warehouse might be changed from lumber to chemical storage (although wood is relatively inert to most chemicals, the fasteners may require special protection or replacement).

A review of factors that affect serviceability (ventilation, condensation, finishes) should be made whenever the environment is altered and a revised maintenance program developed accordingly.

Repairs

Repairs properly designed and properly made will restore a structure or element to its original or better condition. Repairs improperly made will not solve the original problem and may create worse ones. Records should be kept on all repairs. The date, location, and type of repair should be noted and the repaired area or element carefully monitored until it is certain that the repair is performing satisfactorily. If any servicing, adjusting, or special maintenance to the repair area is necessary, it should be so noted.

Major Events

Records of major events such as seismic activity, high winds, heavy rains, heavy snow loads, and accidental damage should be maintained. Because a structure survived one major event does not insure that it will do so again. The structure may have been weakened by prior exposure, events preceding the major one might vary (large snowfalls in short periods of time with little melt), or load distributions (drifting) might differ. For example, there are a lot of lessons to be learned from failures under record snow loads during the winters of 1978 and 1979.

Dimensions and Elevations

Two rules of construction are that nothing is ever square and nothing is ever plumb. Buildings will move, particularly in the vertical direction. Some movements are of no structural concern whereas others might be serious.

Heavy solid sawn timber and glued-laminated beams may deflect either upward or downward due to changes in moisture content between the top and bottom surfaces. Painted surfaces on vertical pipe guides and duct hangers (when the ducts are supported from below) often show an inch or more of vertical movement.

The truss arching phenomenon is another example of vertical deflection. In some cases, trusses arch upward during the winter heating season and come back down during the summer. Again, this is undoubtedly related to differential moisture content (between the top and bottom chords). Neither of the above movements adversely affect structural capacity, but may produce aesthetic or other undesirable consequence.

Other types of movement may result from foundation settlement, creep deflection, or joint slippage. Interestingly, it is often a custodial person who first notices such movements.

The important thing to know, in assessing the consequence of dimensional changes, is when they occurred and why. Bench mark elevations on principal structural elements and other dimensional data are useful in determining if a structural problem exists.

Inspection and Maintenance Records

An effective maintenance program cannot be conducted in a haphazard manner. To insure that inspection and maintenance operations are performed in a timely and systematic fashion, they should be reported to and reviewed by the responsible parties for compliance with procedures and schedule. These reports should not be discarded but should be maintained as part of the building or structure records.

6.1.5.2 Inventory of Facility or Structure

Before any maintenance program can be initiated, or improved if one does exist, it is first vital to take an inventory of what is present. It is amazing how many people do not know what is below the floor, inside the walls, or above the ceiling. Discovery often follows some problem that creates the need-to-know at a time when the problem is far advanced and difficult to remedy.

The inventory should include identification of the material, condition of the material, and environmental factors. The inventory could logically start outside the structure and proceed from the foundation up through the roof.

Prior to beginning the inventory, the factors which influence serviceability of wood structures (Chapter 2), should be carefully reviewed to serve as a guide in assessing environmental factors.

Some things to look for outside the structure might include: landscaping and drainage, particularly as to the adequacy of discharge of storm water away from the structure; adequacy of clearance between wood and soil (at least 8 inches is recommended) to prevent soil-wood contact; adequacy of vents in terms of number, location, and area and freedom from blocking by shrubbery or other obstruction, and from deliberate closing to increase temperature; adequacy of gutters and downspouts to be sure that they are in place and are working properly; and exposed structural elements to be sure that they have been properly designed and protected, as by flashing.

The foundation is probably the most neglected part of any structure, particularly with crawl space construction. (The area between the ceiling and roof is probably second.) Since people seldom enter these areas, the out-of-sight and out-of-mind principle prevails. Record what the structural elements or components are, their condition and environment. These areas are often characterized by a dark damp environment with inadequate ventilation (see section 6.3).

The floor has two sides, the foundation side (discussed above), and the exposed side. Since the top surface is in constant view, it is generally maintained out of necessity for safety and serviceability.

Walls should be inventoried with respect to interior wall coverings; structural system; presence or absence of insulation; vapor barriers; exterior cladding (type and condition); bracing; windows (caulking and glazing); and caulking and sealants in general. The area in and around windows should be given special attention. Check the condition of sills and look for evidence of leaks into the wall cavities. Wall penetration (e.g. entry points for pipes and conduits) are other potential problem areas and should be noted.

Roof and attic spaces experience the greatest extremes from high temperature in summer to low temperature (and condensation) during winter. The type and condition of both the structural system and sheathing should be well documented on the initial inventory.

The roof covering, of course, protects everything within the structure. It is an essential part of the inventory and is discussed in detail under section 6.4.

6.1.5.3 Evaluation

Once the inventory is completed, an effective maintenance program can be established. At a minimum, the inventory should have provided the following information: What is present? What is the current condition of the building or structure? What is the environment? What needs to be done?

The initial evaluation might conclude that there are areas that require immediate attention, such as needed repairs or other work to bring the building or structure up to standard. This is the time to review other chapters in this publication on inspection, repair, etc., and select those things applicable to the needs of the particular building or structure. Once the building or structure is upgraded, if necessary, an inspection program is necessary to insure that it is maintained properly.

6.1.5.4 Inspection Program

Some of the requirements common to any effective inspection program include: assignment of responsibility, training, reporting, and scheduling.

Assignment of Responsibility

An inspection program cannot be successful unless the responsibilities for conducting and administering the program are well defined. The people who will perform the inspection and/or maintenance must be identified. Because a maintenance program does cost money (albeit much less than not having one) the person having administrative responsibility must be assigned line authority to receive recommendations and to budget funds according to need.

Inspection Training

Training is needed to perform inspection effectively. One needs to know what to look for, where to look, and how. Inspection techniques and tools for inspection are discussed in Chapter 3.

Depending upon the size and complexity of the building or structure, more than one person might require general inspection training. Or it might be better to break the inspection down by categories, and train people for different phases of inspection.

Inspection Reports

As a minimum, inspection reports should include: who conducted the inspection; what was inspected and where; when the inspection was done; what was observed; and recommended action. The inspection reports should then become a part of the permanent building or structure records.

Noting who conducted the inspection will insure that the responsible person is meeting his or her commitments, and suggest whether training is adequate. The what and where must be sufficiently identified so that one can relocate the area quickly in the future. The time of inspection is a check to insure that the schedule is being adhered to. Observations might alert one to changing conditions or potential problems. Recommended actions are possibly the most important and crucial part of the inspection report. This is the decision point on whether the building or structure is in a good state of repair or if corrective actions are necessary.

Inspection Scheduling

Inspection scheduling is important to the success of any maintenance program. The frequency of inspections will vary with structure type, age, and environment, but once established should be adhered to. There are basically three levels of inspection: daily, routine, and timely.

Daily inspections are not formal ones. They depend upon observations during the normal course of work. Once a formal maintenance program is established and people are trained, daily inspections become automatic because people are aware of the structure and alert to what to look for. They should report conditions that merit attention.

Routine inspection is conducted on a sceduled basis. Different structures or components thereof can most effectively be inspected at different times of the year. Riedel (section 6.4) recommends that roofs be inspected twice a year, in the spring and fall. Basement/crawl spaces and roof/attic areas should be inspected during both the summer and winter when the exposures to heat, moisture, and condensation are greatest.

Timely inspections are dictated by extreme or unusual events, such as severe winds, heavy rain or snow, or accidental damage. Potential problems with roof ponding, landscape drainage, snowload, etc., are apparent only if an effort is made to observe these areas when severe conditions prevail. For example, to infer snowloads on roofs based simply on ground cover may greatly underestimate the actual load. Drifting, particularly at changes in roof slope or elevation, can seriously overload the roof structure (3).

6.1.6   References

1. American Association of Port Authorities.
    1970. Port Maintenance, A manual prepared by Standing Committee IV, Construction and Maintenance, Washington, D.C.

2. Chambers, J. H.
    1976. Cyclical maintenance for historic buildings, Interagency Historic Architectural Services Program, Office of Archaeology and Historic Preservation, National Park Service, U.S. Department of the Interior, Washington, D.C.

3. Schriever, W. R.
    1978. Estimating snow loads on roofs, National Research Council of Canada, Division of Building Research, Canadian Building Digest, Ottawa, Canada.

6.2   Condensation*

Cold weather condensation problems can be eliminated by the correct use of vapor barriers, by good ventilation, and by control of indoor humidity. However, when one or more of those details have not been included in an existing building and condensation problems occur, they are often more difficult to solve. Nevertheless, there are methods which can be used to minimize such condensation problems after construction.

6.2.1   Walls

One of the most common evidences of condensation in walls is peeling of exterior paint, or streaking from water running down siding. In cold weather, condensation often occurs as frost behind the siding and goes undetected until spring when outdoor temperatures become warm. This condensation may occur only on walls of rooms having high levels of indoor moisture such as kitchens or shower rooms. It is generally more severe on north walls that do not receive the drying effects of the sun or on other walls that are shaded. Since paint peeling can be caused by many factors, the guidelines stated above can be used to establish whether the problem results from condensation or from some other source.

The best preventative measure is to keep indoor relative humidity at a low level during cold weather. The maximum during the heating season should be 35 percent and it should be reduced when temperatures are

---

*Author: Gerald E. Sherwood, Engineer, Forest Products Laboratory, Madison, Wisconsin.

20° F or lower for long periods of time. It is particularly important to exhaust moisture at sources of high concentrations such as showers or cooking facilities. A general reduction in relative humidity throughout a building can be accomplished by ventilation such as adding outside air to a forced-air heating system or by mechanical dehumidification. If mechanical humidification is being used, make sure it is set at an appropriate level.

Where such humidity-control measures do not solve the condensation problem, some vapor resistance can be added by painting the inside faces of walls with a vapor-resistant paint. Some paint companies have paints formulated for that purpose. Caulking around joints where moisture-laden air may leak into wall cavities will also keep some moisture from walls.

6.2.2   Glass Surfaces

Visible surface condensation on the interior glass surfaces of windows can be minimized by the use of storm windows or by replacing single glass with insulated glass. If condensation still occurs, the indoor relative humidity is too high and there is probably condensation in concealed areas also. Humidity should be reduced by the same measures shown for walls.

6.2.3   Attic Areas

Condensation in attic spaces can usually be controlled by ventilation, which is discussed under the next major heading. Reduction of indoor relative humidity will reduce the amount of vapor flow into the attic, and sealing all joints that might result in air leakage into the attic will further reduce moisture flow into the attic.

6.2.4   Crawl Spaces

Some crawl space moisture can be removed by ventilation, which will be discussed under the next major heading. The largest source of moisture in a crawl space is often from the soil. The amount of moisture can be greatly reduced by a moisture-resistant membrane laid over the soil. Such a soil cover is often a polyethylene film, but other materials, such as roll roofing, can be used. The material should be heavy enough that it will not be easily punctured or torn. If polyethylene is used, at least a 6-mil thickness is recommended. Some caution should be used where the crawl space is under a wood floor. Under extreme conditions, the addition of a soil cover may reduce the moisture level so much that flooring boards shrink excessively. It may be best to begin by covering only one-half the soil for awhile and observing shrinkage. Then, additional cover may be added in increments while observing shrinkage until it approaches an excessive amount.

Where insulation is added to an existing floor, a vapor barrier on the warm side of the floor is desirable; however, water vapor passing through the insulation can usually be vented to the outside.

6.2.5 References for Section 6.2

1. American Society of Heating, Refrigerating, and Air-Conditioning Engineers.
   1977. ASHRAE Handbook of Fundamentals, Ch. 20. New York, N.Y.

2. Anderson, L. O., and G. E. Sherwood.
   1974. Condensation problems in your house: Prevention and solution. U.S. Dep. of Agric. Inf. Bull. No. 373.

3. Boner, A. H.
   1977. Status report: Vapor barrier research in the United States. Prepared by the Dep. of Energy, State of Oregon, Salem, Oreg.

4. Busching, H. W., R. G. Mathey, W. J. Rossiter, Jr., and W. C. Cullen.
   1978. Effects of moisture in built-up roofing: a state-of-the-art literature survey. Nat. Bur. of Stand. Tech. Note 965. Washington, D.C.

5. Duff, J. E.
   1974. Comparative effects of brick and wood siding on the moisture conditions in wood walls. USDA For. Serv. Res. Pap. SE113. Southeastern For. Exp. Stn., Asheville, N.C.

6. Duff, J. E.
   1971. The effect of air conditioning on the moisture conditions in wood walls. USDA For. Serv. Res. Pap. SE78. Southeastern For. Exp. Stn., Asheville, N.C.

7. Duff, J. E.
   1968. Moisture distribution in wood-frame walls in winter. For. Prod. J. 18(1), Madison, Wis., Jan. 1968.

8. Duff, J. E.
   1972. Vapor barrier decreases moisture conditions in wood walls exposed to air conditioning and heating. USDA For. Serv. Res. Pap. SE-98. Southeastern For. Exp. Stn., Asheville, N.C.

9. Latta, J. K.
   1976. Vapour barriers: What are they? Are they effective? Nat. Res. Counc. of Can. Can. Build. Dig. CBD 175. Ottawa, Can.

10. Powell, F. J., and H. E. Robinson.
    1971. The effect of moisture on the heat transfer performance of insulated flat-roof constructions. Nat. Bur. of Stand. Build. Sci. Ser. 37. Washington, D.C.

11. Sherwood, G. E.
    1978. Paint as a vapor barrier for walls of older homes. USDA For. Serv. Res. Pap. 319. For. Prod. Lab., Madison, Wis.

12. Sherwood, G. E., and C. C. Peters.
    1977. Moisture conditions in walls and ceilings of a simulated older home during winter. USDA For. Serv. Res. Pap. FPL 290. For. Prod. Lab., Madison, Wis.

6.3 Ventilation*

Ventilation is generally considered under two major areas: occupied spaces and structural cavities. Relative humidity in occupied spaces of existing buildings is usually not high during cold weather because there is usually adequate air leakage to dissipate moisture to the outdoors unless mechanical humidification is being used. The exception is in major moisture-producing areas such as shower rooms or kitchens. If humidity is a problem, exhaust fans should be installed to remove the moisture at its source. The ventilation of structural cavities is more involved and the remainder of this section will address such ventilation.

6.3.1 Attics and Roof Spaces

Inadequate ventilation of attics or roof spaces may result in cold weather condensation, excessively high temperatures, or ice dams. The same ventilation techniques are used to solve all three problems. The most effective ventilation occurs when there are outlet vents at least 3 feet above the inlet vents. The temperature of the air in the roof cavity is raised by heat loss from the building. The lighter, heated air rises and passes through the outlets, drawing outside air in at the lower level. If vents are all at one level, wind is required for air exchange. Recommended ratios of inlet and outlet areas to ceiling area to remove water vapor escaping into the roof cavity from occupied space are shown in table 6.3.1. More ventilation may be desirable for summer comfort. For good moisture control, the ventilation should be well distributed throughout the roof cavity.

Good ventilation is particularly important for control of ice dams (fig. 6.3.2). Ice dams are caused by snow over a warm attic melting, running down the roof, and freezing on the colder roof overhang. This ice forms a trough that catches future melt water. The water backs up under the shingles and leaks down through the wall or to the inside near the wall. Good eave inlet vents together with outlets at higher levels are essential to keep the roof near the outdoor temperature, and adequate ceiling insulation contributes to the solution. When reroofing is required, roll roofing placed at the eaves and extending above the point where the roof is over the outer wall will prevent water damage if an ice dam does occur.

6.3.2 Crawl Spaces

Crawl space ventilation is required to dissipate moisture from the soil and from vapor entering from the occupied space. The main moisture source is ground water. Excessively high humidities in the crawl

---

*Author: Gerald E. Sherwood, Engineer, Forest Products Laboratory, Madison, Wisconsin.

Table 6.3.1--Recommended good practice for loft and attic ventilation[a]

Flat Roof--Slope 3 in. in 12 in. or Less
Condensation Zone I[c]: Total net area of ventilation should be 1/300th[b] distributed uniformly at the eaves plus a vapor barrier in the top story ceiling. Free circulation must be provided through all spaces.
Condensation Zones II and III: Same as for Zone I.

Gable Roof--Slope over 3 in. in 12 in.
Condensation Zone I: Total net area of at least 2 louvers on opposite sides located near the ridge to be 1/300th[b] plus a vapor barrier in the top story ceiling.
Condensation Zone II: Same as for Zone I.
Condensation Zone III: Same ventilation as for Zone II. A vapor barrier is not considered necessary.

Hip Roof
Condensation Zone I: Total net area of ventilation should be 1/300th[b] with 1/600th[b] distributed uniformly at the eaves and 1/600th[b] located at the ridge with all spaces interconnected. A vapor barrier should also be used in the top story ceiling.
Condensation Zone II: Same as for Zone I.
Condensation Zone III: Same ventilation as for Zone II. A vapor barrier is not considered necessary.

Gable or Hip Roof--With Occupancy Contemplated
Condensation Zone I: Total net area of ventilation should be 1/300th[b] with 1/600th[b] distributed uniformly at the eaves and 1/600th[b] located at the ridge with all spaces interconnected. A vapor barrier should also be used on the warm side of the top full story ceiling, the dwarf walls, the sloping part of the roof, and the attic story ceiling.
Condensation Zone II: Same as for Zone I.
Condensation Zone III: Same as for Zone I except that a vapor barrier is not considered necessary if insulation is omitted.

[a] It is recognized that in many areas increased ventilation may be desirable for summer comfort. For winter comfort, insulation is recommended between a living space and a loft or attic ventilated at these rates.
[b] Refers to area enclosed within building lines at eave level.
[c] The zone numbers refer to figure 6.3.1.

Figure 6.3.1--Condensation zones in the United States.

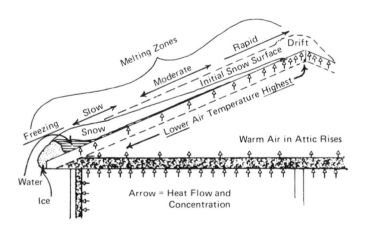

Figure 6.3.2--Schematic representation of roof-snow melting.

space may create conditions that support decay in wood framing. The moisture may also enter the occupied space and create a high indoor relative humidity. In some cases, crawl space moisture has entered wall cavities and caused excessive condensation. To control the moisture, at least four openings should be provided, with one near each corner. The total net area of openings may be calculated by the formula: $a = \frac{2L}{100} + \frac{A}{300}$

where: L = the perimeter of the crawl space, linear feet.

A = the area of the crawl space, square feet.

a = the total net area of all vents, square feet.

Where a soil cover is placed over the entire crawl space, the area of vent openings may be reduced to 10 percent of that calculated above.

6.3.3 Basements

Most basements receive some moisture from the soil and that must be removed by mechanical dehumidification or by venting it to the outdoors. Ventilation should be done at night when the air is coolest and therefore holds the least total amount of moisture. As this cool air is warmed in the basement, relative humidity is reduced. The basement should be closed during the day to deep out warm, moist air. If possible, ventilating should be done during periods when the outdoor humidity is low. Where a dehumidifier is used, the basement should be kept closed, except for brief periods of ventilation for odor control.

6.3.4 References for Section 6.3

1. American Society of Heating, Refrigerating, and Air-Conditioning Engineers.
    1977. Handbook of Fundamentals, Ch. 21, ASHRAE, New York, N.Y.

2. Baker, M. C., and C. P. Hedlin.
    1976. Venting of flat roofs. Nat. Res. Counc. of Can., Can. Build. Dig. CBD 176. Ottawa, Can.

3. Grange, H. L., and L. T. Hendricks.
    1976. Roof-snow behavior and ice-dam prevention in residential housing. Agric. Ext. Serv., Univ. of Minnesota, St. Paul, Minn.

4. H. C. Products Company.
    1976. Fundamentals of residential attic ventilation. H. C. Products Co., Princeville, Ill.

5. Home Ventilating Institute.
    1972. Home ventilating guide. Home Ventilating Inst., Chicago, Ill.

6.4 Maintenance and Repair of Roofs on Wood Structures*

6.4.1 Introduction

Since all roofs can be expected to leak eventually, and leakage leads to damage to insulation, decking, and framing members, maintenance is a very important requirement for all types of roofs.

Roofs that are poorly designed or installed can actually be leaky from their very beginning, while good roofs may soon become leaky when damaged by falling objects such as antennas, by severe weather, etc. Therefore, routine inspection of roofs is important to determine maintenance needs. Twice a year, flat or built-up type roofs should be inspected--once in spring and again in fall.

Common routine maintenance needs include: (1) cleaning out the roof drains or downspouts so that the water will flow freely off the roof; (2) check for damaged or deteriorated composition flashings and damaged or missing metal copings and flashings; (3) look for accidental damage; (4) note blisters that are increasing in size and are a threat to the integrity of the membrane; (5) check on condition of the surfacing material.

6.4.2 Emergency Repairs

Inspections and emergency repairs are necessary, of course, any time a leak appears, because allowing a leak to continue simply compounds the problem of roof, insulation, and deck damage. Deterioration of structural members from prolonged leakage can easily be determined by probing with an ice pick, comparing the depth of penetration into the suspected area with that of a similar member in a known sound area.

Emergency repairs are usually difficult because the need for them usually manifests itself during very inclement weather. Small defects can be sealed with plastic roofing cement and pieces of roll roofing, felt, or asphalt-saturated fabric. The cement should be applied under and over the ply of repair material as smoothly as possible. If the defect includes a depression into which the patch may sag, more than one ply should be used. Cement alone, without the reinforcing ply, cannot be depended upon.

Splits are a common source of emergency action on built-up roofs. They are caused by contraction of the BUR (Built-Up Roof) membrane because of contraction due to low temperature or to shrinkage, when the insulation is not held immovable, or the decking is not sound. Repairs to splits require strength to resist further splitting, so that not less than 2 layers of fabric and cement are necessary.

Cement for emergency repairs should be chosen with care. Some manufacturers market tar-base cement especially for emergency work. They have excellent ability to seal under wet conditions, but should be used only on tar-based BURs. If used on asphalt BURs, they tend to destroy

---

*Author: Roger G. Riedel, P. E., Roofing Consultant, Portage, Wis.

the asphalt. For asphalt-base roofs, special asphalt-base cements are available which contain additives which increase their sealing ability under wet conditions. If the tar-base cement has been used on an asphalt BUR, it should be thoroughly scraped off as soon as possible and replaced with proper asphalt materials.

To distinguish between tar and asphalt, place a small lump of the bitumen from the roof surface into a glass of mineral spirits or gasoline (do not use xylol or xylene). If the bitumen is coal tar pitch, a faint greenish color will develop in the liquid surrounding the sample within a few minutes. If the bitumen is asphalt, the surrounding liquid will immediately begin to turn brown and in a few minutes become opaque with brown to black coloration.

6.4.3 Graveled Roofs

Graveled roofs of either type must be smoothly degraveled and cleaned before applying a patch of any kind, as gravel and its associated dirt will prevent a seal from occurring. A smooth area at least 4 inches wide is necessary for a reliable seal.

After the emergency repairs have been made, they should, when warm dry weather permits, again be checked and properly completed. Sometimes a hot-mopped patch is desirable to replace or overlay the temporary repair, as the cement patch is not as durable as the hot patch.

Of particular importance in the case of graveled roofs in following up on emergency repairs is to re-apply a flood and gravel coat to any area that has been scraped while searching for defects. If this is not done, the exposed felts will deteriorate. This may be done with either hot or cold methods but the material should match the bitumen of which the roof is made.

Well-built roofs last a long time--15 to 20 years and more without maintenance--but poorly built roofs will require major maintenance in as little as 5 years. In addition to the previously mentioned maintenance, the protective surfacing materials will need renewing. When this is needed, it will be evidenced by displacement of the gravel or absence of a complete coating.

Renewal of the surface coat is best done with a bitumen having a solvent base compatible with the original because its fluid consistency and solvent content penetrates through residual dust, soaks into porous felt and blends into the aging original bitumen to partially renew its original properties.

Two grades of solvent-base asphalt are available. The softest and most durable is typified by ASTM D 2823 Type I while the firmer, for smooth roofs and steeper slopes, is ASTM D 2823 Type II.

Tar-based re-saturants are also available from a number of roofing and maintenance materials manufacturers, as are the tar-base repair cements which should always be used on tar-base roofs and never on asphalt-base roofs. The Federal Specification for the tar-base re-saturant is

WOOD STRUCTURES

SS-C-540 b and for the cement is SS-C-153 Type II. There is no ASTM specification for this material.

In the resurfacing process, all loose gravel, dirt and unsound material is removed, all defects in the membrane and flashings repaired, then the new surfacing material is applied. If the roof is the graveled type, new gravel should be applied.

Many built-up roofs require premature resurfacing because the improper type of asphalt was used in its construction. Steep-grade asphalt should not be used on graveled flat roofs because it is not weather-resistant. Within 2 or 3 years, this type of surfacing material will become brittle and cracked (alligatored) and moisture will then soak into the exposed felts and deteriorate them. Although this kind of deterioration can never be fully corrected and the roof made as good as it would have been, a double thick application of D 2823 Type I re-saturant may arrest its degradation for many years.

The wrong grade of asphalt often is used where a roof has flat areas or valleys in conjunction with steep areas. Here, premature deterioration of the flat areas can also be slowed if re-saturation is carried out as soon as the condition is discovered.

6.4.4 Ungraveled Roofs

Ungraveled roofs of all kinds of felt benefit greatly from a coat of aluminum roof paint. This reduces the temperature extremes which the roof develops and protects the roof from weathering just as paint protects other products. If the surface coat is deteriorated, a prime coat that is somewhat harder than a re-saturant (ASTM D 2823 Type II) should first be applied and allowed to cure thoroughly before being painted.

Contrary to recommendations of most manufacturers, the best aluminum roof paint is the premium grade pigmented non-fibrated type. It is smoother and resists soiling better and appears to weather as well as the fibrated, or better, even though it is thinner. Because only half as much is required per square, the non-fibrated material costs only half as much to cover a given area and is easier to apply. Flashings, after being repaired, should also be painted aluminum, whether on a smooth or a graveled roof.

Roll roofing with mineral granules is commonly found applied directly to wood decks. Ordinary weathering causes the granules to come off and this in turn exposes the thin coating of asphalt to the weather which causes the roll roofing to deteriorate. Probably both the granule loss and coating loss occur simultaneously.

Cosmetic corrections can be made by cold methods and aluminum roof paint. Fibrated aluminum coatings are also available in such colors as brown, red, green, etc., which may be preferred.

Recently, a system to spray-coat this type of roof with cold materials and to apply new granules into the coating has been gaining favor.

This process can, of course, be used on all kinds of steeper roof materials, including the repairing of the old roof and even building up a complete new roof.

Splits in roll roofing or in built-up roofs are often found where roof decks are made of boards. This will be found to be caused by accumulation of moisture within the building and its cure is to provide adequate ventilation; otherwise re-splitting will re-occur as the boards shrink and swell from condensation.

6.4.5 Shingled, Tiled, and Slate Roofs

Shingles and other nail-on types of roof coverings such as tiles and slates do not respond well to maintenance procedures. Leaks are caused by irreparable deterioration in the case of asphalt shingles and this requires replacement.

Ice dam leakage in this type of roof is a well-known phenomenon which can be prevented by following manufacturers' recommendations to install a proper underlayment at the eaves when the roof is installed. Details are explicit and they do work if followed. If not followed, there is no maintenance procedure that will correct this problem. Some relief can be had by heating tapes, shoveling snow off the roof and by cooling the attic space, but there is no substitute for a proper eaves underlayment. This subject is discussed also in section 6.3.

6.5 Water Repellents*

6.5.1 Introduction

In many ways, water is one of wood's worst enemies. Water plays a key role in the rapid weathering of wood exposed outdoors, in the performance of exterior finished wood, and in the decay or rotting of wood. Properly seasoned wood that stays dry is not subject to decay, to premature failure of paints and finishes, or to many of the other serious problems associated with weathering (see section 2.6).

Fortunately, there are some relatively simple wood treatments that can be used to slow down the pickup of water and help keep wood dry. These treatments are called water repellents (WR). When a preservative is added to a WR, it is called a water-repellent preservative (WRP). The composition of these two treating materials is very similar; both contain a substance that repels water (usually paraffin wax or related material), a resin or drying oil, and a solvent such as turpentine or mineral spirits. Addition of a preservative such as pentachlorophenol or copper naphthenate to a water repellent helps to protect wood surfaces against decay and mildew organisms.

For more than 40 years, millwork manufacturers have been treating exterior millwork with water-repellent preservatives (9,19,23). They have dip-treated the wood without pressure with resultant reduced

---

*Author: William C. Feist, Chemist, Forest Products Laboratory, Madison, Wisconsin.

swelling, shrinking, and warping of millwork caused by changes in moisture. These treatments also have reduced attack by wood decay and stain organisms that may occur on wet wood (9). In addition, the WRP treatments have improved paint performance (1).

In 1936, the National Woodwork Manufacturers Association (NWMA) initiated a "Seal of Approval" program and proposed a "Code of Minimum Standards for Millwork Preservatives," and in 1938 announced adoption of the program (19). In 1948, the "Seal of Approval" for water-repellent treatments was adopted. In 1950, the NWMA announced its "Standard for Water-Repellent Preservatives for Woodwork." The current Standard, adopted in 1970, resulted from several revisions of the original 1950 version (23).

A large proportion of the damage done to exterior woodwork (paint defects, deformation, decay, leakage, etc.) is a direct result of moisture changes in the wood and subsequent dimensional instability (2,6,8,9,26,29,34). Water generally enters wood through open cracks, unprotected end-grain surfaces, and defects in surface treatments. Although the negative effects of such problems can be avoided or at least reduced by proper design or correct choice of materials, it is extremely difficult to stop checks or cracks from appearing where woodwork is subjected to harsh long-term exposure. Even a preservative coating, which is itself functionally unimpaired, often loses its protective qualities because it cannot take up the movement of the woodwork, especially around joints. Eventually, the coating or treatment gives way.

A great store of information has been accumulated on the effectiveness of WRP's in protecting exterior wood (1-18,20-22,24,25,31-33). WRP's inhibit water from penetrating wood by capillary action and reduce fluctuations in moisture content. The treatments can be applied by immersion (which is preferred), by brush, or by spray application. They improve the performance of many finishes applied over them and add greatly to the durability of exposed wood. Even chipboard or particleboard, which is very susceptible to moisture, can be quite effectively protected against the effects of climatic exposure by using a WRP pretreatment followed by a vapor-resistant coating (27,28). The WRP can, with frequent reapplication, serve as a natural finish for wood (9,30,34,35).

The WRP treatments give wood the ability to repel water, thus denying fungi that cause decay and stain the moisture they need to live. Wood surfaces that remain free of mildew retain an attractive "natural" appearance (9). A WRP reduces water damage to the wood, and helps protect applied paint from blistering, peeling, and cracking that often occurs when excessive water penetrates wood.

A WRP usually contains a fungicide as a preservative that kills any surface mildew living on the wood. Recent research indicates that, in many medium- to low-hazard decay situations, a preservative may not be needed for successful performance of the water repellent (8). That study supports the general feeling that restriction of water from wood is of prime importance in improving durability of exposed wood.

6.5.2 Applying WR or WRP to Wood

Applying WR or WRP solution to the surface of unfinished wood by brushing or by dipping is an effective treatment for siding and exterior millwork (doors, window sash, door and window frames, sills, moldings, fascia).

The following steps are suggested for application to new wood:

(1) If treated siding or millwork is purchased, only freshly cut surfaces need to be brush or dip treated.

(2) Wood that has not been factory treated can be treated by either brushing or dipping. Dipping is more effective. Care should be taken to treat ends of boards and joints between boards. Open joints should be caulked after treating and priming.

(3) Freshly treated wood must be allowed to dry. If the treatment is applied with a brush, 2 days of warm favorable drying weather must be allowed before painting. If dipped for 10 seconds or more, a week of favorable drying weather is necessary before painting. If enough time is not allowed for most of the solvent to dry from the wood, the paint applied over it may be slow to dry, or may discolor or dry with a rough surface that looks like alligator leather.

When applying WR or WRP to previously painted wood, loose paint must be removed, the WR or WRP should be brushed into the joints only, and excess solution wiped from the paint surfaces with a rag. Two days of favorable warm drying weather must be allowed before repainting.

Whether treatment is to new wood or to previously painted wood, particular care should be taken to apply the solution well at the ends of boards, at joints between boards, and to all newly exposed wood such as drill holes. Water will climb by capillary flow up the back of untreated bevel siding from the lap joints; WR or WRP applied to lap joints of the siding does a good job of preventing capillary flow. Accordingly, places that should be well treated include the butt and lap joints of horizontal siding, edges and top and bottom ends of vertical siding, and the edges and corner joints in window sash, sills, window frames, doors, and door frames. Often bottoms of doors and window sash are overlooked. These are areas where water can penetrate deeply and cause extensive damage if not treated. Treatment with WR or WRP will eliminate many such problems later.

6.5.3 Typical WR and WRP Solutions

WR and WRP solutions are widely made and distributed commercially and are available in most paint and lumber stores. Typical formulas for these wood treatments are:

Ingredient	Water repellent (WR)	Water-repellent preservative (WRP)
Pentachlorophenol[a]	0	5%
Boiled linseed oil[b]	10%	10%
Paraffin wax	1%	1%
Solvent (turpentine, mineral spirits, or paint thinner)	89%	84%

[a] Other preservatives used commercially include copper naphthenate, copper-8-quinolinolate, and bis (tri-n-butyltin) oxide. Recent Forest Products Laboratory exposure studies show that pentachlorophenol is a somewhat more effective mildewcide than copper naphthenate, which in turn is better than copper-8-quinolinolate or bis (tri-n-butyltin) oxide. In some states, PCP concentrate may be a restricted pesticide and unavailable.

[b] Exterior-grade varnish can be used in place of boiled linseed oil. If so, twice the volume shown for linseed oil should be used.

6.5.4 References for Section 6.5

1. Anderson, L. O.
    1963. Water repellents improve performance of drop siding. USDA For. Serv. Res. Pap. FPL 4. For. Prod. Lab., Madison, Wis.

2. Borgin, K.
    1961. The effect of water-repellents on the dimensional stability of wood. Nor. Skogind. 11:507-521.

3. Borgin, K., and K. Corbett.
    1970. The stability and weathering properties of wood treated with various oils. With various waxes. With various resins. Plast., Paint Rubber 15:1/2, 3/4, 5/6.

4. Borgin, K.
    1973. A comparison of the effect of weathering of paints, varnishes, and water repellents as coating systems for wood. So. African For. J. 84:10-15.

5. Cooper, G. A.
    1967. Hardwood siding performance. USDA For. Serv. Res. Pap. NC 16. No. Central For. Exp. Stn., St. Paul, Minn.

6. Dooper, R.
    1970. Some aspects of the protection of exterior joinery. J. Oil Col. Chem. Assoc. 53(8):653-668.

7. Dost, W. A.
   1959. Attempts to modify the weathering of redwood. For. Prod. J. 9(3):18A-20A.

8. Feist, W. C., and E. A. Mraz.
   1978. Protecting millwork with water repellents. For. Prod. J. 28(5):31-35.

9. Feist, W. C., and E. A. Mraz.
   1978. Wood finishing: Water repellents and water-repellent preservatives. USDA, For. Serv. Res. Note FPL-0124. For. Prod. Lab., Madison, Wis.

10. Ferruci, A. P., Jr.
    1965. Some aspects of the effect of petroleum waxes on the performance of wood preservatives. Proc. Am. Wood-Preserv. Assoc. 67:97-108.

11. Grantham, J. B., T. B. Heebink, J. M. Black, and E. A. Mraz.
    1976. Natural exterior finishes for wood in the Pacific Northwest. For. Prod. J. 26(8):21-27.

12. Gray, V. R.
    1959. The durability of exterior clear finishes for timber. Res. Rep. C/RR/4m Timber Res. and Devel. Assoc., High Wycombe, Buckinghamshire.

13. Heebink, T. B.
    1970. Performance of exterior natural wood finishes in the Pacific Northwest. For. Prod. J. 20(3):31-34.

14. Hill, R. R.
    1973. Water repellent preservative finishes. J. Oil Colour Chem. Assoc. 56(6):251-258.

15. Hill, R. R.
    1975. Natural finishes for exterior timber. Timber Rev. 158 (33):49-51.

16. Hill, R. R.
    1974. The weathering performance of water repellent stain finishes on various timber and plywood substrates under different exposure conditions. Res. Rep. WT/RR/12. Timber Res. Devel. Assoc., High Wycombe, Buckinghamshire, England.

17. Hill, R. R.
    1975. Effect of preservative pretreatment on the exterior durability characteristics of exterior stain finishes applied to European redwood and whitewood. Res. Rep. WT/RR/13. Timber Res. Devel. Assoc., High Wycombe, Buckinghamshire, England.

18. Hill, R. R.
    1975. Exterior stain finishes--methods of test and assessment on exterior exposure. Res. Rep. WT/RR/14. Timber Res. Devel. Assoc., High Wycombe, Buckinghamshire, England.

19. Lance, O. C.
    1958. History and development of wood preservation for millwork. For. Prod. J. 8(10):61A-65A.

20. Laughnan, D. F.
    1954. Natural finishes for exteriors of houses. For. Prod. J. 4(5):343-345.

21. Levi, M. P., C. Couple, and J. Nicholson.
    1970. Distribution and effectiveness in *Pinus* sp. of a water-repellent additive for waterborne wood preservatives. For. Prod. J. 20(11):32-37.

22. Miniutti, V. P., E. A. Mraz, and J. M. Black.
    1961. Measuring the effectiveness of water-repellent preservatives. For. Prod. J. 11(10):453-462.

23. National Woodwork Manufacturers' Association.
    1970. Industry Standard for Water-Repellent Preservative, Nonpressure Treatment for Millwork, IS 4-70. NWMA, Chicago, Ill.

24. Panek, E.
    1968. Study of paintability and cleanliness of wood pressure treated with water-repellent preservatives. Proc. Am. Wood-Preserv. Assoc. 64:178-186.

25. Richardson, S. A.
    1971. Biological deterioration of ancient buildings. Commonw. For. Rev. 50(1):55-61.

26. Sell, J.
    1975. Basic requirements for the surface treatment of exterior wood: A review of pertinent literature. Holz Roh- Werkst. 13(9):336-340.

27. Sell, J., and U. Krebs.
    1975. Studies on weathered particle and chip boards. Part II. Combating moisture with water repellents and surface coatings. Holz Roh- Werkst. 33(6):215-221.

28. Sell, J., and U. Meierhofer.
    1974. Studies on weathered particle and chip boards. Part I. Field test on treated boards. Holz Roh- Werkst. 32(10):390-396.

29. Tack, G. W.
    1967. The role of water-repellent preservatives (WRP) in the Australian timber industry. Austr. Timber J. 33(7):41-47.

30. Timber Research and Development Association.
    1965. Maintaining timber exposed to the weather. Timber Res. Devel. Assoc., High Wycombe, Buckinghamshire, England.

31. Verrall, A. F.
    1961. Brush, dip, and soak treatments with water-repellent preservatives. For. Prod. J. 11(1):23-26.

32. Verrall, A. F.
    1963. Water-repellent preservatives on exterior woodwork of buildings. For. Prod. J. 13(10):460-462.

33. Verrall, A. F.
    1965. Preserving wood by brush, dip, and short-soak methods. USDA Tech. Bull. 1334. U.S.D.A., For. Serv., Washington, D.C.

34. U.S. Department of Agriculture, Forest Service, Forest Products Laboratory.
    1974. Wood Handbook: Wood as an engineering material. USDA Agric. Handb. No. 72. Washington, D.C.

35. U.S. Department of Agriculture, Forest Service, Forest Products Laboratory.
    1975. Wood finishing: Weathering of wood. USDA For. Serv. Res. Note FPL-0135. For. Prod. Lab., Madison, Wis.

36. U.S. Government General Services Administration.
    Federal Specification TT-W-572B. May 28, 1969. Wood preservative: Water-repellent. Washington, D.C.

6.6   Finishes*

6.6.1   Introduction

The primary function of any wood finish is to protect the wood surface from the natural weathering process and to help maintain appearance. Weathering erodes and roughens unfinished wood. Different finishes give varying degrees of protection, so the type of finish, its quality and quantity, and the application method must be considered in selecting and planning the finish or refinish job. And the finish is the final touch on any building (figs. 6.6.1 and 6.6.2).

Generally, problems associated with outdoor weathering of wood, in the absence of decay, are inclined to be related to aesthetics more than to anything else (see section 2.6). Wood can be left unfinished to weather naturally (fig. 6.6.1). Surface coatings for wood are required to provide protection against weathering for mainly cosmetic reasons, by reflecting light or by absorbing it (13).

On the other hand, however, light and water protection provided to a wood surface by the surface treatment itself will be affected by the

---

*Author: William C. Feist, Chemist, Forest Products Laboratory, Madison, Wisconsin.

Figure 6.6.1--House with western redcedar siding which has been allowed to weather naturally without finish.

weather resistance of the bonding agents of the finish (drying oils, synthetic resins, latexes, etc.) These bonding agents are subject to photolytic degradation to some degree. The mechanism of failure of paints and other finishes has been described in great detail (6,11, 13,22,26, and many others) and will not be discussed further here.

Performance of paints on wood and wood-based materials (in addition to those described in section 2.6, Weathering) has also been studied extensively (6,7,35,41,42). Sell (38) has thoroughly reviewed the basic requirements for surface treatment of wood for outdoor exposure. Protection of wood exposed outdoors by various finishes, by construction practices, and by design factors, to compensate for effects of weather has been addressed in great detail (1-4,8,14,21,32,34,36,43,44).

There are two basic types of finishes (or treatments) used to protect wood surfaces during outdoor weathering: (1) those that form a film, layer, or coating on the wood surface, and (2) those that penetrate the wood surface leaving no distinct layer or coating. Film-forming materials include paints of all description, varnishes, lacquers, and also overlays bonded to the wood surface. Penetrating finishes include preservatives, water repellents, pigmented semitransparent stains, and

Figure 6.6.2--A 100-year-old house, re-sided in 1930 and maintained with good painting practices.

chemical treatments. Various finishes for exterior wood are summarized in table 6.6.1.

6.6.2 Film-Forming Materials

Paints.--Film-forming finishes such as paint have long been used to protect wood surfaces (fig. 6.6.2). Of all the finishes, paints provide the most protection for wood against erosion by weathering and offer the widest selection of colors. A nonporous paint film will retard penetration of moisture and reduce problems of paint discoloration by wood extractives, paint peeling and checking, and warping of the wood. Proper pigments will essentially eliminate UV degradation of the wood surface. <u>Paint, however, is not a preservative; it will not prevent decay if conditions are favorable for fungal growth.</u> The durability of paint coatings on exterior wood is affected by variables in the wood surface and the type of paint (<u>45</u>).

Paints are commonly divided into the oil-base or solvent systems and the latex or water-based paints. Oil-base paints are essentially a suspension of inorganic pigments in an oleoresinous vehicle which binds the pigment particles and the bonding agent to the wood surface.

Table 6.6.1.--Exterior wood finishes: Types, treatment, and maintenance[1]

Finish	Initial treatment	Appearance of wood	Cost of initial treatment	Maintenance procedure	Maintenance period of surface finish	Maintenance cost
Preservative oils (creosotes)	Pressure, hot and cold tank steeping	Grain visible. Brown to black in color, fading slightly with age	Medium	Brush down to remove surface dirt	5-10 yr only if original color is to be renewed; otherwise no maintenance is required	Nil to low
Waterborne preservatives	Brushing	......do.......	Low	......do.....	3-5 yr	Low
	Pressure	Grain visible. Greenish in color fading with age	Medium	Brush down to remove surface dirt	None, unless stained, painted or varnished as below	Nil, unless stains, varnishes or paints are used. See below.
Organic[2] solvents preservatives	Diffusion plus paint	Grain and natural color obscured	Low to medium	Clean and repaint	7-10 yr	Medium
	Pressure, steeping dipping, brushing	Grain visible. Colored as desired	Low to medium	Brush down and re-apply	2-3 yr or when preferred	Medium
Water repellent[3]	One or two brush coats of clear material, or preferably, dip applied	Grain and natural color visible becoming darker and rougher textured	Low	Clean and apply sufficient material	1-3 yr or when preferred	Low to medium

Table 6.6.1—Exterior wood finishes: Types, treatment, and maintenance[1]--con.

Finish	Initial treatment	Appearance of wood	Cost of initial treatment	Maintenance procedure	Maintenance period of surface finish	Maintenance cost
Stains	One or two brush coats	Grain visible. Color as desired	Low to medium	.....do......	3-6 yr or when preferred	....do......
Clear varnish	Four coats (minimum)	Grain and natural color unchanged if adequately maintained	High	Clean and stain bleached areas, and apply two more coats	2 yr or when breakdown begins	High
Paint	Water-repellent, prime, and two top coats	Grain and natural color obscured	Medium to high	Clean and apply top coat; or remove and repeat initial treatment if damaged	7-10 yr [4]	Medium

[1] This table is a compilation of data from the observations of many researchers.
[2] Pentachlorophenol, bis(tri-n-butyltin) oxide, copper naphthenate, copper-8-quinolinolate, and similar materials.
[3] With or without added preservatives. Addition of preservative helps control mildew growth.
[4] Using top quality acrylic latex paints.

One of the more important properties of paint is the ratio of pigment volume to binder volume (6). When this ratio exceeds a critical amount, the volume of binder is inadequate to completely surround all the pigment particles and the paint becomes porous and flat in appearance. Latex paints are suspensions of inorganic pigments and various latex resins in water, and form porous coatings. Acrylic latex resins are very durable, versatile materials. All paint systems work best when applied in the prescribed manner.

Varnishes.--The most attractive initial appearance for wood is no doubt obtained by use of clear varnishes. Other treatments either change wood color or cover it up completely. Unfortunately, clear varnish finishes require frequent maintenance to retain a satisfactory appearance. Despite numerous attempts at using clear finishes having colorless UV light absorbers to help retain the natural color and original surface structure of wood (33,40), it is generally accepted that common pigments provide the most effective and long-lasting protection against light (5,27,28). Even using relatively durable clear synthetic resin varnishes, the weatherproof qualities of the wood/varnish system are still limited because UV light, which penetrates the transparent varnish film, gradually attacks the wood under it (see also 17,19,20,23). Eventually, the varnish begins to flake and crack off, taking with it fibers of the wood which have been degraded photochemically (28,31). Durability of varnish on wood under action of the weather is limited and many initial coats are necessary for reasonable performance (43). Maintenance of the varnish surface must be carried out as soon as signs of breakdown occur. This may be as little as 1 year in severe exposures.

6.6.3 Penetrating Finishes

Water repellents.--It is an accepted fact that a large proportion of the damage done to exterior woodwork (paint defects, deformation, decay, leakage, etc.) is a direct result of moisture changes in the wood and subsequent dimensional instability (12,15). This was discussed in detail in the section on water repellents (6.5). The treatments can also be used as natural finishes for wood (15,18,24,38). Pretreatment of wood with water repellents or water-repellent preservatives is very important in the finishing of wood for exterior uses (2,15).

Stains.--When inorganic pigments are added to water-repellent preservative solutions or to similar transparent wood finishes, the mixture is classified as a pigmented penetrating stain (sometimes referred to as an impregnating paint (9,29,45). Addition of pigment provides color and greatly increases durability of the finish.

The pigmented penetrating stains are semi-transparent, permitting much of the wood grain to show through, and penetrate into the wood without forming a continuous layer. Therefore, they will not blister or peel even if excessive moisture enters the wood. The durability of any stain system is a function of pigment, resin content, preservative, and water-repellent. Their performance during outdoor exposure has received a great deal of attention recently (12,16,18,23-25,28,30,

37,39). Weather resistance of these surface finishes is improved with increasing pigment content, with greater moisture repellency, and with increases in the quantity of material applied.

Penetrating stains are suitable for both smooth- and rough-textured surfaces; however, their performance is markedly improved if applied to rough-sawn, weathered, or rough-textured wood (9,16,18). They are especially effective on lumber and plywood that does not hold paint well, such as flat-grained surfaces of dense species (45). Stains can be prepared from both solvent-base resin systems and latex systems. Latex systems do not penetrate into the wood, however. Penetrating stains can be used effectively to finish such exterior surfaces as siding, trim, exposed decking, and fences. Commercial finishes known as heavy-bodied or opaque stains are also available, but these products are essentially similar to paint because of their film-forming characteristics. Such "stains" do find wide success on textured surfaces and on panel products such as hardboard.

Preservatives.--Although not generally classified as wood finishes, preservatives in wood do protect against weathering in addition to decay and a great quantity of preservative-treated wood is exposed without any additional finish (table 6.6.1). There are three main types of preservative: (1) the preservative oils (e.g., coal-tar creosote), (2) the organic solvent solutions (e.g., pentachlorophenol) and, (3) waterborne salts (e.g., chromated copper arsenate) (45). These preservatives can be applied in several ways, but pressure treatment generally gives the greatest protection against decay. Greater preservative content of pressure-treated wood generally results in greater resistance to weathering and improved surface durability. The chromium-containg preservatives also protect against UV degradation (10,13,14).

6.6.4   Interactions of Weather, Construction Variables, and Finishes

Satisfactory performance of wood finishes is achieved when full consideration is given to the many factors that affect finishes. These factors include the effect of the wood substrate, the properties of the finishing material, details of application, and severity of exposure to elements of the weather (table 6.6.2).

Wood properties.--Wood surfaces that shrink and swell the least are the best for painting. For this reason, vertical- or edge-grained surfaces are far better than flat-grained surfaces of any species, especially for exterior use where wide ranges in relative humidity and periodic wetting can produce wide ranges in swelling and shrinking (figs. 6.6.3 and 6.6.4).

Also, because the swelling of wood is directly proportional to density, low-density species are preferred over high-density species. However, even high-swelling and dense wood surfaces with flat grain can be stabilized with a resin-treated paper overlay (overlaid exterior plywood and lumber) to provide excellent surfaces for painting. Medium-density, stabilized fiberboard products with a uniform, low-density surface or paper overlay are also a good substrate for exterior use.

Table 6.6.2—Finishing methods for exterior wood surfaces: Suitability[1]

Type of exterior wood surfaces	Water-repellent preservative Suitability	Water-repellent preservative Expected life[2] Yr	Stains Suitability	Stains Expected life[3] Yr	Paints Suitability	Paints Expected life[4] Yr
Siding						
Cedar and redwood						
Smooth (vertical-grain)	High	1-2	Moderate	3-4	High	4-5
Rough sawn or weathered	High	2-3	Excellent	8-10	Moderate	3-4
Pine, fir, spruce, etc.						
Smooth (flat-grain)	High	1-2	Low	2-3	Moderate	3-4
Rough (flat-grain)	High	2-3	High	6-8	Moderate	3-4
Shingles--sawn	High	2-3	Excellent	8-10	Moderate	3-4
--split	High	1-2	Excellent	8-10	--	--
Plywood (Douglas-fir and southern pine)						
Sanded	High	1-2	Moderate	3-4	Moderate	3-4
Textured (smooth)	High	1-2	Moderate	3-4	Moderate	3-4
Textured (rough sawn)	High	2-3	High	6-8	Moderate	3-4
Medium density overlay	--	--	Excellent	8-10	Excellent	6-7
Factory stained	--	--	Moderate	3-4	--	--
Special qualified finishes[5]	--	--	--	--	Excellent	10+
Plywood (cedar and redwood)						
Sanded	High	1-2	Moderate	3-4	Moderate	3-4
Textured (smooth)	High	1-2	Moderate	3-4	Moderate	3-4
Textured (rough sawn)	High	2-3	Excellent	8-10	Moderate	3-4
Factory stained	--	--	Moderate	3-5	--	--
Hardboard, medium density						
Smooth						
Unfinished	--	--	Low to high	3-8	Moderate	3-4
Preprimed	--	--	--	--	Moderate	3-4
Textured						
Unfinished	--	--	Low to high	3-8	Moderate	3-4
Preprimed	--	--	--	--	Moderate	3-4
Factory stained	--	--	Moderate	3	--	--

Table 6.6.2—Finishing methods for exterior wood surfaces: Suitability[1]—con.

Type of exterior wood surfaces	Water-repellent preservative		Stains		Paints	
	Suitability	Expected life[2] Yr	Suitability	Expected life[3] Yr	Suitability	Expected life[4] Yr
Millwork (usually pine) Windows, shutters, doors, exterior trim	High[6]	--	Moderate	2-3	Moderate	3-4
Decking New (smooth) Weathered (rough)	High High	1-2 2-3	Moderate High	2-3 3-6	Low Low	2-3 2-3
Glued Laminated Members Smooth Rough	High High	1-2 2-3	Moderate High	3-4 6-8	Moderate Moderate	3-4 3-4

[1] This table is a compilation of data from the observations of many researchers.
[2] Development of mildew on the surface indicates a need for refinishing.
[3] Smooth, unweathered surfaces are generally finished with only one coat of stain, but rough-sawn or weathered surfaces, being more adsorptive, can be finished with two coats, with the second coat applied while the first coat is still wet.
[4] Expected life of two coats, one primer and one top coat. Applying a second top coat (three-coat job) will approximately double the life. Top quality acrylic latex paints will have best durability.
[5] The American Plywood Association maintains a directory of qualified coatings with exceptionally good performance characteristics on plywood.
[6] Exterior millwork, such as windows, should be factory treated according to Industry Standard 4-70. Other trim should be liberally treated by brushing before painting.

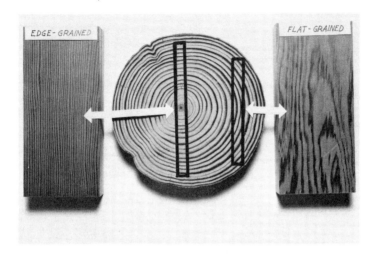

Figure 6.6.3--Edge-grained (or quartersawed) board (left) and flat-grained (or plainsawed) board (right) cut from a log.

Vertical-grained western redcedar and redwood, however are probably the species most widely used as exterior siding which is to be painted. These species are classified as those easiest to keep painted (45). Edge-grained surfaces of all species actually are considered excellent for painting, but most species are generally available only as flat-grained lumber (44,45).

Many wood products of lumber, plywood, shingles, and fiberboard are prepared with a roughsawn and absorptive surface that enhances the durability of stains by providing for better penetration.

Such wood characteristics as high density, flat grain, and tight knots detract from paintability of boards but do not necessarily affect finishing with penetrating preservatives and stains. These finishes penetrate into wood without forming a continuous film on the surface. Therefore, they will not blister or peel even if excessive moisture penetrates into wood.

6.6.3.1 Controlling Effects of Light

Finishes always weather fastest on the sides of a building that receive the most light. In the degradation of surfaces by photo-oxidation, the ultraviolet portion of the light spectrum, which is the high-energy portion, is the most damaging. Clear coatings are embrittled and

Figure 6.6.4--Paint applied over edge-grained boards (top and bottom) performs better than that applied to flat-grained boards (middle).

check. After checks develop, rain and dew have ready access to the wood substrate. Subsequent cycles of wetting and drying produce serious water staining of the wood under the coating and, in time, flaking of the clear coating from the wood. In paint, pigment particles serve as effective ultraviolet absorbers, so paint degrades more slowly than do clear coatings. In addition, very shallow checks develop which gradually loosen and cause chalking or erosion of the paint surface. Chalking of colored and tinted paints is a common cause of fading.

The photodegradation of wood is an important aspect of how it weathers. Wood which has weathered badly before painting will have deep valleys of degraded springwood between sharp ridges of the more photo-resistant summerwood (section 2.6). This kind of surface is not very good for painting, and a paint coating is likely to crack over the ridges of summerwood. However, weathering of the wood surface is beneficial when a penetrating finish is employed, since it permits the wood to adsorb and retain much more of the preservative or pigment stain, so that the finish is more durable.

In addition to light's degradative effects, it also produces heat. Absorption of radiant energy in dark-colored coatings may produce surface temperatures of 100° C or higher under favorable conditions. When the wood surface is hotter than the rest of the board, moisture in the wood substrate moves away from the surface; when the surface is cooler, moisture moves toward it.

Heat from absorbed radiant energy is a problem in the application of paint, particularly dark-colored paint. If paint is applied too thickly on a cool surface which is subsequently warmed by the sun, blisters may form in the paint due to rapid volatilization of paint thinners. This problem is avoided by spreading the paint in only moderate thicknesses and on a surface which has already been warmed by the sun.

By pigmentation.--Adverse effects, particularly the photodegradation effect of sunlight, can be readily retarded in many ways. Addition of pigment to clear coatings effectively protects both wood and coating. This technique is used to great advantage in plastic overlays made of polymers which are transparent to ultraviolet light. Because the polymers are transparent and do not absorb ultraviolet, they have long life out of doors. When applied to wood, such clear polymers permit ultraviolet to attack the wood surface under the coating; this leads to early separation of clear coating from the wood. Addition of pigment to a UV-transparent polymer can protect the wood from breakdown by light.

By construction factors.--The feature most widely employed in construction to protect from sunlight is roof overhang. When a 4-foot-wide overhang is provided, approximately two-thirds of a conventional one-story side wall is protected from exposure to full sunlight. If clear natural finishes are desired, such as on exterior doors, they might be protected by either recessing the entrance or using appropriate loggia construction. Gable-end constructions can be utilized on the north side of the buildings while hip-roof construction provides protection on all sides of a house. A-frame designs of course, also give generally good protection.

Vertical siding patterns may also be beneficial in reducing the effects of light and weathering. Water drains better off vertical boards than off horizontal boards. Vertical siding is also slightly more resistant to sunlight than is beveled horizontal siding because the latter is nearly perpendicular to the incident sunlight.

6.6.3.2 Controlling Effects of Outside Water

Rain and dew account for large volumes of water which come in contact with exterior surfaces. This kind of wetting can produce cracking and peeling failure in paint and clear coatings (fig. 6.6.5).

Water can pass through cracks in the coating and produce peeling (fig. 6.6.5). Leaks in the roof, inadequate drainage of water from valleys on steep roofs, or the formation of ice dams on the roof also allow outside water to enter side walls and damage the paint. Decay can develop where excessive penetration of water occurs in joints of untreated wood that have a low natural resistance to decay.

Figure 6.6.5--Advanced paint failure, cracking, peeling, flaking, due to water getting behind paint film.

The discoloration of paint by the water-soluble extractives in woods such as western redcedar and redwood may be evidence of outside water damage. When the discoloration is a rundown or streaked pattern, it indicates water is getting behind the horizontal siding and running down at breaks in the lap joint. This kind of discoloration is commonly caused by ice dams. An overall diffused pattern of discoloration, on the other hand, indicates penetration of rain and dew through thin layers of porous paint.

Discoloration can further be produced by fungal growth (mildew) on the surface. This failure also is associated with the dew pattern of the house.

By construction factors.--Most exterior water problems resulting from rain are eliminated with properly designed roofs, valleys, gutters, and overhang. Proper use of metal flashing in critical areas is most important. Flashing should be used at the junction of a roof and a wood or masonry wall, at chimneys, over exposed doors and windows, at siding material changes, in roof valleys, and in other areas where rain or melted snow may penetrate into the house. Wide overhangs are most effective in reducing the amount of rain and dew that wet exterior side walls.

Where coating failure is associated with ends of boards and joints between boards, or when extractive discoloration is present, the best

precautionary step is to apply water-repellent preservative before
painting (section 6.5). Window and door trim of a species susceptible
to decay should be factory treated with water-repellent preservative.
If not, they should be treated liberally on the site before painting.
After treating, prime the surface with nonporous paint. Larger cracks
and openings should be calked after treating and priming. Painting
untreated sapwood and species low in decay resistance usually enhances
the possibility of decay.

Formation of ice dams on horizontal roof edges and in valleys of houses
in north temperate zones can be largely eliminated by providing insulation in the attic floor to arrest heat losses. In addition to insulation, the attic space should be ventilated (section 6.3.1).

6.6.3.3 Controlling Effects of Inside Water

Water from inside a building can attack paint on the outside by dif-
fusing through the walls. This water can come from such faulty conditions as leaks in plumbing, or shower spray on a bathroom wall that is
not properly sealed. Or it may result strictly from conditions of high
interior humidity.

Water vapor inside the building can be a source of many gallons of
water daily. In the winter time, this moisture is attracted toward
the cold surfaces of the outer walls. If the outer walls have no vapor
barrier or a poorly installed one, the water vapor passes into the
walls and condenses to liquid in the sheathing and siding. In very
cold weather, it may condense into frost, and later be melted by the
warm spring sun. The condensed water vapor soaks into the siding and
wets the paint. This problem is called cold-weather condensation and
is a common cause of paint blistering and peeling (fig. 6.6.6). Discoloration patterns can also occur from the movement of water-soluble
extractives out to the surface of paint.

By construction factors.--To avoid paint failure by cold-weather condensation, the following procedures are suggested:

(1) Increase the resistance to the penetration of vapor by the use of
a vapor barrier and insulation in side walls.

(2) If paint is peeling on gable ends, increase insulation and ventilation in the attic.

(3) Reduce relative humidity in the house during cold weather.

By penetrating stain.--The most blister- and peel-resistant of all
finishes are the penetrating stains. These finishes are virtually
blister- and peel-proof because they form no coating on the surface.
The penetrating stains are therefore excellent finishes wherever a
serious interior moisture condition is involved but must be applied
to base wood surfaces.

Figure 6.6.6--Blistering of paint due to moisture (interior) behind the paint film.

6.6.3.4 Controlling Effects of Organisms

Fungal organisms are a common cause of discoloration on wood and paint surfaces (13). Such growth usually does no serious structural damage to either the house or the finish system but it does detract from the appearance. It is also evidence that moisture and temperature conditions may be favorable for the growth of other fungi which could cause decay. Many times it is difficult to distinguish between fungal staining and dirt collection. Usually an identification by microscopic examination is required to be certain that fungal growth is present.

Fungi are most likely to grow on surfaces that remain wet for long periods of time. For this reason, it is not uncommon to see fungal growth on exterior house paint which is wet by dew. These areas are usually the surfaces between studs where the surface cools quickly by radiation and the insulation in the wall between the studs prevents warming of the exterior surface with heat from inside the house. The painted area over studs and around nails, however, usually remains free of fungal growth because the area is warmed by heat from within.

Control of fungal staining and of decay problems is achieved through either the selection of wood species that have a high natural resistance to decay or by the use of wood treated with a fungicide or water-repellent preservative. Where greater resistance is needed, liberal brush treatment or even pressure treatment with a preservative is advisable. Exterior wood trim, such as windows and doors, which are commonly made of pine sapwood, should always be treated with water-repellent preservatives before painting or staining. Treatment not only retards fungal growth in the paint but also reduces penetration of water at joints; the water entering the end grain can accelerate paint peeling and decay in wood.

Fungal growth in finishes is controlled also by the addition of fungicides (mildewcides) to the finish. The use of fungicides in latex and flat alkyd-base paints is quite important because these paints, being porous, will hold water on the surface, thus favoring fungal growth. In severe conditions, paints which are pigmented with zinc oxide are recommended to control fungal growth.

The inhibition of fungal growth on wood surfaces is an effective method of achieving an attractive natural finish. Wood treated at regular intervals with a water-repellent preservative solution remains free of graying by fungi and weathers to a light tan color which is natural in appearance.

6.6.4 Summary

Film-forming finishes, such as paint, provide the most protection for wood against sunlight and offer the widest selection of colors. A nonporous paint film is needed to retard the penetration of moisture and to reduce discoloration by wood extractives, paint peeling, and checking. Paint is not a preservative and will not prevent decay if conditions are favorable. Because a coating has been formed on the surface, failure by cracking, blistering, and peeling are possible for a paint finish. To achieve optimum performance, both the substrate and coating must be carefully selected and used on well-designed and constructed structures that reduce the exposure of the coating to a minimum of sunlight, moisture, and fungal organisms.

In contrast to the film-forming finishes are the penetrating-type finishes. Because there is no coating formed on the surface, there is no failure by cracking, peeling, and blistering. Quality of the substrate is not critical; it can be rough, smooth, weathered, knotty, flat grain, dense, porous, and of any species. The penetrating preservative and pigmented oil-base stains are low in cost, easily maintained, essentially troublefree, and durable. Further, because peeling and blistering failures have been eliminated, the dependence of the finish on design and construction factors has been decreased.

6.6.5 Additional Reading

U.S. Forest Products Laboratory

1. 1966. Painting Outside Wood Surfaces, U.S. For. Serv. Res. Note FPL-0123.

MAINTENANCE 293

2. 1966. Discoloration of House Paints by Water-Soluble Extractives in Western Redcedar and Redwood, U.S. For. Serv. Res. Note FPL-0132.

3. 1966. Finishing Exterior Plywood, U.S. For. Serv. Res. Note FPL-0133.

4. 1966. Temperature Blistering of House Paints. U.S. For. Serv. Res. Note FPL-0126.

5. 1966. Intercoat Peeling of House Paints. U.S. For. Serv. Res. Note FPL-0127.

6. 1966. Cross-Grain Cracking of Oil-Base House Paints. U.S. For. Serv. Res. Note FPL-0129.

7. 1970. Blistering, Peeling, and Cracking of House Paints from Moisture, U.S. For. Serv. Res. Note FPL-0125.

8. 1975. Mildew on House Paints. U.S. For. Serv. Res. Note FPL-0128.

9. 1975. Weathering of Wood, U.S. For. Serv. Res. Note FPL-0135.

10. 1978. Water-Repellent Preservatives, U.S. For. Serv. Res. Note FPL-0124.

11. 1979. FPL Natural Finish, U.S. For. Serv. Res. Note FPL-046.

6.6.6 References for Section 6.6

1. American Plywood Association.
    1978. Stains and paints on plywood, APA Pamphlet B407. Am. Plywood Assoc., Tacoma, Wash.

2. Anderson, L. O.
    1970. Wood-frame house construction. USDA For. Serv. Agric. Handb. No. 73. Washington, D.C.

3. Anderson, L. O., and G. E. Sherwood.
    1974. Condensation problems in your house: Prevention and solution. USDA For. Serv. Agric. Info. Bull. No. 373, Washington, D.C.

4. Anderson, L. O., T. B. Heebink, and A. E. Oviatt.
    1972. Construction guides for exposed wood decks. USDA For. Serv. Agric. Handb. No. 432. Washington, D.C.

5. Ashton, H. E.
    1967. Clear finishes for exterior wood, field exposure tests. J. Paint Technol. 39(507):212-224.

6. Banov, A.
    1973. Paints and coatings handbook. Structures Publishing Co., Farmington, Mich.

7. Barnes, H. M.
   1973. Exterior finishing of southern yellow pine wood and wood products. For. Prod. Util. Lab., Information Ser. No. 16. Miss. State Univ., Miss.

8. Black, J. M.
   1971. Finishes, construction factors, and design to compensate for effects of weather on wood surfaces. Symp. World consultation on the use of wood in housing [Vancouver, B.C., July 5-16].

9. Black, J. M., D. F. Laughnan, and E. A. Mraz.
   1975. Forest Products Laboratory natural finish. USDA For. Serv. Res. Note FPL-046. For. Prod. Lab., Madison, Wis.

10. Black, J. M., and E. A. Mraz.
    1974. Inorganic surface treatments for weather-resistant natural finishes. USDA For. Serv. Res. Pap. FPL 232. For. Prod. Lab., Madison, Wis.

11. Browne, F. L.
    1959. Understanding the mechanisms of deterioration of house paint. For. Prod. J. 9(11):417-427.

12. Dooper, R.
    1970. Some aspects of the protection of exterior joinery. J. Oil Col. Chem. Assoc. 53(8):653-668.

13. Feist, W. C.
    1979. Weathering of wood in structural uses. In Structural use of wood in adverse environment. Society of Wood Science and Technology.

14. Feist, W. C.
    1977. Finishing wood for exterior applications--Paints, stains, and pretreatments. In I. S. Goldstein, ed., Wood technology: Chemical aspects. ACS Symp. Ser. No. 43. Am. Chem. Soc., Washington, D.C.

15. Feist, W. C., and E. A. Mraz.
    1978. Wood finishing: Water repellents and water-repellent preservatives. USDA For. Serv. Res. Note FPL-0124. For. Prod. Lab., Madison, Wis.

16. Feist, W. C., E. A. Mraz, and J. M. Black.
    1977. Durability of exterior wood stains. For. Prod. J. 27(1):13-16.

17. Golden Gate Society for Coatings Technology.
    1967. Exterior durability of catalyzed clear coatings on redwood. J. Paint Technol. 39(514):655-662.

18. Grantham, J. B., T. B. Heebink, J. M. Black, and E. A. Mraz.
    1976. Natural exterior finishes for wood in the Pacific Northwest. For. Prod. J. 26(8):21-27.

19. Gray, V. R.
    1959. The durability of exterior clear finishes for timber. Res. Rep. C/RR/4, Timber Res. and Devel. Assoc., High Wycombe, Buckinghamshire, England.

20. Gray, V. R., and A. C. Oliver.
    1958. External clear finishes for timber. Wood 23:237-239, 297-299.

21. Great Britain Forest Products Research Laboratory.
    1975. A new trend in timber protection-exterior wood stains. Proc. Symposium held at Building Research Establishment, Princes Risborough Laboratory, Building Research Station, Gr. Brit. For. Prod. Res. Lab., Aylesbury, Buckinghamshire, England.

22. Hess, M.
    1965. Paint film defects, their causes and cure. 2nd ed. Chapman and Hall, Ltd., London.

23. Hill, R. R.
    1974. Natural finishes for exterior timber. Timber Rev. 158(33):49-51.

24. Hill, R. R.
    1974. The weathering performance of water repellent stain finishes on various timber and plywood substrates under different exposure conditions. Res. Rep. WT/RR/12. Timber Res. Devel. Assoc., High Wycombe, Buckinghamshire, England.

25. Hill, R. R.
    1975. Effect of preservative pretreatment on the exterior durability characteristics of exterior stain finishes applied to European redwood and whitewood. Res. Rep. WT/RR/14. Timber Res. Devel. Assoc., High Wycombe, Buckinghamshire, England.

26. Hoffman, E.
    1971. Weathering of paint films and development of accelerated tests. J. Paint Technol. 43(563):97-106.

27. Janotta, O.
    1975. Transparent exterior coatings on timber. Holzforsch. Holzverwert. 27(4):82-87.

28. Kühne, H., U. Leukens, J. Sell, and O. Wälchli.
    1970. Investigations on weathered wood surfaces. Part I. Scanning electron-microscope observations on mold-fungi causing grey stain. Holz Roh-Werkst. 28(6):223-229.

29. Laughnan, D. F.
    1954. Natural finishes for exteriors of houses. For. Prod. J. 4(5):343-345.

30. Leukens, U., J. Sell, and O. Wälchli.
    1973. Investigations on weathered wood surfaces. Part IV. Tests with an impregnating paint on the "Madison formula basis." Holz Roh-Werkst. 31(2):45-51.

31. Miniutti, V. P.
    1967. Microscopic observations of ultraviolet irradiated and weathered softwood surfaces and clear coatings. USDA For. Serv. Res. Pap. FPL 74. For. Prod. Lab., Madison, Wis.

32. Oviatt, A. E.
    1975. Protecting exposed ends of timber beams in the Puget Sound area. USDA For. Serv. Res. Note PNW-263, Pacific NW For. Range Exp. Stn., Portland, OR.

33. Philadelphia Society for Paint Technology.
    1967. Ultraviolet light absorbers in clear coatings for wood. J. Paint Technol. 39(515):736-751.

34. Rowell, R. M., J. M. Black, L. R. Gjovik, and W. C. Feist.
    1977. Protecting log cabins from decay. USDA For. Serv. Gen. Tech. Rep. FPL-11. For. Prod. Lab., Madison, Wis.

35. Schultz, H., P. Böttcher, and W. Neigenfind.
    1973. Effect of some paints and colors on the moisture condition of naturally weathered wood specimens. Holz Roh-Werkst. 31(3):132-137.

36. Seifert, E.
    1974. The causes of premature deterioration in wooden window frames. Holz Roh-Werkst. 32(3):85-89.

37. Sell, J.
    1973. Weatherproofing of wood with impregnation paints. Holz Roh-Werkst. 31(3):124-127.

38. Sell, J.
    1975. Basic requirements for the surface treatment of exterior wood: A review of pertinent literature. Holz Roh-Werkst. 33(9):336-340.

39. Sell, J., and U. Leukens.
    1971. Investigations of weathered wood surfaces. Part III. Open-air weathering with modern impregnation paints. Holz Roh-Werkst. 29(11):415-424.

40. Tarkow, H., C. F. Southerland, and R. M. Seborg.
    1966. Surface characteristics of wood as they affect durability of finishes. Part I. Surface stabilization. USDA For. Serv. Res. Pap. FPL 57. For. Prod. Lab., Madison, Wis.

41. Teichgräber, R.
    1973. Measurement and evaluation of the interactions between paint systems, wood species, and climate. Holz Roh-Werkst. 31(3):127-132.

42. Thompson, W. S.
    1968. Permeability and weathering properties of film-forming materials used in wood finishing. Res. Rep. No. 4. For. Prod. Util. Lab., Mississippi State Univ.

43. Timber Research and Development Association.
    1965. Maintaining timber exposed to the weather. Timber Res. Devel. Assoc., High Wycombe, Buckinghamshire, England.

44. U.S. Department of Agriculture, Forest Service, Forest Products Laboratory.
    1973. Wood siding, installing, finishing, maintaining. USDA Home and Garden Bull. No. 203. For. Prod. Lab., Madison, Wis.

45. U.S. Department of Agriculture, Forest Service, Forest Products Laboratory.
    1974. Wood Handbook: Wood as an engineering material. USDA Agric. Handb. No. 72, Washington, D.C.

6.7   Heavy Timber (Solid-Sawn Timber)*

Glued-laminated (glulam) timber has replaced solid-sawn structural timbers for most heavy construction. However, there are still times when solid-sawn timbers are preferred, such as for short-span low-profile bridges, beams, stringers, bridge ties, and columns. Additionally many buildings still in service and currently being maintained were constructed with solid-sawn members.

There are differences in the maintenance and servicing of solid-sawn timbers and glulam. For new construction, solid-sawn timbers will generally be installed green because it is not feasible to dry them to service moisture content prior to installation. Existing solid-sawn construction, because of smaller member sections and spans, possesses more and simpler joints that may need attention and servicing.

Many engineers with extensive experience with solid-sawn timber are now retired. In addition, publications on the subject are generally quite old and difficult to retrieve although the information is as valid today as when written. In order to make available some of this older, still-valid information, the following section is excerpted from an article written by Verne Ketchum (deceased), former chief engineer for Timber Structures, Inc. It appeared in the January 1945 issue of Western Construction News, which is no longer being published.

---

*Author: Roger L. Tuomi, Principal Research Engineer, USDA, Forest Service, Washington, D.C.

"In those regions of the United States where timber has long been used extensively as a building material, bolt tightening and the servicing of timber framing after erection is taken as a matter of course.

"In other locations where timber framing has not been used to so great an extent during normal times, owners and engineers are not so familiar with the need of servicing.

"In all regions of the United States during the past few emergency years there has been a large increase in the use of timber framing and a greater proportion of unseasoned timber has been used.

"This article is not to be taken as an alarm that all timber structures require extensive and costly maintenance and repair, but rather to call attention to the need for normal inspection and to outline workmanlike methods of servicing and repair.

"Timber structures built with seasoned timber require little if any servicing after erection.

"Glued laminated members are constructed of kiln dried material and usually no bolt tightening or other servicing is required after the initial checkup at completion of installation.

"Unseasoned lumber needs servicing

"When unseasoned lumber is installed and allowed to season in place, servicing is essential.

"Lumber seasons to a moisture content in balance with the surrounding air, and shrinks about 3 percent or about 1/32 inch per inch in cross sectional dimension from unseasoned to seasoned condition. Lumber does not shrink appreciably lengthwise.

"The amount of shrinkage and the time required for seasoning depends on the moisture content of the wood at time of installation and on the temperature and humidity of the air to which the structure is exposed in use. The higher the temperature and the lower the humidity, the faster the seasoning and the greater the ultimate shrinkage. Also a thick piece will take longer to season than a thinner one even under the same air conditions.

"In a warm dry climate, seasoning may be complete within six months and in a cool damp climate, full seasoning may require a year or more. Servicing can be at relatively longer intervals under slow seasoning conditions but under fast seasoning conditions should be at quite frequent intervals.

"Shrinkage reduces joint strength

"Standards of design for timber connectors require faces of adjacent members to be in contact. If contact faces become separated by shrinkage, a joint no longer conforms to design assumptions, nor is it

practicable to make an assumption as to the reduction in strength which
may develop.

"Access to fabricated trusses built of unseasoned lumber should be
available after erection, and they should be inspected periodically
for separation between adjacent members. When the surrounding air is
warm and dry, the first inspection should be within a month after
erection; under less rapid conditions of seasoning, a longer time can
intervene, but it should not be more than three or four months. Nuts
should be tightened on bolts when necessary, until the wood members
have become thoroughly seasoned, in order to keep adjacent faces in
contact. Early and timely servicing is easier, safer and more economical than tardy servicing.

"Servicing operations should preferably be included in the contract and
specifications, and performed by the fabricator or contractor.

"Servicing personnel

"All servicing and repair should be outlined by an engineer experienced
on timber construction. The engineer should make an inspection of conditions and a report of recommendations in advance of the work. The
servicing should then be carried out under direction of the engineer
and under the supervision of a superintendent assigned full time to
this work.

"Where structures are extensive, experience has shown that servicing is
handled more advantageously by special crews employed for that purpose
than where the job is turned over to regular plant maintenance personnel, who have other duties that take the bulk of their time and
interest.

"Especially should the engineer report on safe or unsafe condition of
the structure as a whole, on location of any critical local potential
or actual failures and on the type and extent of repairs needed to
eliminate them. He should check loss of camber and alignment of members, and look closely for signs of failure or movement.

"Invariably local failures are accompanied by some visible evidence
such as split members, pulled joints, exposed bright wood, broken paint
lines, elongated bolt holes or other visible signs of distress.

"It is so fundamental that where there is no visible sign of distress
there is no movement and where there is no movement there is no failure
that an engineer can safely inspect and report accordingly.

"Servicing operations

"All nuts on all bolts, including stitch bolts, in trusses, columns,
bracings, marquees, catwalks and other framing should be carefully
tightened.

"There will, however, be some locations where extensive separation has occurred, where some warping of the members is present, where bolts may be slightly bent, threads may be rusty, rings may be out of alignment or out of routings, shrinking may have reduced or elongated ring routings, and the nuts will not tighten easily to the extent which will pull adjoining members into proper contact. Where these difficult joints are encountered it will necessary to proceed slowly and carefully, using such tools and expedients as will close the joint without damaging the material. It is suggested that one or more as necessary of the following helps be used and that the experienced operator may have other 'time tried' aids at his disposal which may be employed.

"Any tightening or servicing which does not result in restoring contact between all faces of members in the joint is to be considered as no servicing at all and the structure will still remain in a weakened and dangerous condition. It should be stated here most emphatically that the most dangerous feature of loose connections is the separation of members, and loss of contact between adjoining faces of members in either bolted or bolt and timber connector joints.

"Where threads are dirty or rusted a drop or two of oil placed behind the nut before the start of nut tightening will reduce bolt friction and extend a much greater proportion of wrench pressure to nut movement.

"Should it be found that the pressure on the nut is causing the washer to bite or sink into the wood, it is advisable to remove the present washers and substitute a larger washer during the bolt tightening operation. Such larger washers may or may not be left in place after operations are completed. The oversize washers should be placed under both head and nut. It would be desirable to have a few large washers made up in advance for this type of emergency. Such washers should be about 4 in. square by 3/8 in. thick and have a hole in the center 1/16 or 1/8 in. larger than bolt size. The present malleable or wrought washer may be used over the plate washer and under the nut to provide needed packing.

"A slotted horseshoe washer may be installed under the bolt head without removing the bolt. Use these singly or in pairs, sliding in from the upper side and nailing to prevent rotation or backing out.

"Always install the horseshoe washer between wood and present washer. They should be 1/4 in. or more in thickness and the slot 1/16 to 1/18 in. larger than the bolt size.

"<u>Even tightening</u>

"All nuts at a single joint should be tightened small amounts in rotation and this operation repeated as long as necessary to close the joint. Under no consideration should all tightening be done on one nut while others remain loose and idle.

MAINTENANCE

"Where some timbers in a joint are warped and the joint does not readily close, the work should be attacked rather slowly and given rest periods. Perhaps one-fourth of the slack should be taken out of these difficult joints and the joint then allowed to rest for a few hours or overnight while the crew proceeds with other less difficult joints or works on a difficult joint which has enjoyed a suitable rest period.

"A hand sledge should be used to strike the heads of bolts while the nuts are being tightened.

"A regular spud wrench is the best tool for bolt tightening and it may be found that a short piece of pipe placed over the handle of the wrench and spot welded thereto will prove to be a labor saver, but pipe handle extensions should be used very carefully or stripped threads and broken bolts will result, rather than a better and easier operation.

"It will be found that good steel clamps may assist in bringing warped joints together. The clamps should be tightened in rotation with the joint bolts. It may also be helpful to reverse one or two of the bolts, thus placing the nut tightening pressure simultaneously on both sides of the joint.

"Should it be found that the original panel of bolts in the joint are not sufficient to pull the joint together without excessive or undue strain, it is expedient to install additional bolts and washers at strategic locations, and to tighten all bolts in rotation until the joint is in satisfactory condition. The new bolts may or may not be left in place.

"It is recommended that at least on all structures subject to vibration a 'Palnut' or other suitable type of lock nut be placed on all bolts over the present nut immediately after the nut is tightened. The operator may carry a supply of lock nuts in his pocket or apron and install them as soon as the nut is tightened and before he moves on to the next.

"At final bolt tightening, all bolt heads and nuts and all washers may be given a coat of paint. This painting is highly necessary in locations of high humidity or where salt elements are present in the atmosphere, treatment, or manufacturing process, but optional under other normal conditions.

"Displaced rings

"An inspection of joints will indicate that in most cases the split rings are not out of place, although in extreme conditions the joints may be opened sufficiently between two contact faces to have the ring nearly or entirely pulled out of the routing on one side. The operator should carefully watch joints of extreme separation and be sure, as he tightens the nut, that the ring feeds or leads properly back into the routing. It will be apparent that a joint cannot be pulled together properly where the ring is out of place and remains out of place during the attempted bolt tightening and all such efforts result in labor lost.

"Sometimes shrinkage cracks or minor splits are found in wide and narrow members, such as three by tens, three by twelves, or two by eights and it is desired to install restraint to eliminate further opening of the check or split. Members of this type should be repaired, using bolt and plate collars around the member, or using side plates and cross bolts. Collars should have bolts parallel to and in close contact with the wide faces and steel straps across the narrow face. Side plates should be long enough to throw cross bolts well back from the check, providing good anchorage.

"Repairs on two-leaf members may be made with two steel bars and two bolts. The bars extend across both leaves, one on top and one on bottom. The bolts lie against the inside wide faces of the leaves and extend through. It is desirable to place a malleable washer between the plate and head and nut of the bolt for better cantilever effect. Holes in plates should be oversize to permit close field adjustment. Collar type repairs do not reduce the net section and have very low installation costs.

"Stitch bolts are undesirable in wide and narrow members as holes for their installation may weaken the member below design strength. Stitch bolt repairs may be made in members which do not have too great a depth-over-width ratio, or where unit working stress is well below design strength, so that the stitch bolt hole does not remove too great a percentage of the cross section. Doubtful cases should be decided by an engineer. Stitch bolts except in seasoned lumber must be retightened as the wood shrinks or their value is reduced or lost entirely.

"The maintenance crew should chalkmark all bolts as tightened to avoid skipping and to permit supervisors to check the work.

"In bolt tightening operations, it is fundamental to good practice and good results that operations be carried on slowly and carefully, using methods and expedients which will nurse and not force the separated members back into their proper place.

"Other servicing jobs

"In some types of solid member, steel-rod, timber trusses, servicing adjustment may increase present camber. In most timber connector trusses the camber cannot be increased by the bolt tightening operations.

"Any operation on trusses already in place to eliminate sag or increase camber would be very costly.

"Bolt tightening does act to stiffen up trusses and acts to prevent further loss of camber.

"Some servicing has been carried to the unnecessary extreme of attempting to stitch bolt all checked members and to tighten stitch bolts to the extent of closing up checks. These extreme actions are unnecessary and a waste of labor and materials. Stitch bolts which force checks to close may cause new checks elsewhere or reintroduce stresses which checks have relieved.

"It is wise to avoid instructions and operations which result in the use of large quantities of stitch bolts or other items as a matter of routine regardless of whether or not they are actually needed."

6.8   Other Wood Structures*

A number of structural types have not been covered in this chapter. Many are extremely important structures, and omissions were not due to oversight but rather to the inability to get leading experts to contribute suitable material. Fortunately, there are some recent publications that provide useful information on the maintenance of bridges and poles and piles, along with fairly comprehensive reference lists. These will be discussed briefly.

6.8.1   Bridges

There are several thousand wood bridges in the Federal, state, county, and municipal road systems, and about 1,500 miles of timber bridges and trestles within the railroad network.

As of 1971, it is mandatory that all highway bridges in the Federal-aid system be safety-inspected at least every 2 years. The American Association of State Highway and Transportation Officials (AASHTO) has published two manuals, one on bridge maintenance (1) and another on maintenance inspection (2). Neither is very detailed nor explicit on how to inspect and maintain wood bridges.

The railroads, because of their long experience with wood bridges and trestles have established and practice sound maintenance and inspection programs. The American Railway Engineering Association's Manual for Railway Engineering (3) mentions maintenance but provides little information on how to do it. The chief engineer for one company indicated that it was company policy to conduct inspections every 3 months at the local level plus one annual inspection by the division engineer. They rely upon on-the-job training more than on written instructions.

In 1979 Eslyn and Clark prepared a handbook to aid bridge inspectors and bridge maintenance crews (4). It is equally applicable to both highway and railroad structures.

6.8.2   Poles and Piles

Annually, about 6 million poles and 40 million linear feet of piles are installed in the United States. There is a considerable amount of ongoing research on new inspection techniques (sonic testing and tomography) and in-place treatments for decay control.

---

*Author: Roger L. Tuomi, Principal Research Engineer, USDA, Forest Service, Washington, D.C.

Graham and Helsing (5) published a wood pole maintenance manual in 1979, and Highley (6) completed a study on in-place treatments for waterfront structures in 1980.

6.8.3 References for Section 6.8

1. American Association of State Highway and Transportation Officials.
    1976. AASHTO Manual for Bridge Maintenance, Washington, D.C.

2. American Association of State Highway and Transportation Officials.
    1978. Manual for Maintenance Inspection of Bridges,
       Washington, D.C.

3. American Railway Engineering Association.
    1973. AREA Manual for Railway Engineering, Chicago, Ill.

4. Eslyn, W. E., and J. W. Clark.
    1979. Wood bridges-decay inspection and control. U.S. Dep.
       Agric. Handb. No. 557, U.S. Gov. Print. Off.,
       Washington, D.C. 20402.

5. Graham, R. D. and G. G. Helsing.
    1979. Wood pole maintenance manual: Inspection and supplemental
       treatment of Douglas-fir and western redcedar poles, Res.
       Bull. 24, For. Res. Lab., Oregon State Univ.,
       Corvallis, Oreg. 97331.

6. Highley, T. L.
    1980. In-place treatments for control of decay in waterfront
       structures, For. Prod. J. 30(9):49-51, For. Prod. Res. Soc.,
       Madison, Wis.

CHAPTER 7 DESIGNING TO AVOID PROBLEMS*

7.1 Loadings**

In general, design loadings are set out by codes or by the unique requirements of the structure being designed. That is, most building codes contain minimum loadings for which floors, roofs, and other structural systems must be designed. In some cases, the specified minimums are not likely to be exceeded and may well be used directly as the basis for design. In other cases, however, the anticipated usage of the structure will be such as to require the designer to give special thought to the loads which he will use in design. For example, storage of loose granular material may well impose especially high lateral loads on the walls of a storage building. Or the anticipated loads on a warehouse floor may be substantially above the code minimum and will require special consideration.

In such instances, where there is reasonable prospect of anticipating service loadings, design is straightforward. In other instances, the loadings which may be encountered in service may not be so obvious and may, if not considered, lead to distress or failure of the structure. It is these which tax the ingenuity of the structural designer. A few examples may illustrate the point.

In some areas, roof loads in service are, with rare exceptions, rather low, and roof beams and girders of relatively small section may provide ample strength and stiffness and be in compliance with code requirements. If, however, architectural or other considerations dictate a flat roof, problems may result. That is, irregularities in the roof surface may permit the accumulation of water in depressions of greater or lesser area. If the stiffness of the roof-support elements is low, ponds which are initially small may enlarge as the roof system deflects and, if conditions are right, serious if not catastrophic ponding may occur. Obviously, one solution is to impart to the roof a slope sufficient to prevent initial ponding in roof irregularities. If other considerations prevent sloping the roof, good judgment suggests that the possibility of serious ponding be investigated and, if necessary, the roof-support elements be redesigned to prevent it. Methods for analysis of the possibility of occurrence of ponding are available.

Excessive roof loads from water may occur under other circumstances as well. Roof forms may be such as to require a drain to prevent accumulation of water behind a parapet, in a valley, etc. But what happens

---

*Chapter coordinator: Alfred E. Oviatt, Jr., Research Architect, Forest Products Laboratory, Madison, Wisconsin.

**Author: Alan D. Freas, formerly Assistant Director, Forest Products Laboratory, Madison, Wisconsin.

if the drain becomes clogged? This can happen if maintenance is not thorough. Is there some way to assure that this cannot happen; can the roof be redesigned to limit the depth of the pond? Or, possibly, the designer might anticipate the amount of additional loading which might occur and provide strength and stiffness in the roof-support elements to compensate for some or all of the possible added load.

In areas subject to snow loads, care must be exercised in anticipating actual snow loads and the possibility of nonuniform loading. In fact, nonuniform rather than uniform loading may be the norm. That is, the effects of wind may be to sweep certain areas bare and to deposit the snow from that area in another place. For example, the snow might be cleared by the wind from most of a roof and deposited on a roof at a lower level, leading to an extremely high load at that level. Some illustrations of drifting and load concentration of this type are given by W. R. Schriever in his publication "Estimating Snow Loads on Roofs" (Canadian Building Digest 193, February 1978).

Sliding of snow accumulations on roofs may result in drifts and thus concentrations on lower levels--or perhaps even injury to pedestrians if the snow slides onto a walkway. Melting of snow and subsequent refreezing can cause significant concentration of load. On flat roofs, for example, the melt might run to the lowest level of the roof, such as at midspan, freeze in that location and be augmented by subsequent melting and refreezing to the point where ponding and serious overload results. If a roof is used as a parking area, snow removal may cause problems if the snow removed is allowed to accumulate in large piles.

It is not uncommon to find appendages hung from beams and girders-- monorails for a light conveyor system, for example, or unit heaters, or some other type of equipment. And one wonders whether those were a part of the original design or whether they were afterthoughts. If the latter, how much have they reduced the factor of safety? Could the designer have anticipated the possibility that such things might be added? Should he? This suggests that the structural designer might want to know something about the probable usage of the structure and the possible need for light conveyors or other equipment in connection with the manufacturing or other process planned for the building.

Perhaps, in some instances, there is no way to anticipate what might occur. In one building with minimum roof slope, flooding of the roof was common with each rainstorm and water ran into machinery enclosures on the roof. The solution was simple--raise the level of parts of the roof by the addition of several inches of lightweight concrete! It solved the problem but created a new one when, after a time, the roof beams began to fail. There seems to be no way to outguess such an occurrence, but at least one would hope that, if such a solution was used, someone--the owner, the manager, the contractor who installed the concrete--would wonder whether the roof system was sufficiently strong and stiff to support the added load and would require an analysis. Insofar as is known in the case of this building, no analysis of the effect of the added load was done.

One thing that can be prevented is a design change without reanalysis before or during construction. As an example, a certain building was designed with long-span laminated wood roof beams and a specific type of skylight between the beams. A different type of skylight than that originally contemplated was installed. The one actually installed was such that a portion could be moved longitudinally (parallel to the length of the roof beams) so as to leave a portion of the room beneath open to the sky. As a consequence, with the skylight in the open position, the load was no longer uniform. There was, however, no evidence that any consideration had been given to the effect of this change in loading. A combination of this with deterioration (decay) led to a serious hazard.

Without question, many more instances of problems resulting from unanticipated loadings could be cited. We believe, however, that this discussion is sufficient to alert structural designers to the possibility of loadings beyond those normally considered in design and to the possibility of problems if such loadings occur. We hope, too, that it has alerted them to wonder about such possibilities and to attempt to determine what they might be or, if this is not possible, to do a little imagining and, where feasible, to allow a little margin in the structural design to reduce the probability of the occurrence of problems. In other words, the designer should always keep in mind the function of the structure and find out all he can about its intended use.

7.2   Weather Exposure*

7.2.1 Introduction

When wood is exposed in service to rain and sun, alternate wetting and drying can cause checking, staining, cupping, twisting, paint failure, and loosening of joints. In wet and humid climates, high moisture contents may persist long enough for decay to develop. Factors such as orientation or shading can significantly affect the persistence of moisture, as can design details which favor capillary retention or prevent free drainage and air drying (3).

Orientation of a building or other structure with respect to the sun affects the drying time of exposed members. Adjoining buildings or trees may also be factors in providing shade, and prevailing winds may sometimes be used to hasten drying. Thus, if exposed elements such as wood decks or screens can be located toward the south side, exposed to prevailing breezes or free of major shading by trees or buildings, the duration of exposure to moisture is shortened.

Capillary attraction is the molecular force which causes a liquid to cling to a solid. Small cracks or spaces between members thus can retain a considerable quantity of water for long periods, allowing it to be absorbed by the wood. This kind of wetting can be minimized by wider spacing of members and wider separation of end joints, as in wood

---

*Author: Alfred E. Oviatt, Jr., Research Architect, Forest Products Laboratory, Madison, Wisconsin.

decks or fences. Other devices such as flashings, sealing the joint with an elastomeric adhesive or with a water-repellent coating are also effective. This can be very important at end-grain surfaces, which absorb moisture rapidly. End-grain areas occurring at connections are equally vulnerable. Wood performs best when it is dried to the approximate moisture content it will attain in service (Section 7.5, table 7.5.1) and is maintained at or near that moisture content. Lumber may be kiln dried or air dried, but the former is preferred because the temperatures used in the kiln disinfect the wood, killing any decay organisms which may be present, and better control of final moisture content is possible.

7.2.2 Treatments

Pressure preservative treatment with toxic chemicals may be needed for exposed wood elements. Primary structural framing elements like glulam columns and arches exposed to high humidity or free water are examples. Members imbedded in or in contact with the ground, or other framing subject to excessive wetting and which are costly to replace generally require treatment.

Treatments can effectively preserve wood from decay organisms, but chemical preservation does not solve all exposure problems. Most preservatives which are suitable for architectural work do not serve to exclude moisture from wood or to diminish its other effects. An exposed laminated beam end or arch, for example, still suffers the effects of rain and sun, which cause wide swings in moisture content and severe stresses due to dimensional changes. Such stresses in a large section always result in checking and an unsatisfactory surface condition of paint failure or unsightly stains. Increasingly deep checks or splits may even penetrate the treated surface regions of the timber, allowing moisture and spores to penetrate beyond the treatment and effectively bypass it. Oil-borne treatments do help to seal the surfaces of members used in bridges, towers, and waterfront structures, and a high retention of creosote may be effective for many years by bleeding to the surface and repelling moisture.

Moisture has similar, though lesser effects on smaller wood members: checking, staining, paint failure, cupping, twisting, and loosening of joints. It is important that the designer recognize these problems of dimensional change in exterior exposures, for they cannot be prevented by treatment. Design methods to control them within acceptable limits and careful detailing can be effective. These may involve details which shed water quickly, provide air circulation, or cover critical joints.

Many exposed constructions of wood can be so designed that adequate performance may be expected without the use of pressure treatment. Such light structures as exposed wood decks above the ground, covered walkways, and screens are examples. The choice of treatment by the designer may depend upon the availability of treatment and upon the cost-benefit relationship, a factor which will vary with climate and design (2).

7.2.3 Flashings and Covers

A familiar example of cover for exposed wood is the old covered bridge, where a complete roof was thrown over the vital structural elements. Cover for wood construction may take many forms, such as wide overhangs which effectively protect structure, siding, sash, and trim from frequent wetting and the high sun. Flashings and covers, limited to vulnerable surfaces or joints, are effective protective devices when properly detailed and applied (fig. 7.2.1). These may be of metal, wood, plywood, or plastic, and may be either decorative features or blended to match wood surfaces.

Figure 7.2.1.--Plywood covers protect exposed beam ends.

Exposed ends of timber beams which are not covered by overhangs are particularly vulnerable (fig. 7.2.2). Properly designed flashings of metal or plywood have been shown to be effective in excluding excessive moisture (1) (fig. 7.2.3). Glulam arches exposed to the weather can become badly checked on the top surfaces, admitting water to the interior if not fully flashed at the tops. Exposed base connections are a special problem, requiring special provisions for drainage or packing with resilient caulking.

310 WOOD STRUCTURES

Figure 7.2.2.--Decay in an unprotected beam end.

7.2.4 Designing for Long Life

Critical points for moisture absorption are:

1. All exposed end-grain surfaces.

2. Closely spaced members, side by side or end to end, where water may be held by capillary attraction.

3. Connections.

4. Pockets, splits, and checks in side-grain surfaces.

Following are some of the means which have been successful in providing long life to fully exposed structures:

>   1. Slope end-grain surfaces away from direct rain and sun exposure. Cap the ends of posts and other verticals. Seal all end grain, including miters, by dipping 5 minutes or more in a water-repellent preservative. Brush-flood all fabricated ends and holes with water-repellent preservative. This type of treatment is most effective on end-grain surfaces, where it penetrates deeply enough to seal against water.

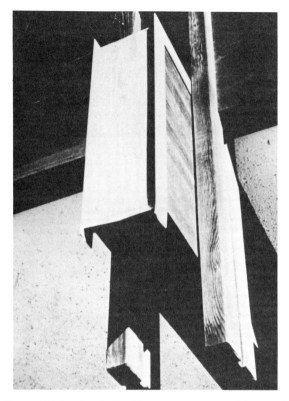

Figure 7.2.3.--Metal flashing protects exposed beam end.

2. Provide natural drainage and allow for air circulation. When joints or other details are designed to shed water and to permit rapid drying after a rainfall, the duration of conditions favorable to decay can be greatly lessened. Slope railings, stairs, and tops of deck stringers or joists. Space members to avoid capillary voids which retain water. Space butt joints in railings and decks to allow air circulation.

3. Use wood grades and species which stay in place well. Clear grades with limited slope of grain, if obtainable, are best. Species such as western redcedar, redwood, spruce, poplar, and the soft pines are stable. Choose sections which are more nearly square rather than thin, wide boards, for example: A 3 by 4 inch instead of a 2 by 6 inch or a 1 by 8 inch. This minimizes dimensions

for swelling and shrinkage and limits changes in grain pattern. Use quartersawn or edge-grain lumber, if available.

4. Use the heartwood of decay-resistant species for maximum durability in the weather. Western redcedar is an excellent choice for permanence and stability. Redwood and the light cedars found in various regions are also good choices on both counts. Commonly found hardwoods such as white oak also have decay-resistant heartwood.

5. Fasten assemblies well with rust-resistant fasteners or waterproof adhesives. Aluminum nails are now common, and stainless steel is available at a higher price.

6. Well-designed and well-maintained finishing systems, with prior dipping of all ends and fabricated surfaces in a water-repellent preservative, can extend the performance of exterior wood by slowing the rate of moisture penetration. Surface coatings, however, depend heavily on the effectiveness of end seals, since water absorbed in end grain will eventually loosen or blister them. The essence of good performance in any finishing system is adequate maintenance, which is all too often neglected.

7.2.5 References for Section 7.2

1. Oviatt, A. E.
   1975. Protecting exposed ends of timber beams in the Puget Sound area. Pac. Northwest For. and Range Exp. Stn., Portland, Oreg. Nov.

2. Scheffer, T. C., and Verrall, A. F.
   1973. Principles for protecting wood buildings from decay. USDA For. Serv. Res. Pap. FPL 190, For. Prod. Lab., Madison, Wis.

3. Schein, Edward W.
   1968. The influence of design on exposed wood in buildings of the Puget Sound area. Pac. Northwest For. and Range Exp. Stn., Portland, Oreg.

7.3 High-Humidity Environments*

7.3.1 Introduction

Buildings of special occupancy may be exposed to interior environments where unusual humidity or temperature conditions exist. Among these are interior swimming pools, shower rooms, skating rinks, freezer rooms, food processing and storage spaces, laundries, and many others.

---

*Author: Alfred E. Oviatt, Jr., Research Architect, Forest Products Laboratory, Madison, Wisconsin.

The severity of these environments varies with many factors, among them the moisture source, temperature, air circulation, air conditioning, and the character of the building enclosure. Humidity and its companion, condensation, can be equally damaging to nonwood materials, but we shall consider here only their possible effects on timber components.

It is known that extremes of humidity above 90 percent relative humidity, if maintained for extended periods, can result in wood moisture content of 20 percent or more (6). The continuous occurrence of condensation on wood surfaces has been observed to be a more severe hazard, causing even higher moisture contents. Continuous moisture conditions of 20 percent or more in structural members cause significant weakening of the wood, deterioration of many adhesives, and troublesome dimensional changes. Decay may also result when temperatures are favorable.

Condensation can most often be prevented by adequate insulation and ventilation. Under some conditions dehumidification or local heating may be needed. Vapor barriers can be used to prevent condensation within a building assembly, such as a wall or roof, where hidden deterioration might occur.

There is little published information on surveys of damage due to high interior humidities, and these conditions vary considerably in different occupancies. However, the U.S. Forest Service has made some field surveys of moisture in wood members in use in some highly humid occupancies (3). These studies offer some guides to the control of moisture in certain building types.

7.3.2 Enclosed Swimming Pools

With typical water temperatures of 75° to 80° F and interior air temperatures of 78° to 83° F, relative humidities as high as 60 percent may be expected. However, this humidity level does not result in an equilibrium moisture content in wood members high enough to cause significant weakening or any hazard of decay (fig. 7.3.1).

Condensation on wood members or on adjacent surfaces must, however, be avoided. Adequate insulation of the enclosure and a nominal amount of ventilation are usually sufficient to prevent this (5). In most climates, less insulation is needed to prevent condensation than would normally be used for energy conservation, and the number of air changes required is most often no more than would be needed for healthful human occupancy.

Conditions of high outside temperature and humidity may require a greater degree of ventilation, perhaps calling for two-speed fan control. In the exceptional case, when outside temperature and specific humidity exceed inside design objectives, cooling of the ventilating air may be indicated (1).

314   WOOD STRUCTURES

Figure 7.3.1.--A community swimming pool.

Poorly insulated building components, such as windows and skylights, can be expected to accumulate condensation. This can usually be drained by adequately designed gutters and drains, and stainless or plastic-encased window frames can be provided.

If gutters are to be used to collect condensate from skylights, windows, and the like, care must be taken in installation and maintenance. In one instance, gutters were installed on the side of a large laminated beam supporting a skylight. The beam length was such, however, as to require two lengths of gutter, with the joint between them near the center of length of the beam. This joint was not sealed, and a large waterspot was found below each joint. In some instances, decay was present in the waterspot area. It is greatly preferable, therefore, to use single lengths of gutter, if available, and, if not, to be sure that the joints are sealed.

Similarly, the gutter must have a grade sufficient to insure that it will drain to an outlet and that the grade connot be overcome by deflection of the beam. Care must be taken that the outlet (tubing, pipe, etc.) does not empty onto wood surfaces, walkways, etc. Draining onto wood surfaces can result in high moisture contents and decay, while draining onto walkways can result in unpleasant consequences for users of the walkway, from wetting of clothing to icing of walks and danger of falls. The outlet should be inspected periodically to be sure that it has not become obstructed by lime deposits or by dirt or debris from the gutter.

Vapor barriers are essential near the interior surfaces of walls or ceilings of the enclosure to prevent vapor passage into wall cavities, attics, or other spaces. These are of critical importance in this type of building, since interior vapor pressures can be very high. Therefore it is very important that they be continuous and well sealed at electrical and plumbing penetrations. Careful field supervision is needed.

7.3.3 Enclosed Ice Rinks

Indoor ice rinks range from large arenas for such spectator events as hockey and ice-show productions to more modest buildings for neighborhood skating and curling (2). Air humidity conditions are high, with vapor generated directly from the ice surface and the respiration of active participants and spectators. Air temperatures and the temperatures of interior surfaces are low, encouraging condensation (fig. 7.3.2).

Figure 7.3.2.--An indoor skating rink.

When the outside air is sufficiently dry, adequate insulation and ventilation may prevent condensation. In regions where humid seasons occur, however, seasonal dehumidification may be needed. Under more severe conditions, heating and forced circulation of the dehumidified air or local heating of surfaces may be required for full control.

Marked radiation from the interior surfaces of the enclosure to the ice surface occurs in a rink building. Ice temperatures range from 22° F for hockey to 26° F for other skating. This is usually well below the temperatures of other interior surfaces, which are further cooled by radiation to the ice. Temperatures at the soffits of members directly above the ice may thus be lower than those of similar areas further away, causing local condensation and a significant increase in wood moisture content. Measurements in one building (4) indicated average soffit moisture content in a laminated beam over the ice surface near 22 percent, while the average in the same beam over spectator areas was 4 percent less. These levels were later reduced by the circulation of heated air by unit heaters. Care must be taken not to direct air from unit heaters at the beam itself, since excessive drying can cause checking of the surfaces. A similar result could have been obtained by using resistance heaters attached to the beam soffits.

Skating rinks in colder climates, often used only seasonally, are sometimes protected by a roof structure but otherwise largely open to natural ventilation. These are largely at the mercy of local weather, and condensation on the roof members is likely under humid air conditions. These structures are not well suited to warmer and more humid climates, where condensation may sometimes occur on all ceiling areas. Low temperatures may inhibit any decay in the members, but dripping onto the ice may make it unsuitable for any precision skating or curling. Such a structure should be inspected frequently for moisture damage which may impair strength, particularly at connections.

7.3.4 Food-Processing Buildings

The processing of foods produces a wide variety of interior environments, and the quality of building enclosures varies greatly. Moisture effects resulting from several processes have been observed in a canning plant, where fruits and vegetables are seasonally stored, sorted, and sometimes cooked prior to canning (3).

Refrigerated storage is provided for short periods prior to processing in a fully insulated building. Temperatures range from 35° F to -10° F, depending on the product stored. Between loads, the interior is hosed down and fumigated, usually every 2 or 3 months. Measured moisture contents in the wood structure were highly variable over a 2-year period, ranging from 14 to 25 percent at different times. Operation of this space and the frequent washing obscured clear conclusions, except the general indication of relatively high moisture contents in this use.

The sorting operation used a wet conveyor which washed and spread the produce for hand sorting. It is operated seasonally, and the entire area is hosed down after each shift. When in operation, this process resulted in high moisture contents in a large timber beam over the sorter. Condensation dripped from the soffit and from nearby wood plank ceiling areas, and moisture contents in the lower portion of the beam reached 25 percent. The building is unheated and uninsulated. Special ventilators over the sorting line would reduce the humidity, but heating and roof insulation might also be required to prevent seasonal condensation.

Steam generated by cooking vats enveloped a laminated beam nearby, causing condensation on its surfaces and moisture contents as high as 20 percent. These receded to 10 percent when the equipment was not in operation, but for continuous operation such a unit would require either complete enclosure or forced ventilation to the exterior (fig. 7.3.3).

Figure 7.3.3 Cooking vats in canning plant.

7.3.5 Frozen-Food Storage

Frozen-food storage buildings are widely used as local depots in the distribution of frozen food. They enclose environments at high relative humidities as a result of high percentages of moisture in the produce and temperatures as low as -20° F. Moisture generated by workers as well as the frequent opening of loading doors to load frozen produce in or out also affect interior conditions.

These conditions have sometimes raised doubts among designers about using timber structures. However, some preliminary observations by the U.S. Forest Service and some measurements made by a competent private group indicate that resulting moisture contents in wood structures are very low. In addition, the low temperatures involved increase the strength of wood somewhat and prevent the growth of decay fungi.

In these studies a section of glued-laminated timber with moisture sensors installed was placed in one frozen-food storage building for a period of over 3 years and observed every 6 months, winter and summer. The section, when placed, had a moisture content of 11 percent; during the whole period of observation, the moisture content varied only between 11 and 12 percent. Observed interior temperatures varied between -3° and -10° F.

Two other laminated samples were conditioned before placement, one to 16 percent moisture content and the other to 7 percent. Placed in a second storage building, moisture contents converged to an average of 11 percent in a 6-month period, at a temperature of -15° F. Readings in a heavy timber-framed storage building in operation were reported by others to be in a similar range.

The refrigeration process maintains the air at a very low moisture content in this type of building, and moisture contained in the frozen goods apparently is frozen before it can affect the wood. Moisture from other sources is quickly drawn off.

Where humid interior conditions similar to these are expected in design, a competent mechanical engineer familiar with the particular building type should be consulted in the early stages of design.

7.3.6   References for Section 7.3

1. Daily, Robert E., and Bishop, Walter P.
    1969. Swimming pool design. Air Cond., Heat., and Vent. May and October.

2. Handegard, G. O., and Crocker, C. R.
    1962. Control of condensation in curling rinks. Can. Bldg. Dig. No. 35, Natl. Res. Counc., Ottawa, Ontario.

3. Oviatt, Alfred E., Jr.
    1968. Moisture content of glulam timbers in use in the Pacific Northwest. Pac. Northwest For. and Range Exp. Stn., Portland, Oreg.

4. Oviatt, Alfred E., Jr.
    1972. Moisture in wood structures for ice skating. For. Prod. J. Mar.

5. Oviatt, A. E.
    1973. Wood is ideal in pool natatorium roof construction. Swimming Pool Weekly/Age Data and Ref. Ann.

6. U.S. Forest Products Laboratory.
    1974. Wood handbook: Wood as an engineering material. USDA Agric. Handb. No. 72, rev. Superintendent of Documents, Washington, D.C. 20402.

7.4　Fire*

7.4.1　Structural Systems and Fire

Designing to minimize a building's contribution to death or to its destruction by fire must be incorporated early in the architectural planning process. Under control of the architect are such fire safety considerations as: building height; floor and compartment areas; location and design of stairways, exits and corridors; interior finish materials; general materials of construction; use of automatic sprinklers; private fire protection; and access for public fire fighting. The design engineer must consider the same factors if they influence the structural design.

Limitations are imposed for all of the above factors by building codes. Unfortunately, all are interrelated, and attempts to express them in specific rules, as is done in building codes, necessarily cause some oversimplification. Best results are obtained only by a fire protection engineering analysis of the structural problems, occupancy, and fire protection in a building rather than by a literal application of rules which are bound to have exceptions.

Fundamental to the problem of incorporating fire safety in structures is the manner in which fire, heat, and smoke spread. Table 7.4.1 gives the principal building design deficiencies which contribute to the spread of fire, heat, and smoke throughout a building or from one building to another.

It has been confirmed that open doors and stairways, together with lack of firestopping in concealed spaces are responsible for spread of fire and smoke in a high percentage of fatal dwelling fires. Combustible finish material is suspected of contributing to death by fire spread in more than half of all fatal fires (12).

The design for fire integrity of a completed building is predicated on equal fire resistance for all structural assemblies. Any single weakness may negate all other protection features under fire conditions and result in extensive fire damage and danger to occupants and firefighters. It can be assumed that a wall will offer the degree of fire resistance indicated by its rating only if it is constructed and installed properly and has been joined with other assemblies in such a way that the fire resistance of the combination will not be weakened. As a result, a wall with door and conveyor openings and open expansion joints may become ineffective if the openings and joints are not properly protected. Not infrequently the protection of structural elements and the fire integrity of division walls are endangered by the installation of building service equipment, such as ducts and lighting fixtures, and by piercing the walls for pipe cases.

The methods for correction of many of these fire safety shortcomings are obvious, but are much less so for others. The following sections

---

*Author: Erwin L. Schaffer, M. ASCE, Supervisory Research Engineer, USDA Forest Service, Forest Products Laboratory, Madison, Wis.

discuss design considerations that deal with the problems described above.

Table 7.4.1 Design deficiencies responsible for spread of fire, heat, and smoke (17)

1. Throughout a building due to:

Lack of or inadequate vertical and/or horizontal fire separations, such as unprotected or inadequately protected floor and wall openings for stairs, doors, elevators, escalators, dumbwaiters, ducts, conveyors, chutes, pipe holes, and windows.

Concealed spaces in walls and above ceilings without adequate firestopping or fire divisions.

Combustible interior finish including combustible protective coatings.

Combustible structural members (beams and girders, joists) framed into fire walls.

Improper anchorage of structural members in masonry bearing walls.

Lack of adequate explosion venting where required.

Inadequate protection of framing, resulting in weakening or destruction of floors and walls used as fire barriers.

Lack of roof vents to ventilate fire gases.

2. From one building to another due to:

Lack of or inadequate spacing and/or fire division walls between adjoining buildings.

Unprotected or inadequately protected openings in fire division walls between adjoining buildings or in fire walls between detached buildings.

Inadequate fire resistance of exterior walls.

Use of combustible roofs, roof coverings, roof structures, eaves, trim, etc.

Lack of protection at openings to passageways, pipe tunnels, conveyors, ducts, etc., between detached buildings.

Inadequate separation or explosion venting between adjoining or detached buildings.

Inadequate fire resistant design resulting in collapse of exterior wall.

7.4.2 Firestopping and Draftstopping

As mentioned previously, fire spread from one area of a building to another through open concealed spaces has become a major problem. In response to this, the National Forest Products Association has prepared a guide recommendation on where and how such firestopping and draftstopping should be used (22). Such stopping is needed in all buildings, whether or not a building or its contents are classified as noncombustible, since combustibility of a building's structural components is often not a factor in fire spread. Firestopping prevents movement of flame and gases in concealed passages in building floors, walls, and stairs. Draftstopping is employed for the same purpose in large concealed passages such as attic spaces and floor assemblies with suspended ceilings or open-web trusses.

Acceptable firestopping may consist of 2-inch nominal lumber or two thicknesses of 1-inch nominal lumber with lap joints, or one thickness of 3/4-inch plywood with joints backed by 3/4-inch plywood. Typical firestopping details in wood frame constructions are illustrated in figures 7.4.1 to 7.4.3.

Draftstopping materials typically accepted are 1/2-inch gypsum board or 3/8-inch plywood; both must be adequately supported. In floor-ceiling assemblies, draftstop partitioning into horizontal areas not exceeding 1,000 square feet is recommended (see a typical detail in figure 7.4.4). Draftstopping is critical in multifamily dwellings, motels, and hotels, when tenant separation walls do not extend to the roof sheathing above open roof spaces (such as attics, mansards, overhangs) (fig. 7.4.4). In these and other buildings, draftstopping that partitions attic spaces into horizontal areas of 3,000 square feet or less is recommended. Draftstopping is not required where approved sprinklers are provided. (Note: To prevent moisture condensation problems from occurring in concealed roof spaces, ventilation should be provided in each partitioned space.)

Canadian code requirements (11) for firestopping and draftstopping are similar to American practice. However, 1/2-inch-thick plywood backed with 1/2-inch-thick plywood at joints is acceptable for a fire or draftstop, and partitioning into at most 3,000-square-foot areas is recommended.

7.4.3 Fire Classifications of Structures

The fire resistance of various structural assemblies (such as walls, floor-ceilings, doors, etc.) has been defined as the ability of the component to restrict passage of fire through it. For load-bearing components (including members and assemblies), the term fire endurance is used to characterize the ability to restrict passage of fire and to continue its load-carrying function. To determine, or "rate," a component's fire resistance or fire endurance, the component is subjected to a standard fire exposure (4), during which time its behavior is observed and recorded. A "rating" is given in terms of the minutes or hours the component can be expected to restrict passage of fire and maintain its integrity. Code authorities employ this "rating" to judge

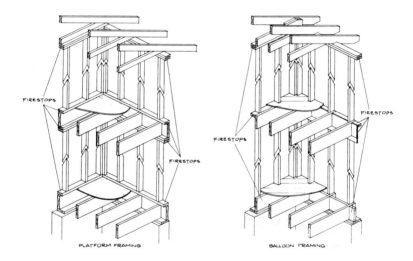

Figure 7.4.1--Typical firestopping in concealed spaces of stud walls and partitions, including furred spaces at ceiling and floor levels. (a) Platform framing; (b) Balloon framing (23).

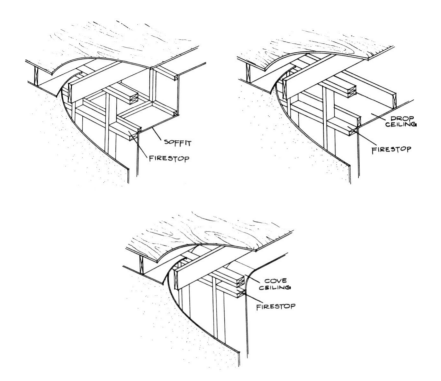

Figure 7.4.2--Firestopping at interconnections between concealed vertical and horizontal spaces such as at soffits (a), drop ceilings (b), and cove ceilings (c) (23).

324   WOOD STRUCTURES

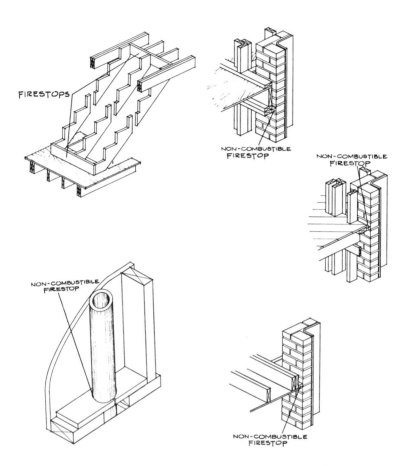

Figure 7.4.3--Firestopping in concealed spaces between stringers at the top and bottom of stairs (a) and at openings around vents, pipes, ducts, chimneys, and fireplaces at ceiling and floor levels with noncombustible materials (b and c) (23).

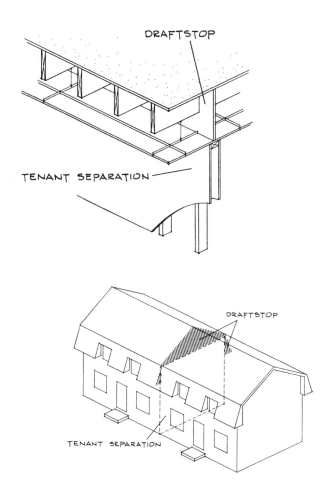

Figure 7.4.4--Draftstops in multifamily buildings. (a) In the floor-ceiling assemblies and (b) in attics, mansards, overhang, or other concealed roof space above and in line with tenant separation when tenant separation walls do not extend to the roof sheathing above (23).

a given construction's ability to meet a general rating requirement for components in a building type.

The design of components to meet code requirements for fire-safe barriers can be done in either of two ways and be acceptable to most code authorities: (1) through employment of previously fire-tested components known to have the requisite rating level or (2) through analytically-based design of the component using engineering principles to establish the rating. The first method is the traditional code-acceptable means to meet building requirements. The second method, however, is being increasingly recognized and will be universally used by designers in the future.

7.4.3.1 Heavy Timber and Glued-Laminated Constructions

Before the advent of glued-laminated construction, the sizes of solid-sawn timbers that were available limited the versatility of heavy timber construction. Even so, heavy timber or mill-type construction was used extensively in multistory buildings. These have exterior walls of masonry, and interior columns, beams, and floors of wood, with straight members of relatively short span and a minimum of surface or projections exposed to fire.

Glue-laminating techniques have since provided the means of manufacturing solid wood structural members with extremely long spans and a variety of shapes. Thus, laminating has permitted the construction of heavy timber buildings with larger unobstructed areas and has given the architect a group of wood products that he can use in ways never before possible. Because of this new versatility, the term "mill-type" has properly been replaced and this type of construction is now universally referred to as heavy timber construction.

Heavy timber construction is defined in building codes and standards by the following minimum sizes for the various members or portions of a building:

	Inches, nominal
Columns--supporting floor loads	8 by 8
supporting roof and ceiling loads only	6 by 8
Floor framing	
Beams and girders	6 wide by 10 deep
Arches and trusses	8 in any dimension
Roof framing--not supporting floor loads	
Arches springing from grade	(6 by 8 lower half
	(6 by 6 upper half
Arches, trusses, other framing springing from top of walls, etc.	4 by 6
Floor (covered with 1-in. nominal flooring, 1/2-in. plywood, or other approved surfacing)	
Splined or tongue-and-groove plank	3
Planks set on edge	4
Roof decks	
Splined or tongue-and-groove plank	2
Plank set on edge	3
Tongue-and-groove plywood	1-1/8

The overall fire resistance of members in heavy timber construction obviously increases with increased cross-sectional dimensions. Building codes, however, recognize that heavy timber construction has performance similar to that of noncombustible construction with a 1-hour fire resistance, and with one exception permit its use in all fire districts for all types of occupancy. The exception (20) is a Fire Zone No. 1 area--normally downtown areas of larger cities. This acceptance is based on experience with the performance of heavy timber construction in actual fires, the lack of concealed spaces, and the high fire resistance of walls in this type of construction.

Roof arches and truss members may have spaced members which must be 3 inches thick and blocked solidly throughout the intervening spaces, or closed by a continuous wood cover plate (2 inches thick) secured to the underside of the members.

To maintain the fire integrity of heavy timber structures, special care must be given to the design details for the many connections. It is recommended that the designer obtain both AITC[1] 108-69 "Standard for Heavy Timber Construction" (2) and NFPA[2] "Heavy Timber Construction Details" (19). The NFPA reference (19) contains many detail drawings of a variety of approved connections between heavy timber components.

Although building code requirements vary somewhat for walls in heavy timber construction, they generally require that exterior and interior bearing walls be of 2-hour fire-resistive noncombustible construction.

---

[1] American Institute of Timber Construction.
[2] National Forest Products Association.

Also, use of fire-retardant-treated wood in exterior 2-hour, fire resistive bearing walls is permitted in some codes (UBC) (15). However, 3-hour fire-resistive non-combustible construction is required for exterior walls when the distance from other buildings or the property line is 3 feet or less. The 3-hour requirement also applies to nonbearing walls, except that for building or property line separations between 20 and 30 feet the fire resistance may be reduced to 1 hour and beyond 30 feet no fire resistance is required. Consequently, today there is much heavy timber construction in combination with glass or other nonrated exterior wall material when there is adequate separation from other buildings. Such construction is widely used for educational, religious, supermarket, and other buildings not built close to property lines.

The fire resistance of glued-laminated structural members, such as arches, beams, and columns, is approximately equal to the fire resistance of solid members of similar sizes. Available information indicates that laminated members glued with phenol, resorcinol, or melamine adhesives are at least equal in fire resistance to a one-piece member of the same size, and laminated members glued with casein glues have only slightly less fire resistance.

Available data (26,32) indicate that the char zone in a casein-glued laminated member will be as much as 10 percent deeper at glue lines than in a solid beam of the same size after exposure to an ASTM E 119 fire for 1 hour. The appearance of the casein-glued joints and the results of shear tests indicate, however, that little if any weakening of the glue joints as a result of the fire exposure occurs beyond the pyrolysis zone (fig. 7.4.5), or about 1/4 inch below the char base (26).

When the fire endurance required of a wood member is less than the time required for the zone of char to penetrate through the outer laminations, the type of adhesive used is unimportant. On the basis of a rate of penetration of char of 1-1/2 inches per hour of exposure to fire, laminated members with outer laminations not less than 1-1/2 inches thick (such as nominal 2-in. lumber laminations) would be equivalent in fire resistance to solid members of the same actual size and having a fire resistance rating up to 1 hour. For use in the heavy-timber-construction classification, laminated members glued with phenol, resorcinol, or melamine adhesives or laminated members glued with casein adhesive and having nominal 2-inch outer laminations are considered equivalent to solid-sawn members of the same actual size.

Because the remaining uncharred wood in large members after fire exposure retains much of its original strength (due to the very short duration of exposure to relatively low temperature beyond the char zone), these members can be rejuvenated by simple removal of the charred wood. Load-carrying capacity after char removal is calculated from the residual cross section. It should be noted here that, in designing a structure, increasing the cross sectional dimensions of members beyond those required to carry the design load in order to counteract the

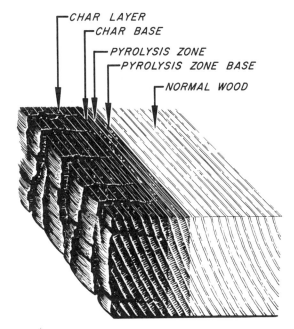

Figure 7.4.5--Degradation zones in a wood section exposed to fire on one surface.

results of possible future severe fire exposure is an accepted practice. However, as sections are exposed to fire on two or more adjacent sides, one must recognize that the corners develop deeper char depth (24). This is shown in figure 7.4.6.

7.4.3.2 Ordinary and Light-Frame Construction

The term "ordinary construction" defines buildings with exterior walls of masonry and interior wood framing with members not less than 2 inches (nominal) thick. This type of construction has been widely employed in commercial or public buildings up to five or six stories in height. Ordinary construction differs from heavy timber construction in that exterior walls generally are not as heavy and interior framing is of lesser cross-section. These differences are reflected in smaller building heights and floor areas being allowed. It differs from light-frame construction in its larger allowable heights and areas, its self-supporting masonry walls, and in a number of interior requirements appropriate to the occupancy. There are detailed code requirements in ordinary construction for firestops (nominal 2-in. thick wood or the equivalent) in concealed spaces in walls or ceilings through which fire might spread.

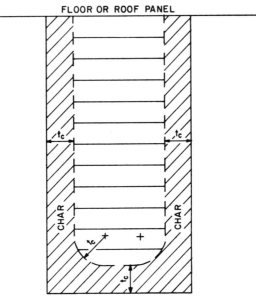

Figure 7.4.6--Corner charring behavior of large timber structural members exposed to fire; t_c = char depth (14).

Most residential and some commercial institutional, industrial, and assembly buildings of wood are of light-frame or, as it is more commonly known, frame construction. Originally restricted to the conventional type of building with stud walls, joisted floors and ceilings, and raftered roofs, light-frame construction has been diversified by the introduction of prefabricated, panelized, or stressed-skin structural elements.

A type of wood construction known as "protected light frame," in which elements are designed to have a fire resistance of 1 hour, is commonly used. Based on the areas allowed, codes rate the fire performance of this type as intermediate between ordinary and heavy timber construction. There are many recognized assemblies involving wood-framed walls, floors, and roofs that provide a 1-hour, and even a 2-hour, fire resistance (30). Fire resistance ratings of 1 hour have been obtained through test for several dozen floor-ceiling assemblies having wood joists and plywood floors. The ceilings usually consist of at least 1/2-inch gypsum wallboard or acoustical tile; the floor is usually two layers of plywood, although single layers have also been rated (30). A double-layer ceiling of 5/8-inch gypsum wallboard and two layers of plywood for the floor are needed to achieve a 2-hour rating.

Unprotected light-frame wood buildings do not have the natural fire resistance of the heavier wood frames. In these, as in all buildings, attention to good construction details is important to minimize fire hazards. Of particular importance are firestops, separation of wood from masonry around chimneys and fireplaces, and design of walls, ceilings, floors, roofs, stairways, and doors. The fire resistance of single-layer plywood floors without protective ceilings has been investigated and forms the basis for HUD requirements in one- and two-family housing (28). The fire resistance of these floors is about 10 minutes. The fire resistance of other light frame constructions can be found in references (13,17,18,30,31). A conventional light-frame bearing wall has at least a 20-minute fire endurance, for example (13).

7.4.4 Reducing Surface Flammability: Fire Retardants

7.4.4.1 Interior Finish Restrictions

The interior finish commonly referred to for building constructions includes the exposed interior surfaces where the surface is an integral part of the building or is affixed thereto, such as the materials for walls and ceilings, interior partitions, interior trim, paint, and wallpaper. Decorations and furnishings which are not affixed to the structure are not considered interior finish, and are not limited by building codes, even though they may furnish the primary source of fuel to an incipient fire.

The standard test for determining the flame spread rating of materials is that described in ASTM E84-79 (5). In this method, the rate of travel of flame along a material under test is compared with asbestos-cement board which does not spread flame and is rated 0 and red oak flooring which is rated at 100.

Materials are classified into groups based on their flame-spread index values--in class A from 0 to 25, class B from 26 to 75, class C from 76 to 200, class D from 201 to 500, and class E over 500.

The requirements for surface flammability of interior finish generally prescribe class A in the exitways of unsprinklered buildings intended for large assembly and institutional purposes, and class B for school, small assembly, mercantile, and hotel buildings. In general, the next higher class, C, (greater flammability) is permitted for the interior finish used in other areas of these building which are not considered exitways. This flame-spread classification is permitted when the areas are protected by automatic sprinkler devices, except that class C finish is usually the highest permitted in any of these buildings. Class D and E finishes are not allowed in buildings of this type, but are permitted in 1- and 2-family dwellings.

These requirements frequently exempt interior trim or permit up to 10 percent of the total wall and ceiling surface areas in any use area or occupancy group to be materials with flame-spread classification as high as class C. The exposed portions of structural members of heavy timber construction are also exempt from these flame-spread requirements in several types of occupancies. Also, wallpaper, paint, and floor coverings may be exempt from these requirements, unless they are judged to be unusual fire hazards. Generally, the common paints and varnishes have only a slight effect on the flame-spread index values of wood, usually lowering the values.

Most wood species have flame-spread index values of 90 to 160 by the ASTM E 84 method, and therefore are accepted for interior finish only for those applications permitting class C interior finish. A few species (western cedar, redwood, hemlock, and northern spruce) have flame-spread index values of slightly less than 75, and these can be used for class B applications. The Underwriters Laboratories, Inc., Card Data Service C 60, U.L. 527 (29), lists the flame-spread index for various wood species.

E 84 flame-spread tests have been conducted by Underwriter's Laboratories on a variety of plywood panels manufactured under Product Standard PS 1. Variables studied included effects of species groups, thickness, glue type, and surface texture. Flame-spread values ranged from 75 to 225, but for all common untreated, unfinished plywood grades were under 200, or within class C.

Therefore, fire-retardant treatments or coatings are necessary for wood interior finish when class A, and sometimes class B, flame-spread performance is required.

7.4.4.2 Fire-Retardant Treatments and Coatings

Two general methods are available for improving the fire performance of wood by the use of fire-retardant chemicals. One method consists of impregnating the wood with waterborne salts, using conventional vacuum-pressure methods, such as used in the wood-preserving industry. The second method involves the application of fire-retardant chemical paint coatings on the wood surface. The impregnation methods are usually the more effective and lasting and are intended for use on new wood construction. For wood in existing constructions, the surface application of the fire-retardant paints offers the only means for increasing fire-retardant characteristics.

The proper fire-retardant treatment of wood improves fire performance by greatly reducing the amount of flammable products released, thus reducing the rate of flame spread over the surfaces. Treatment also reduces the amount of heat available or released in the volatiles during the initial stages of fire, and also results in the wood being self-extinguishing once the primary source of heat and fire is removed or exhausted. Fire-retardant coatings and treatments do not prevent the wood from decomposing and charring under fire exposure, and the rate of fire penetration through treated wood is approximately the same as for untreated wood (27). The effect of fire-retardant treatments on strength is discussed in Section 7.4.4.3.

In the impregnation treatments, wood is pressure impregnated with water-soluble chemical solutions using full-cell pressure processes similar to those used for chemical preservative treatments. Retentions of the fire-retardant salts must be fairly high (2-1/2 to 5 lb of dry salt per cu ft of wood) to be effective. The salts used in the current fire-retardant formulations are principally the same ones which have been known for their fire-retardant characteristics for over 50 years--monoammonium and diammonium phosphate, ammonium sulfate, zinc chloride, sodium tetraborate, and boric acid. These salts are combined in various formulations to develop optimum fire performance characteristics and still have acceptable characteristics with regard to hygroscopicity, strength, corrosivity, machinability, surface appearance, gluability, paintability, and cost. See references (6) and (23) for typical formulations.

Fire-retardant-treated wood and plywood is currently being used for interior finish and trim in rooms, auditoriums, and corridors where codes require materials with low surface flammability. In addition, many codes, including the model building codes, have recently accepted the use of fire-retardant-treated wood and plywood in fire-resistive and noncombustible constructions for the framing of nonload-bearing walls and roof assemblies, including decking. Fire-retardant treatment for all wood used in buildings over 150 feet high is also prescribed in New York City. Some building codes also permit the use of increased floor area limits in heavy timber, ordinary, and wood-frame constructions when the structural wood members have been given fire-retardant treatment. Fire-retardant-treated wood products should be in a "dry" state (less than 19 pct and preferably less than 15 pct moisture content) before installation. AWPA specifications (7) contain a restriction on drying methods and attained moisture content to satisfy this need. The use of fire-retardant-treated wood products in areas of a building where the relative humidity can exceed 80 percent requires special consideration. In such areas, the treatment should be no more hygroscopic than untreated wood, or a nonleachable fire retardant (exterior type) should be prescribed, to minimize corrosion of metallic fasteners.

Effective leach-resistant fire-retardant treatments for severe-environment applications, such as exterior exposure or high humidity, have been needed for some time. Several commercial treatments or treated products are available on the market which appear to be equally effective (9,16,25). Currently, the following companies supply durable fire retardant treated wood products:

>Koppers Company ("Non-Com X")
>Pittsburgh, PA
>
>J. H. Baxter Company
>San Mateo, CA
>
>Hoover Universal ("Fire-X")
>Thomson, GA

There are many commercial paint-coating products available to provide varying degrees of protection of wood against fire. These paint coatings generally have low surface-flammability characteristics and "intumesce" to form an expanded low-density film upon exposure to fire, thus insulating the wood surface below from pyrolysis reactions. They have ingredients to restrict the flaming of any released combustible vapors.

Fire-retardant paints include those based on water-soluble silicates, urea resins, carbohydrates and alginates, polyvinyl emulsions and oil-base alkyd, any of which may be clear or pigmented. In many of the water-soluble paints, ammonium phosphate or sodium borate is used in the formulation to obtain fire-retardant characteristics. The oil-base paints frequently make use of chlorinated paraffins and alkyds plus antimony trioxide to limit the flammability of any pyrolysis products produced. Inert materials, such as zinc borate, mica, kaolin, and inorganic pigments are also used in these formulations. Intumescence is obtained by the natural characteristics of some of the organic ingredients or special materials, such as isano oil, may be used. There are a limited number of clear fire-retardant finishes available. Generally they do not have the durability of the fire-retardant paints as pigmentation and opaque chemical additives are usually necessary to provide durability.

7.4.4.3 Treatment-Related Properties of Fire-Retardant-Treated Wood

Strength--The combined treatment and drying process, with <u>interior inorganic</u> fire-retardant chemicals can have a deleterious effect on some strength properties. Kiln dried clear wood experiences a reduction of 13 to 20 percent in modulus of rupture, and a higher loss (to 34 pct) in shock or impact resistance (1,10,14). The stiffness (as reflected by modulus of elasticity) is reduced only 5 percent or so (1,14). The strength losses are reduced significantly by air-drying the treated wood, while increasing the drying temperature results in greater losses (1). As a result, both the impregnation and redrying processes need control. The American Wood Preservers' Association (AWPA) recommends that wood treated with inorganic fire-retardant salt solutions not be kiln-dried at dry bulb temperatures exceeding 160° F (71° C) in order to minimize strength loss (7).

As a result of these researches, the National Forest Products Association (21) and the American Institute of Timber Construction (3) currently recommend that the allowable design values for lumber pressure-impregnated with fire-retardant chemicals and the allowable loads for fastenings in such wood shall be reduced 10 percent. For structural glue laminated timber similarly treated, the effect on strength is dependent upon the species and treatment combinations involved, and the manufacturer of the treatment should be consulted for specific information on adjustment of design allowables.

There are commercial fire-retardant treatments based on chemicals other than inorganic salts that are effective in both interior and exterior exposures because they are polymerized within the wood. One treatment (1,25) results in a slight to moderate increase in bending strength with no significant change in stiffness and impact resistance. No adjustment in design values would apparently be required for wood treated by this process.

In addition to the reduction in the strength of wood after fire-retardant treatments with inorganic salt, progressive loss of strength with time can be expected in those treatments involving phosphates and sulfates (10). Such salts gradually generate acidic residues in the treated wood which may continue to degrade strength even under ordinary use conditions.

Corrosivity--Wood normally undergoes changes in absorbed moisture in response to the changing relative humidity and temperature of the environment it is in. If the wood is treated with an inorganic fire-retardant treatment, the wood can be expected to absorb more moisture than does untreated wood. (The polymerized resin fire-retardant treatment mentioned previously does not evidence this increase in hygroscopic behavior (25)).

The increase in moisture content is greatest above relative humidities of about 65 percent, but most inorganic-salt-based commercial formulations are developed for use in conditions not exceeding 80 percent relative humidity.

Commercial inorganic salt treatments also contain corrosion inhibitors or are formulated to a near-neutral pH (31) in order to decrease the corrosive attack of metal fasteners. Such attack is recognized as a function of the salts used and how humid the environment. Not only can attack and deterioration of fasteners occur, degradation of the wood adjacent to the fasteners can occur as well (8,10). For example, it is reported that fire-retardant treatments, involving ammonium sulfate, of lumber used in metal-plate-connected trusses, corroded the zinc and galvanized steel plates in humid conditions (10).

A designer or fabricator of fire-retardant-treated structural members or components should use fasteners composed of metals recommended by the commercial treater to minimize such events.

Durability--Fire-retardant-treated wood is durable and nonhygroscopic under normal exposure conditions (temperatures not exceeding 150° F and relative humidities not exceeding 80 pct RH). Treatments using inorganic water-soluble salts, however, are not recommended for exterior exposures to rain and weathering unless the treatment can be adequately protected by a water-repellent coating.

7.4.5 References for Section 7.4

1. Adams, E. H., G. L. Moore, and J. D. Brazier.
   1979. The effect of flame retardant treatments on some mechanical properties of wood. BRE Inf. Pap. IP 24/79. Princes Risborough Lab., Aylesbury, Bucks, U.K.

2. American Institute of Timber Construction.
   1969. Standard for heavy timber construction. AITC 108-69. AITC. Englewood, Colo.

3. American Institute of Timber Construction.
   1974. Timber Construction Manual, Second Edition. Wiley Interscience, John Wiley and Sons, New York.

4. American Society for Testing and Materials.
   n.d. Standard methods of fire tests of building construction and material. ASTM Stand. E 119. Philadelphia, Pa.

5. American Society for Testing and Materials.
   n.d. Standard methods of test for surface burning characteristics of building materials. ASTM Stand. E 84.

6. American Wood Preservers' Association.
   n.d. Standards for fire-retardant formulations. Stand. P10. Bethesda, Md.

7. American Wood Preservers' Association.
   1974. Pressure fire retardant treatment of lumber (Standard No. C-20). Pressure fire retardant treatment of plywood (Standard No. C-27).

8. Baker, A. J.
   1974. Degradation of wood by products of metal corrosion. USDA For. Serv. Res. Pap. FPL 229. For. Prod. Lab., Madison, Wis.

9. Bescher, R. H.
   1967. A new class C treatment for wooden shingles and shakes. Fire J. 61(5). Sept.

10. Brazier, J. D. and R. A. Laidlaw.
    1974. The implications of using inorganic salt flame-retardant treatments with timber. BRE Inf. Pap. IP 13/74. Princes Risborough Lab., Aylesbury, Bucks, U.K.

11. Canadian Wood Council.
    1978. "Fire stopping." CWC Fire Protective Design Data File FP-5 Can. Wood Counc., Ottawa, Can.

12. Christian, W. J.
    1974. The effect of structural characteristics on dwelling fire fatalities. Fire J. 68(1):22-28. Jan.

13. Eickner, H. W.
    1975. Fire endurance of wood frame and sandwich wall panels. J. of Fire and Flammability 6:155-190.

14. Gerhards, C. C.
    1970. Effect of fire retardant treatment on bending strength of wood. USDA For. Serv. Res. Pap. FPL 145. For. Prod. Lab., Madison, Wis.

15. International Conference of Building Officials.
    1970. Uniform Build. Code. 651 pp. Whittier, Calif.

16. Juneja, S. C.
    1972. Stable and leach-resistant fire-retardants for wood. For. Prod. J. 22(6). June.

17. National Board of Fire Underwriters.
    1964. Fire resistance ratings of less than 1 hour.
18. National Fire Protection Association. 1969. Fire Protection Handbook 13th ed. NFPA. Boston, Mass.
19. National Forest Products Association.
    1961. Heavy timber construction details. Wood Construction Data No. 5. NFPA. Washington, D.C.
20. National Forest Products Association.
    1965. Fire limits as they relate to growth of a community.
21. National Forest Products Association.
    1977. National design specification for wood construction.
22. National Forest Products Association.
    1980. Improved fire safety: design of firestopping and draft-stopping for concealed spaces.
23. National Safety Council.
    1960. Fire-retarding treatments for wood. Data sheet 372. Revised. 425 North Mich. Ave., Chicago.
24. Ollis, W.J.B.
    1968. Standard tests and calculation of fire-resistance in timber. J. of Municipal Eng. June 14. U.K.
25. St. Clair, W. E.
    1969. Leach resistant fire-retardant treatment for outdoor exposure. Proc., Am. Wood Preserv. Assoc.
26. Schaffer, E. L.
    1968. A simple test for adhesive behavior in wood sections exposed to fire. USDA, For. Serv. Res. Note FPL-0175. For. Prod. Lab. Madison, Wis.
27. Schaffer, E. L.
    1974. Do fire-retardant treatments affect charring rate of wood? Fire Retardant Chemistry Supplement J. of Fire and Flammability. May.
28. Son, B. C.
    1971. Fire endurance tests of unprotected wood-floor constructions for single family residences. NBS Report 10320, Natl. Bur. of Stand. Washington, D.C.
29. Underwriters' Laboratories, Inc.
    1971. Card data service C 60. Wood-fire hazard classification. Chicago.
30. Underwriters' Laboratories, Inc.
    1974. U. L. fire resistance index. Underwriters' Lab. Inc., Northbrook, Ill.

31. U.S. Department of Agriculture.
    1974. Wood Handbook. For. Prod. Lab., Madison, Wis. U.S. Dep. Agric., Agric. Handb. No. 72. Available from U.S. Gov. Print. Off., Supt. of Doc., Washington, D.C.

32. Yuill, C. H.
    1963. An evaluation of performance of glued-laminated timber and steel structural members under equivalent fire exposure. Southwest Res. Inst. Rep. 1-923-38.

7.5 Decay Fungi, Insects, Marine Organisms*

7.5.1 Fungi and the Importance of Moisture Control

The most effective method of protecting wood in buildings from decay fungi, and from many insects, is to use properly seasoned wood and to assure that it remains dry after construction (table 7.5.1, fig. 7.5.1). Wood in a properly designed, constructed and maintained building rarely would have a moisture content above 15 to 16 percent.

Table 7.5.1--Recommended moisture contents for wood at the time of construction

Use	Moisture content, percent of ovendry weight					
	Southwest[1/]		Coastal areas[1/]		Other areas[1/]	
	Average	Range[2/]	Average	Range[2/]	Average	Range[2/]
Interior finish	6	4-9	11	8-13	8	5-10
Softwood flooring	6	4-9	11	8-13	8	5-10
Hardwood flooring	6	5-8	10	9-12	7	6-9
Siding, exterior trim, sheathing, framing[3/]	9	7-12	12	9-14	12	9-14

1/ See figure 7.5.1 for area involved.
2/ Range is generally more important than average. If pieces fall within the range, condition will be satisfactory regardless of average.
3/ Framing with moisture contents up to 18 percent may be used if lumber with lower moisture contents is not available, but this is not recommended as a general practice.

---

*Author: Terry L. Amburgey, Professor, Forest Products Utilization Laboratory, Mississippi State University, Mississippi State, Miss.

DESIGNING TO AVOID PROBLEMS

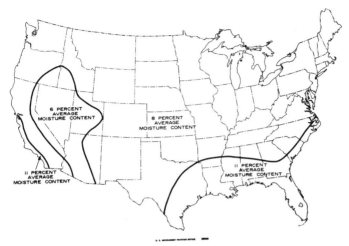

Figure 7.5.1--Recommended average moisture contents for interior use of wood products in various areas of the United States.

Since decay fungi cannot colonize wood with a moisture content below 28 to 30 percent, and mold and sapstain fungi cannot colonize wood with less than about 20 percent moisture, it is important to understand how design and construction practices influence wood moisture content.

7.5.2   Moisture--Sources and Control

Water necessary for the growth of fungi may come from (1) the original water in green wood, (2) rainwater, (3) ground water, (4) condensation, and (5) plumbing leaks. Design and construction details, and maintenance procedures that eliminate these sources of moisture assure that a structure will not fail prematurely due to decay. It cannot be assumed, however, that design features and construction practices that have caused no problems in one geographic region can be used successfully in another. As illustrated in figure 7.5.2, the decay hazard to structures in any region is dependent on both temperature and rainfall. Special precautions to avoid decay must be used in hot, humid regions such as the southeast (tables 7.5.2 and 7.5.3).

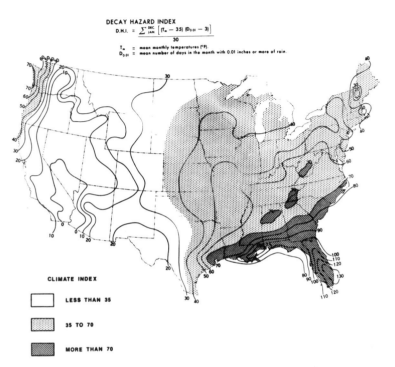

Figure 7.5.2--Relative decay hazard of wood exposed above ground as indicated by climate index. Dark areas, with an index of more than 70, are the wettest areas most conducive to decay. Lightest areas, with an index less than 35, are the driest and have a low decay hazard. Gray areas, with indexes of 35 to 70, are moderately wet and have moderate hazard.

Table 7.5.2--Amount of protection needed for foundation and substructure items in different climate zones[1,2]

Item[3]	Climate index[1]		
	More than 70	35-70	Less than 35
Flooring on concrete slab	D	D	D
Foundation: Pole, lumber, plywood in ground contact or separated from soil by a moisture-proof membrane	A	A	A
Furring strips in basements[4]	A	A	A
Joists, girders, or beams with ends imbedded in concrete or masonry[5]	A or D	A or D	A or D
Piers in crawlspace[5]	A	A	D
Posts set in ground (fence, etc. not a part of the building)	A	A	C
Sills or plates imbedded in concrete slab	A	A	A
Sills and plates on concrete slab or foundation wall	A	C	D

[1] Climate zones are shown in figure 7.5.2.
[2] Needed protection is designated by: A. Maximum protection required, i.e., pressure treatment or equivalent. See table 7.5.4. B. Moderate protection required, i.e., nonpressure treatments. See section on nonpressure treatments. C. All heartwood of species having high natural decay resistance. Small amounts of sapwood sometimes is acceptable. D. No treatment needed.
[3] It is assumed that foundation and substructure designs will meet recommendations given in the text.
[4] B or C may be adequate in well-moisture-proofed basements but A is recommended.
[5] D is permissible if all wood is at least 8 inches (20 cm) above the soil.

Table 7.5.3--Amount of protection needed for outside above-above-ground items subject to rainwetting in different climate zones[1,2]

Item[3]	Climate index[1]		
	More than 70	35-70	Less than 35
Access panels, doors, and frames to basementless space	A	B	B
Arches and other major load-bearing items	A	A	A
Columns (porch and carport)	B or C	B or C	B or C
Fence framing and panels	A or C	A or C	D
Frames, sash, trim (window, screen, door)	B	B	B
Gutters, wood	A	A	B or C
Porch, patio, deck, balcony framing and decking	A	A or C	A or C
Roof edge (fascia, rake boards)	A or C[4]	C	D
Shutters	B	B	D
Siding and trim	C[4]	C	D
Treads, stringers (step, stair)	A	B or C	D

[1] Climate zones are shown in figure 7.5.2.
[2] Needed protection is designated by: A. Maximum protection required, i.e., pressure treatment or equivalent. See table 7.5.4. B. Moderate protection required, i.e., nonpressure treatments. See section on nonpressure treatments. C. All heartwood of species having high natural decay resistance. Small amounts of sapwood sometimes are acceptable. D. No treatment needed.
[3] It is assumed that practices recommended for restricting rainwetting will be followed.
[4] It is assumed that naturally resistant wood be given a nonpressure treatment.

7.5.2.1 Rainwater--Foundation Area

The foundation and roof areas are where the greatest danger of rain seepage occurs. Too often, well-designed structures become badly decayed because improper site drainage causes water to accumulate under them or adjacent to foundation walls. It is important that water drain away from structures on all sides or be prevented from accumulating next to the foundation. This is particularly important in homes built above crawlspaces where the crawlspace grade is below the exterior grade. If a low-profile structure is desired, it is safer to build up the exterior grade around a crawlspace than to achieve a low profile by excavation. Whenever possible, the grade in a crawlspace should be at or above the exterior grade.

On sloped lots, proper site drainage may entail constructing flower beds, retaining walls, swales, or drains on the upslope side to deflect water around structures. The outside surface of all below-grade portions of foundation walls should be waterproofed (fig. 7.5.3). The

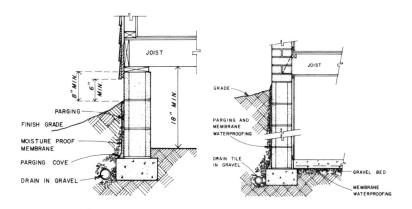

Figure 7.5.3--A crawlspace below outside grade (A) and basement (B) showing wood clearance, waterproofing of outside wall, and footing drain.

surfaces of all porches, patios, carports, or walkways should slope away from the structure, and the bases of untreated wooden porch columns should not be in direct contact with the deck (fig. 7.5.4).

Figure 7.5.4--Porch columns can be protected from rainwetting (A) by placing them on elevated supports; columns in direct contact with porch surfaces (B) frequently become decayed.

7.5.2.2 Rainwater--Building Envelope

A pitched roof and wide overhang are very important in deflecting rainwater away from siding and millwork. Roofs with little or no pitch or overhang, or with water-trapping systems such as recessed gutters, greatly increase the risk of decay (fig. 7.5.5). Flashing should be used to separate siding from roof surfaces in structures with multiple roof levels and to protect trim at the eave and rake (fig. 7.5.6). Roofing shingles should extend about an inch (2.5 cm) beyond the flashing. Gutters may be omitted in stuctures with a wide overhang, but are necessary to protect siding, doors, and trim from rain splash in structures with an overhang less than 12 inches (30 cm) or wherever discharge from the roof is directed toward concrete or a similar hard surface.

Figure 7.5.5--Water flowing from roofs with little pitch or overhang can result in (A) decay of trim or (B) wetting of walls. (C) decay following a hidden leak in a recessed eave gutter.

Figure 7.5.6--Siding should be separated from roof surfaces by at least 2 inches of flashing to avoid problems with decay and paint peeling.

Snow can indirectly wet walls and ceilings. Heat escaping from the attic melts snow on the roof; the water runs down the roof to the cooler overhang area where it freezes. Additional water backs up behind this dam and soaks under the shingles. The result often is damage to ceilings or walls. To prevent ice dams from forming, it is desirable to increase attic insulation and ventilation (to keep snow from melting), and to insert flashing such as 55-pound roll roofing under the shingles to completely cover the overhang area (to keep water from soaking the sheathing).

Roofing shingles of all-heart, decay-resistant woods formerly gave long-time service, but reports of relatively early decay (5 to 7 years) of western redcedar shingles and shakes are now increasing. In older types of construction, shingles were laid, without paper, on narrow, widely separated wooden strips. This permitted rapid inward as well as outward drying after rains. In modern construction, wood shingles and shakes are laid on solid sheathing, often plywood, which restricts drying from below. Also, modern roofs frequently have lower pitches than were used in older structures and, as a result, drain more slowly and have more of a tendency for leaves to collect on the shingles. Accumulation of organic matter favors wetting and slows drying.

Early failure of wood shingles and shakes is occurring in both moderate- and high-rainfall areas. Therefore, in areas where the climate index is over 70 (fig. 7.5.2), it seems best that wood shingles and shakes, regardless of wood species, be preservative-treated (table 7.5.4) if laid on low-pitched roofs with solid sheathing.

Table 7.5.4--**Preservatives and their retentions for use under high hazards**

Product and service condition	Minimum retentions in pounds per cubic foot[2/]								
	Coal-tar creosote	Creosote-coal-tar solution	Creosote-petroleum solution	5 Percent pentachlorophenol			Acid copper chromate	Ammoniacal copper arsenate	Chromated copper arsenate
				In heavy oil AWPA P9A	In light petroleum oil AWPA P9C	In volatile solvent AWPA P9B			
I. Lumber, plywood, structural timbers									
1. In ground contact									
All-weather wood foundation	[3/](15)	--	--	[3/](0.75)	--	--	--	0.60	0.60
Other	10	10	10	.50	0.62	0.62	0.50	.40	.40
2. Above-ground use									
Lumber[6/]	8	8	8	.40	.40	.40	.25	.25	.25
Laminates prior to gluing	6	--	--	.30	.30	.30	.25	.25	.25
Glued laminates	6	6	--	--	--	--	--	--	--
II. Poles, building	15	--	--	.75	--	--	--	.70	.70
III. Posts									
1. Building, round	15	--	--	.75	--	--	--	.70	.70
2. Fence	6	6	7	.30	.38	.38	.50	.40	.40
IV. Piling									
1. In coastal waters	[4/]20	[4/]20	--	--	--	--	--	[5/]	[5/]
2. In fresh water or on land									
Douglas-fir, western hemlock, lodgepole pine	17	17	17	.85	--	--	--	1.0	1.0
Other pines	12	12	12	.60	--	--	--	.80	.80

1/ Specification also requires minimum penetrations: Southern and ponderosa pines mostly 2.5 inches (7.5 cm) or 85 percent of sapwood, Douglas-fir, western hemlock, western larch mostly 0.5 inch (1.3 cm) and 90 percent of sapwood.
2/ Retentions of waterborne salts are pounds of dry salt on an oxide basis; others as pounds of liquid.
3/ Not in specifications but will give good service if odor and bleeding are not objectionable.
4/ 20 to 25 pounds depending on species of marine borers present.
5/ Permissible at high retentions when no pholads are present.
6/ Also acceptable for this use is fluorchrome arsenate phenol at 0.22 pcf.

Sometimes wood is used on the exterior for purely decorative reasons (fig. 7.5.7). Balustrades are used to relieve the severe appearance of a roof, or wood strips are embedded in brick and stucco or are laid on the surface of panel siding. In moderate-rainfall areas (climate index 35 to 70, fig. 7.5.2), such decorative features should be made of naturally decay-resistant woods if fully exposed and preferably given a 3-minute dip in a water-repellent preservative as well. In high-rainfall areas (climate index over 70) items which are difficult or expensive to replace, such as strips embedded in brick or stucco, should be made from wood given a deep-penetrating preservative treatment (table 7.5.4). Unless well protected by roof overhang, no surface strips on panel siding should be used except the vertical battens over joints.

Many log houses now are being constructed throughout the country using untreated, unfinished wood or wood given only a superficial treatment. Deep checks and cracks that commonly develop in these logs hold large amounts of water and thus are prime areas for colonization by decay fungi and wood-boring beetles. Also, many of these structures have

Figure 7.5.7--Decorative exterior woodwork often is exposed to severe rainwetting.

little or no roof overhang; rainwater flows over the sides and ends of logs. Almost no research has been done to determine the best method of protecting log structures from deterioration in high-hazard areas of the country (fig. 7.5.2), but literature of a general nature is available (14,20).

All joint areas and the ends of structural timbers should be protected from excessive wetting. Decay usually is most prevalent at joints because water is absorbed more readily by the end grain of wood than by other surfaces and is retained longer (fig. 7.5.8). All design features on building exteriors should promote rapid drainage of water (fig. 7.5.9).

7.5.2.3 Ground Water

Ground water can wet any wood directly in contact with the soil. If the wood is untreated, it almost always is colonized by decay fungi or subterranean termites; soil-wood contacts are the common points of entry for the very destructive water-conducting fungi. As with ordinary decay fungi, the water-conducting decay fungi and subterranean termites are most prevalent in the hot, humid southeast.

Figure 7.5.8--Decay usually starts where wood end grain is exposed to rainwetting. (A) Untreated beam extending past the roof line. (B) Decay in the bottom rail of a screen door where condensate and rainwater ran down the wire and collected in a joint.

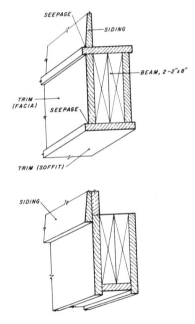

Figure 7.5.9--Boxed porch beam (top) has horizontal ledges that trap rainwater; boxed beam (bottom) with facia extending below the soffit is less subject to seepage.

Many foundation systems have been designed to separate untreated wood from the soil, but construction practices such as those listed below continue to create decay and termite hazards. The HUD Minimum Property Standards recommend that structural wood members be at least 8 inches (20 cm) and siding 6 inches (15 cm) above grade. However, wood-soil contacts such as the following are quite common: (1) Unprotected sills or headers in contact with soil under earthfilled porches, patios, flower planters, carports, etc.; (2) forms left on concrete foundations; (3) basement supports that extend through the concrete floor; (4) wood piers without adequate concrete footings; and (5) siding and wood step bases that become covered with soil.

Ground water can also wet wood indirectly by migrating through concrete or cinder block foundations. Any wood resting on moist concrete will absorb water (fig. 7.5.10). Metal flashing (termite shields) often

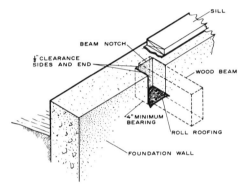

Table 7.5.10--Beams resting in notches in foundation walls can be kept dry by placing them on waterproof membranes and providing ventilation around them.

used to cap masonry walls is very effective in preventing sills from becoming wet even though it cannot be relied upon to protect a structure from subterranean termites. However, proper site drainage and waterproofing of the below-grade portions of foundation walls will greatly reduce the moisture reaching the sills. With slab-on-grade construction, pouring the footings and slab as one unit significantly reduces the danger from subterranean termites, and eliminates joints where water can penetrate a structure.

Earthfilled Appendages.--Many earthfilled porches, flower planters, patios, and other appendages are constructed directly against house foundations and with the level of the fill above that of the subfloor framing behind them. With or without a protective flashing between the soil and the wood, this type of construction is very hazardous and frequently leads to colonization by decay fungi and subterranean termites.

The fill under a porch may be separated from wood by removing the fill from the area adjacent to the wood members of the house before pouring the concrete slab, so that the slab extends as a wedge against the wood (fig. 7.5.11). Flashing must be installed at the top and bottom of the

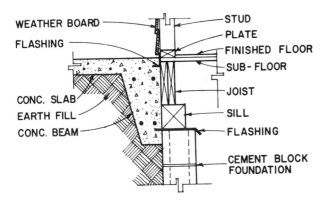

Figure 7.5.11--Porch slab designed to separate the earthfill from the framing timbers.

wedge to prevent rain seepage into the wood and the possible entry of fungi or termites from the fill. If the porch is next to a basement wall, the foundation dampproofing must extend to the top of the fill.

The best method of constructing earthfilled structures (fig. 7.5.12) is to make them completely independent units separate from the structure. By separating the appendage from the foundation, wood-destroying fungi and insects that may be present in the earthfill cannot enter the building. This design also decreases the chance of moisture accumulation in the wood adjacent to porches. The space between the porch and structure must be kept free of debris. Self-supporting concrete slabs are safer than those over earth fills. However, forms used for pouring the slabs must be removed because they form a direct contact between the soil and the wood in the house proper. Access to the areas under suspended slabs is essential for the removal of forms and for periodic inspections.

Wood Foundations.--Although foundations have traditionally been designed to separate wood from soil, wood-soil contact is an integral part of some foundation systems. With these systems, wood cannot be kept dry. Protection from decay fungi and subterranean termites is afforded by treating the wood with high retentions of preservative chemicals (table 7.5.4).

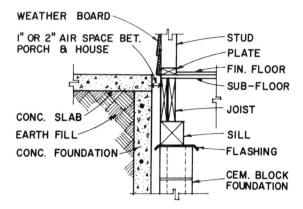

Figure 7.5.12--An earthfilled porch constructed as a unit separate from the house removes the decay and termite hazard.

Pole-Type Foundations.--Poles must be treated with one of the most effective preservative treatments suitable for use in the ground (table 7.5.4). The preservative retentions recommended are higher than those commonly recommended for ground contact but are considered necessary for long-time service under conditions where hazard is high and replacement difficult. In most cases the ACA and the CCA treatments are preferred because they leave a clean, paintable surface. Aside from being smelly and dirty, and having low paintability, the oil treatments listed will give equally satisfactory service; they are not approved for use in habitable spaces.

Round poles usually are superior to squared poles, particularly with such refractory woods as Douglas-fir. Squared material has less of the more easily treated sapwood and may have exposed heartwood which treats poorly and yet is decay-susceptible. Where squared poles are desired, it may be safer to bolt together two or more previously treated 2-inch (5-cm) planks or to use glued-laminated members fabricated from pressure-treated material. If a pole must be trimmed, it should be on the upper end where any untreated wood exposed at the cut will be protected by the roof; trimmed ends that are exposed to rainwetting should be given a brush treatment with water-repellent preservative.

All-Weather Wood Foundations.--All-weather and similar types of foundations in which footings, studs, and all other components of the foundations are composed of wood, now are widely accepted building designs. Extreme care is needed in selecting materials for such foundations as well as in designing and fabricating them.

The preservatives recommended for all-weather type foundations are chromated copper arsenate (CCA) and ammoniacal copper arsenite (ACA) at 0.6 pound per cubic foot retention (table 7.5.4). This is an appreciably higher retention than is usually recommended for wood in ground contact, but is justified by the importance of foundations and their vulnerability to colonization by fungi and subterranean termites. Creosote and pentachlorophenol (table 7.5.4) have given satisfactory service in all-weather wood foundations but can create such problems as bleeding through inside wall finishes and creating odor or toxic fumes in basement houses and where crawlspaces are used as a plenum for heating or cooling systems.

All lumber and plywood used in all-weather or similar type foundations should be labeled with the AWPB-FDN quality mark. The American Wood Preservers' Bureau, Arlington, Va. can furnish names of treaters who provide FDN-grade treated lumber and plywood.

Because of the exacting requirements of wood foundations, lumber and plywood used in them should be cut to size before treating. If this is not possible, retreating of cut surfaces must be well done and consist of several brushings or sprayings with the preservative.

Information on designs and construction methods for all-weather wood foundations is contained in (1) American Wood Preservers' Bureau, "FDN Standard, quality control program for softwood lumber and plywood pressure treated with waterborne preservatves for ground contact use in residential and light commercial foundations, Appendix E"; and in (2) National Forest Products Association, "All-weather wood foundation system. Design, fabrication, installation manual."

7.5.2.4 Condensation

Condensation occurs when warm, moist air contacts a cold surface and becomes supersaturated; the excess water vapor condenses as liquid water on the cold surface. Conditions favoring condensation in structures usually arise from cold weather or from air-conditioning by refrigeration. Winter condensation is a problem in regions where the average January temperature is 35° F (2° C) or below (fig. 7.5.13). Winter condensation seldom leads to appreciable decay, except around the periphery of crawlspaces, but often results in paint failure, loss of the effectiveness of thermal insulation (wet insulation readily transfers heat), and water stains or mold growth on inner wall and ceiling coverings when frozen condensate within walls, ceilings or attics melts in the spring and soaks through plaster or sheet coverings.

Condensation resulting from air-conditioning may cause widespread decay in floor joists, subflooring, and finished floors in structures over a crawlspace. Floor buckling, peeling and blistering of paint, loosening of wallpaper, or growth of mold fungi on walls and ceilings may also occur. Whenever the moisture content of the wood in a crawlspace increases, the danger of attack by wood-boring beetles is increased also.

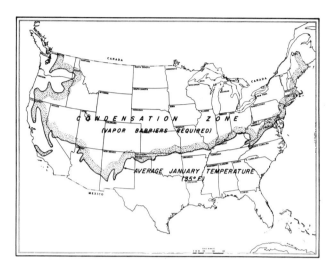

Figure 7.5.13--Winter condensation occurs mainly where the average January temperature is 35° F (2° F) or lower.

The amount of condensate forming on a surface will depend on: (1) The duration of a temperature at or below the dewpoint temperature, (2) the amount of water vapor present in the atmosphere adjacent to the surface in question, (3) the vapor permeability of any material between the free, warm atmosphere and the condensing surface, and (4) the vapor pressure gradient. The vapor permeability of materials between the condensing surface and the cold atmosphere will determine how long the condensate will persist.

The basics of condensation control are to: (1) Use vapor barriers to prevent water vapor from contacting a surface which is at or below the dewpoint temperature, (2) use thermal insulation to insure that the dewpoint temperature remains on the cold side of the vapor barrier, (3) restrict the release of water vapor into the indoor atmosphere, (4) use ventilation to remove excess water vapor or to prevent the accumulation of condensate and, (5) control temperature.

Crawlspaces.--If a structure is designed to eliminate problems associated with rainwater and ground water, damage resulting from condensation is much less likely to occur. However, crawlspaces, especially those under air-conditioned structures, require additional protection. All crawlspaces should be vented and, except in very dry regions, have a soil cover (e.g., 6-mil polyethylene). At least 1 square foot of vent per 150 square feet of crawlspace area should be provided if no soil cover is present; this can be reduced to a minimum of 1 square foot per 1,500 square feet if a soil cover is used. There should be a minimum of 2 vents located opposite one another. Some vents should be located as near corners as possible to eliminate dead air spaces.

Often vents are closed in winter to conserve heat and are not reopened in spring. This problem can be solved by using the relatively inexpensive vents now available that automatically close at approximately 40° F (4° C) and reopen at approximately 70° F (21° C) (e.g. Temp-Vent, Shelby, N.C.).

Live-steam pipes in crawlspaces can raise the dewpoint temperature to 80° F (27° C). Serious condensation may occur rapidly with even moderate cooling if a steam leak occurs. Live-steam pipes should be avoided under air-conditioned spaces.

Attics.--The attic space also should be vented, with intake vents located at the soffits and exhaust vents located near the peak. Mansard (fig. 7.5.14) and flat roofs require continuous soffit vents. As with crawlspaces, there should be a minimum of 1 square foot of vent area for every 150 square feet of attic space; 1 square foot of vent per 300 square feet is adequate if a vapor barrier is present in the ceiling. Requirements for ventilation are much greater for insulated than for noninsulated structures. Care must be taken that insulation does not block soffit vents.

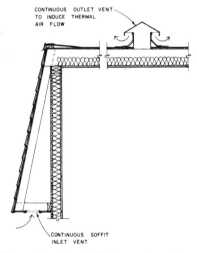

Figure 7.5.14--Mansard roof ventilation. A continuous ridge vent is needed because each inter-rafter area needs separate ventilation.

Interiors.--Excessive moisture generated within a structure should be vented to the outside. Exhaust fans should be installed over kitchen stoves and in bathrooms. Clothes dryers should be vented to the outside. Closets and other small, enclosed spaces should have louvered doors.

Insulation.--Insulation should be installed only after all potential sources of excessive moisture have been eliminated.

Most moisture problems associated with the improper installation of insulation are caused by lack of vapor barriers or by vapor barriers placed in the wrong locations. Most batt or blanket insulation has a vapor barrier on one side, but loose fill, foamed in-place, or rigid insulation usually does not have a barrier. When insulation has a vapor barrier, the barrier should be placed on the living-space side of the insulation and located just behind the interior wall covering. In walls, the vapor barrier should be oriented inward, in attics downward, and below the floor upward. It is recommended that a soil cover be used whenever insulation is placed between the floor joists in a structure built over a crawlspace.

In many instances, insulation without a vapor barrier is added to attic spaces or pumped into wall cavities. Except in warm, moist areas such as Gulf Cost region, vapor barriers should be added to the interior wall and ceiling surfaces where high indoor humidities are expected. Paint-on vapor barriers, such as aluminum primer placed under decorative wall or ceiling coverings or paints are effective. If interior wall coverings can be removed, polyethylene film stapled directly to the interior stud surfaces provides a good vapor barrier. If insulation without a vapor barrier is used in air-conditioned structures in warm, moist regions, any moisture that condenses from the outside air will do so on interior wall covering where it can be readily evaporated into the air within the structure. On the other hand, if a barrier is located next to the living space, moisture from the outside air may condense on the vapor barrier and be trapped within the walls.

Little research has been done to evaluate the effectiveness of foamed-in-place insulations in wood-frame walls. These materials have been used primarily to fill the voids in masonry construction. Care should be taken when adding urea-formaldehyde foams to wood-frame walls with water-impervious surface coatings, since it is not known how long it takes for the water injected into the walls along with the foams to dissipate. If the water becomes trapped, it could result in decay within the walls or blistering of paint on siding. In the winter, trapped water could freeze and cause walls to buckle.

Materials that act as both vapor barriers and insulation (e.g., styrofoam and aluminum-foil-backed foam panels) now are being used as sheathing directly behind exterior wall coverings. The hazard of decay associated with these materials has yet to be determined, but condensation is likely to occur on the back side of wood siding placed over them.

7.5.2.5 Exterior Structures

Wood is often used in structures that are fully exposed to the environment. Observation platforms and towers, bridges, open stadia, playground equipment, and roller coasters are just a few structures where virtually all of the wooden components are constantly subjected to weathering.

Poles, piles, or squared timbers supporting exterior structures should be treated with preservative chemicals to retentions recommended for ground contact (table 7.5.4). Whenever ordering treated wood, be sure to specify that it be marked with the appropriate AWPB quality stamp (1). Wood treated with either waterborne (e.g., chromated copper arsenate, acid copper chromate) or oil-borne (e.g., creosote, pentachlorophenol) preservatives may be used. If the members are to be stained or painted, a water-borne preservative or pentachlorophenol in a volatile solvent (e.g., Cellon process) should be specified. If supporting timbers must be cut to size after treatment, the uncut end should be placed in the soil; all cut surfaces should be given a brush treatment with a preservative such as 5 percent pentachlorophenol in mineral spirits. All fasteners should be of similar metals to prevent electrochemical activity and should be rust-resistant. It is particularly important to use either galvanized or nonferrous fasteners in wood treated with water-borne preservatives.

Pressure-treated wood is preferred for use as decking, trim, or other nonstructural components, but the heartwood of decay-resistant species such as redwood or western redcedar may be used. Where strength is critical or replacement difficult (e.g., bridge decking), pressure-treated wood should be specified. Remember that pressure treatment will not raise the grade of wood, so pressure-treated wood of the proper grade should be specified. As with structural members, all fasteners should be of similar metals and rust-resistant.

Whether made of wood that is naturally resistant or pressure treated, all decking should be placed "bark side up" (fig. 7.5.15). This will minimize cupping or splitting and thereby reduce maintenance costs. In general, using narrower decking boards (e.g., two 2 x 6's rather than one 2 x 12) is preferable because wider boards have a greater tendency to cup and split. Decking also should be treated with a good water repellent to further reduce splitting and to aid in dimensional stabilization. Maintenance programs should include periodic retreatment of deck surfaces with a water-repellent preservative (e.g., 5 percent pentachlorophenol and water repellent in mineral spirits) to treat wood exposed by splits or checks and to maintain water repellency.

Wood treated with a waterborne preservative (e.g., CCA) or the heartwood of a naturally decay-resistant species is preferable for use in picnic tables, stadia seats, park benches, decks around residences or swimming pools, or playground equipment. As was discussed above, horizontal members should be installed "bark side up." CCA-treated wood used in picnic tables or playground equipment should be scrubbed with detergent before use to remove any unfixed chemicals from the surface.

Non-film-forming finishes (e.g., stains) rather than paints should be used whenever possible on exterior wooden structures. Painted surfaces require much more maintenance than do those which are stained, but paint peeling and blistering can be minimized by first treating the wood with a water repellent. Stains can be formulated with water repellents and preservatives, so that several maintenance operations can be performed at once.

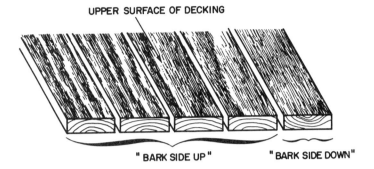

Figure 7.5.15--Decking and other horizontal members should be placed "bark side up" to minimize cupping and splitting.

7.5.3 Insects

Controlling wood moisture content is very effective in preventing damage by wood-boring beetles in crawlspaces, but is less effective in preventing infestation by termites.

7.5.3.1 Wood-Boring Beetles

Anobiid beetles (commonly referred to as powderpost beetles) feed on seasoned hardwoods and softwoods and usually colonize crawlspace timbers in structures 10 or more years old. Anobiid eggs will not hatch on wood with a moisture content less than about 12 percent, and the feeding activity of the larvae is slower in drier wood. Lyctus powderpost beetles colonize recently seasoned hardwood sapwood, and thus usually are found in new homes. Lyctus damage has begun to increase since tropical hardwood veneer has been used as the inner plies in many plywood products. Old house borers colonize only recently seasoned softwoods; they usually are found in the framing timbers of structures less than 10 years old.

All of the wood-boring beetles that colonize wood products require unfinished wood for egg laying and larval feeding. Applying a finish to crawlspace or basement timbers would prevent egg laying and hatching by these beetles. However, unfinished wood in dry, vented crawlspaces and basements is not likely to be colonized. The increased use of untreated, unfinished wood resulting from the current popularity of log houses likely will be accompanied by a greater incidence of beetle damage.

7.5.3.2 Subterranean Termites

Subterranean termites also require moist wood to feed successfully, and they feed most actively in wood wetted by rain seepage, plumbing leaks, or soil. Wood-soil contacts or earthfilled appurtenances afford easy access of termites to a structure. As was discussed in preventing damage from decay fungi, any wood that must be used in contact with the soil should be pressure treated (table 7.5.4); earth-filled structures should be constructed as units separate from the foundation.

Subterranean termites may also gain access to a building by constructing earthen shelter tubes on foundation walls or within voids in foundation materials. Termite shields may limit the areas accessible to termites, but they cannot be relied upon to exclude termites from a structure.

The most effective way of excluding subterranean termites is to apply a toxic barrier around a structure. The soil under all concrete slabs (including porches, patios, carports, etc.) and adjacent to foundation walls and piers (down to the footings) should be treated by a qualified pesticide applicator with an insecticide that provides long-term protection against subterranean termites. Even though pressure-treated wood is toxic to termites, they can construct shelter tubes across the surface of treated wood to reach untreated wood above it. Soil adjacent to pressure-treated wood also should be treated. Slab and foundation perimeters should be retreated after all grading and landscaping has been completed.

Concern now is being expressed about how to exclude subterranean termites from structures that use the crawlspace as an air-flow plenum. It is feared that chemicals applied to the soil in plenums will enter into the air stream and be deposited within the living space. This problem has not been resolved.

7.5.3.3 Drywood Termites

Drywood termites, unlike subterraneans, do not maintain connections with the soil and, therefore, cannot be excluded from a building by soil treatments. As the name implies, these termites require less wood moisture than do subterraneans. Damage by drywoods occurs primarily in attic and crawlspace framing of structures in the deep south, in southern Texas, New Mexico, and Arizona, and along the Pacific coast in California.

In the past, screening of vents in attics and crawlspaces has been recommended to exclude drywood termites. However, screening will not exclude all drywoods, and the small size of the mesh (18 to 20 mesh) restricts air movement through the vents and causes other problems. Drywoods can enter a structure under siding, through wood shingles, or underneath and between tiles or shingles and sheathing.

Ebeling and Wagner (11) described a system of controlling drywood termites that has merit. Fluoridated silica aerogel dusts (e.g.,

Dri-die 67) are applied during construction in attics, crawlspaces, and wall voids. These dusts absorb a portion of the waxy layer covering the exoskeleton of drywoods and cause a lethal rate of water loss.

7.5.4 Marine Borers

Wood exposed as fixed or floating objects in marine aquatic environments is subject to attack by species of mollusks (e.g., the "shipworm" Teredo) and crustaceans (e.g., Limnoria). These organisms require the presence of salt or brackish water to survive, and rarely attack structures in fresh water or water with very low salinity.

The use of pressure-treated wood is the only reliable means of protecting against marine borer damage. Heavy treatment with coal-tar creosote or creosote-coal tar solution is adequate where borer activity is moderate. However, wood must be treated with both CCA (chromated copper arsenate) and coal-tar creosote where borer hazard is severe. Portions of treated wood that will be exposed below the water line should not be cut. All bolt holes or notches that must be made after treatment should be given a thorough brush application of creosote. The cut ends of all piles exposed above water also should be given a thorough brush treatment to protect them against decay fungi. Maintenance programs should include retreatment of pile ends to provide continued protection from decay fungi.

Jacketing the below-water portions of treated piles to provide additional protection also has been successful, providing that the jacket can be maintained intact. Heavy polyvinyl plastic, as well as metal such as copper, has been used to wrap piles.

7.5.5 References for Section 7.5

1. American Wood-Preservers' Association.
    n.d. Book of Standards. (Includes standards on preservatives, treatments, methods of analysis, and inspection.) (American Wood Preservers' Bureau official quality control standards.) (see current issue.)

2. American Plywood Association.
    1976. The all-weather wood foundation: Why, what, and how. 35 p.

3. American Wood-Preservers' Institute.
    n.d. Pole house construction. Am. Wood-Preservers' Inst.

4. Anderson, L. O.
    1970. Wood-frame house construction. U.S. Dep. Agric., Agric. Handb. 73. 223 p.

5. Anderson, L. O., T. B. Heebink, and A. E. Oviatt.
    1972. Construction guides for exposed wood decks. U.S. Dep. Agric., Agric. Handb. 432.

6. Anderson, L. O., and G. E. Sherwood.
   1974. Condensation problems in your house: Prevention and solution. U.S. Dep. Agric. Inform. Bull. 373.

7. Arsenault, R. D.
   1975. CCA-treated wood foundations. A study of permanence, effectiveness, durability, and environmental considerations. Am. Wood-Preservers' Assoc. Proc. 71:126-149.

8. Black, J. M., and E. A. Mraz.
   1974. Inorganic treatments for weather-resistant natural finishes. U.S. Dep. Agric., For. Serv. Res. Pap. FPL-232. 41 p.

9. Brown, W. S.
   1959. Protection from moisture for slab-on ground construction and for habitable spaces below grade. Natl. Acad. Sci., Natl. Res. Counc. Publ. No. 707. 66 p.

10. Clapp, W. F., and R. Kenk.
    1963. Marine borers. An annotated bibliography. Office of Naval Res., Dep. of Navy, Rep. ACR-74. 1135 p. (U.S. Gov. Print. Off. No. 1963-0-679485).

11. Ebeling, W., and R. E. Wagner.
    1964. Built in pest control. Pest Control 32(2):20-32.

12. Johnston, H. R., V. K. Smith, and R. H. Beal.
    1972. Subterranean termites: Their prevention and control in buildings. U.S. Dep. Agric., Home and Garden Bull. 64. 30 p.

13. Koenigshof, G. A.
    1978. Preventing basement leakage in treated-wood-foundations. U.S. Dep. Agric., For. Serv. Res. Pap. SE-195. 13 p.

14. Look, D. W.
    1975. Selected bibliography on log structures. U.S. Dep. Interior, Heritage Conservation and Recreation Service, Office of Archaeology and Historic Preservation. 3 p.

15. Mampe, C. D. (ed.).
    1973. Approved references procedures for subterranean termite control. Natl. Pest Control Assoc., 8150 Leesburg Pike, Vienna, Va. 168 p.

16. National Forest Products Association.
    1973. All-weather wood foundation system. Design, fabrication, installation manual. 76 p.

17. Oviatt, A. E.
    n.d. Optimum insulation thickness in wood-framed homes. U.S. Dep. Agric., For. Serv. Gen. Tech. Rep. PNW-32. 37 p.

18. Rice, J. T., and R. F. Blomquist.
    1969. Upgrading joint sealant performance in residential construction. For. Prod. J. 19(11):11-14.

19. Rossiter, W. J., Jr., R. G. Mathey, D. M. Burch, and E. T. Pierce.
    1977. Urea-formaldehyde based foam insulations: An assessment of their properties and performance. U.S. Dep. Commer., Natl. Bur. Stand., Tech. Note 946. 92 p.

20. Rowell, R. M., J. M. Black, L. R. Gjovik, and W. C. Feist.
    1977. Protecting log cabins from decay. U.S. Dep. Agric., For. Serv. Gen. Tech. Rep. FPL-11. 11 p.

21. Scheffer, T. C., and A. F. Verrall.
    1973. Principles for protecting wood buildings from decay. U.S. Dep. Agric., For. Serv. Res. Rep. FPL 190.

22. Smith, R. S., and G. W. Swann.
    1975. Colonization and degradation of western red cedar shingles and shakes by fungi. Mater. u. Organ., Suppl. 3, Organismen und Holz, p. 253-262.

23. Snyder, T. E.
    1969. Control of nonsubterranean termites. U.S. Dep. Agric., Farmers' Bull. 2018. 16 p.

24. U.S. Federal Supply Service.
    n.d. Wood preservatives: Water-repellent. Fed. Spec. TT-W-572. (See current issue.)

25. U.S. Federal Supply Service.
    n.d. Wood preservation treating practices. Fed. Spec. TT-W-571. (See current issue.)

26. U.S. Forest Products Laboratory.
    1974. Wood handbook: Wood as an engineering material. U.S. Dep. Agric., Agric. Handb. 72 rev. 421 p.

7.6 Mechanical Damage*

The strength of wood structural members is frequently impaired by cutting or drilling for piping, ducts, or wiring. This usually occurs after completion of the framing and in the process of roughing in for mechanical equipment. It may be the result of poor layout, inadequate on-the-job inspection, or both (fig. 7.6.1).

Planning is an important factor in avoiding conflicts between structure and mechanical equipment. Space for ducts and pipes must be anticipated early in the design process, often dictating the direction of framing or the location of building elements (fig. 7.6.2). Common plumbing walls, when possible, simplify piping runs and reduce costs.

---

*Author: Alfred E. Oviatt, Jr., Research Architect, Forest Products Laboratory, Madison, Wisconsin.

Figure 7.6.1.--Cutting of framing members to accommodate plumbing.

The use of 2- by 6- or 2- by 8-inch plates in these walls facilitates lateral pipe runs without damage to structure, using double studs placed flatwise (fig. 7.6.3).

Notching of floor joists in light framing should be avoided, but when it must be done, it should be limited to one-sixth of the depth and confined to the end quarters of the span. The effect is less severe on the top edge. Holes for pipes or wiring should be no more than 2 inches in diameter, and the edge of the hole should be at least 2-1/2 inches to 3 inches from top or bottom edges (fig. 7.6.4).

Holes for wiring are generally small but should be confined to the neutral plane of bending members. In general, holes which do not exceed the allowable knot size in the lumber grade are permissible, provided they are not made near maximum knots.

7.6.1   References for Section 7.6

1. American Institute of Timber Construction.
    1977. Timber construction manual, third edition.

2. Anderson, L. O.
    n.d. Wood frame house construction. USDA Agric. Handb. No. 73. Superintendent of Documents, Washington, D.C. 20402.

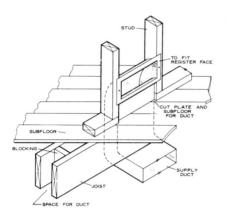

Figure 7.6.2.--Spaced joists for supply ducts.

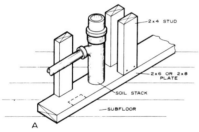
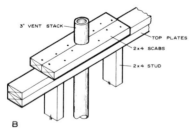

Figure 7.6.3.--Installation of plumbing stack.

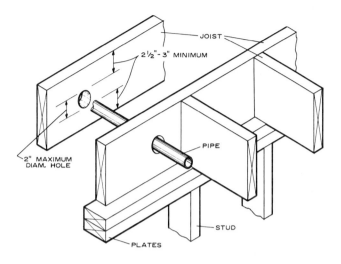

Figure 7.6.4.--Drilled holes in joists.

CHAPTER 8   RECYCLING OLD STRUCTURES (CASE HISTORIES)*

8.1   Introduction**

Today, renovation and recycling are frequently the chosen alternatives to the complete demolition and rebuilding of a structure. There are several reasons for this: economics, quality of life, and preservation. Material costs are rising faster than those for labor-intensive repair, so that renovation is often the more economical choice. The high-density housing units built in the United States in the 50's and 60's have led to a myriad of social problems. People are opting for the improved quality of life which they feel comes from rehabilitated low-rise, low-density, housing. A search for national roots has given additional impetus to the movement to preserve structures of historic interest.

Many structures scheduled for renovation, recycling, or repair are timber-framed. The case studies which follow discuss problems and solutions encountered during design and construction. Included are an industrial building which was repaired after an explosion, a warehouse recycled into shops, a historic courthouse renovated for use for a variety of civic functions, a church repaired and remodeled, several covered bridges structurally strengthened, and a center for the performing arts which suffered fire damage while under construction. Repair, renovation or recycling was the choice in each instance because of the economics, the quality of life obtainable, or for historic preservation.

8.2   Reconstruction of Wooden Cathedral***

8.2.1   Introduction

Holy Trinity Cathedral is a historic church noted in architectural guidebooks to San Francisco. The cathedral was built in 1909 to replace an earlier church destroyed in the 1906 earthquake. A fire damaged the interior in 1934. At that time the building was restored and enlarged to the west of the tower and main entrance. By 1970, reconstruction was considered necessary--in fact, urgently needed--for the following reasons:

---

*Chapter coordinator: Ross M. Lanius, Jr., Chairman, Dept. of Civil and Environmental Engineering, University of New Haven, West Haven, Connecticut.

**Author: Ross M. Lanius, Jr.

***Author: Constantine C. Chekene, Structural Engineer, Rutherford and Chekene, San Francisco, Calif.

1. Known decay and termite damage in parts of the bearing wall system required immediate attention.

2. The front exterior stairs had deteriorated due to continuous exposure to the elements until they were no longer safe and really constituted a hazard to the users.

3. Except for the exterior walls, the cathedral had no pre-engineered lateral resisting system. The interior partitions provided only the degree of stability inherent in plastered and wood-sheathed walls.

4. Should a seismic event of medium or major intensity occur, the generally deteriorated condition of the building constituted a real hazard, not only to the congregation of the church but also to the valuable historical treasures of the cathedral. In addition to old icons, vestments and historical church items and records, Holy Trinity Cathedral possesses a set of unique Russian bells cast in Moscow in 1888.

5. The existing general circulation in the building had become confused and complicated by minor additions and reconstruction through the years, and needed complete reorganization.

8.2.2 Investigation

The cathedral community engaged an architect; members of the parish themselves provided structural, electrical and mechanical engineering. Nearly 4 years passed from the inception of the project to the beginning of construction because of the time required to mount a fund drive, to secure parish approval of proposed alterations, and especially to give careful consideration to proposed architectural changes to this historic structure. Drawings were prepared to show the existing construction in as much detail as was practical, and an evaluation was begun to establish the architectural and structural concepts of reconstruction.

The existing building is a single-story church with a full basement partially below grade along the east and south elevations. The foundation of the 1909 portion is of brick laid in lime-cement mortar. The 1934 addition is on a foundation of reinforced concrete. The superstructure is of wood; studs in the exterior walls are rough, full-size, 2-inch by 6-inch (51-mm by 152-mm), with 7/8-inch (22-mm) redwood siding on the exterior. Plaster on wood lath covers the inner surfaces. Concentrated loads from roof framing beams, balcony framing beams, bell tower framing, and other load-carrying elements are all transferred by full-size 6-inch by 6-inch (152-mm by 152-mm) posts directly to the sill plates at the foundation. Most of these posts were found to be single-length timbers, some exceeding 30 feet (9.1 m) in length.

WOOD STRUCTURES

Studs of two widths were used to construct the bell tower: the outer layer was made of 2-inch by 6-inch (51-mm by 152-mm) studs, and the inner layer, placed immediately behind the outer, of 2-inch by 4-inch (51-mm by 102-mm) studs. In general, the construction may be described as "balloon" framing braced by tightly fitted compression blocking at random locations and next to the openings.

Floor framing consists of 2-inch by 12-inch (51-mm by 305-mm) members at 16 inches (406 mm) on center, with 1-inch (25.4-mm) diagonal sheathed subfloor and 1-inch (25.4-mm) hardwood floor over, and plastered ceiling on wood lath below. There is a considerable amount of decorative wood trim on the exterior of the cathedral; the interior has some decorative plaster and wood trim.

8.2.3 Solutions

Because of the high cost to remove and restore existing architectural finishes, the parish decided to limit structural upgrading to partial code compliance. It can be seen from figure 8.2.1 that the cathedral has a limited number of walls capable of providing lateral stability in either direction.

Determination of lateral forces based on the requirements of the San Francisco Building Code yielded the following loads:

Seismic load for K = 1.33   V = 10 kips   (44,480 N) to roof
  in both directions   V = 10 kips   (44,480 N) to floor

Wind load for W = 15 pounds per square foot (720 N/m^2) to 30 feet above grade and W = 20 pounds per square foot (960 N/m^2) above 30 feet

North-South direction   W = 16.97 kips (75,490 N) to roof

East-West direction   W = 13.14 kips (58,450 N) to roof
                             W = 6.48 kips (28,820 N) to main floor

The cathedral floor plan was divided into three areas designated "A," "B," and "C," and the work was organized accordingly (fig. 8.2.1).

8.2.3.1 Area "A"

The necessary reorganization of existing and internal circulation in Area "A" (fig. 8.2.1) provided a good opportunity to add lateral resistance capacity to the existing walls in an east-west direction.

RECYCLING OLD STRUCTURES 369

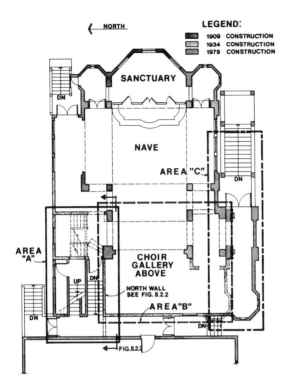

Figure 8.2.1--Plan of the main church showing areas of reconstruction.

The north wall was selected to receive a 3/8-inch (9.5-mm) plywood sheathing full height from foundation sill plate to roof, since the north face of this wall could be sheathed in a continuous way without interruption by floors and stairs, which were to be installed later. Figure 8.2.2 shows the approximate arrangement of tie-downs and lateral ties. The wall was designed to resist one-half the total lateral force (13.14/2 = 6.57 kips) (29,220 N) at the roof level and (6.48/2 = 3.24 kips) (14,410 N) at floor level generated by wind in an east-west direction. Wall nailing was 10d at 4 inches (102 mm) on centers at all edges of all panels. Tie-downs were installed at only two points (fig. 8.2.2). Special strapping and blocking at openings were omitted and there was no continuous tying of lateral load collectors at floor and roof levels due to the difficulty of access and the cost of finish restoration. The general character and quality of the original construction, however, give reasonable assurance that the lateral load could be transferred to the shear wall on both levels. New sill bolts were installed to provide additional shear transfer from the existing sill to the concrete foundation (fig. 8.2.3).

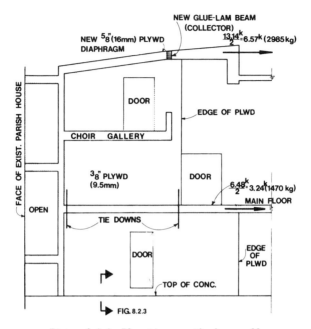

Figure 8.2.2--Elevation, north shear wall.

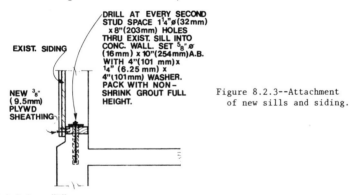

Figure 8.2.3--Attachment of new sills and siding.

8.2.3.2 Area "B"

To improve the general acoustics in the church, it was necessary to construct a new roof over the choir gallery and part of the nave. The new roof was built with a slope of 42 inches (1,070 mm) in 18 feet (5.5 m). A new glue-laminated beam with a clear span of 20 feet

RECYCLING OLD STRUCTURES 371

(6.1 m) and a 5-foot (1.5-m) cantilevered end extending toward Area "C" was placed at a new elevation to replace the existing (solid wood) beam. The new beam was supported on existing 6- by 6-inch (152- by 152-mm) columns with column extensions fabricated from 5-inch-square (127-mm) standard tubes. This beam was ultimately connected to the braced steel framing of the new bell tower in Area "C" and became a lateral load collector in a north-south direction. Building a new roof allowed installation of new 5/8-inch (16-mm) plywood in the area of new construction, thereby considerably improving the lateral stability of the rear area of the church. Roof framing was nominal 2- by 12-inch (actual 1-1/2- by 11-1/2-inches (38- by 292-mm)) beams at 16 inches (406 mm) on centers (figs. 8.2.4 and 8.2.5).

8.2.3.3 Area "C"

This area includes the entrance stairs (concrete, with steps dressed in granite), a new concrete base for the extension of the nave, and a new bell tower to support a complement of seven church bells. In this area, the following procedure for construction was determined:

    1. Existing front wooden stairs and landing were demolished.

    2. New foundations free from interference with existing footings were excavated and poured to accommodate positioning of the new tower to the south of the existing tower.

    3. The small bells were removed and steel needle beams were installed to support the main bell (2,361 kg) and the upper part of the existing tower. The north ends of the needle beams were supported by the existing wood wall and the south ends were supported by a temporary braced frame built of 5-inch (127-mm) diameter pipe columns and bracing L's. Columns were bearing on existing foundations left after removing the front stair (figs. 8.2.6 and 8.2.7).

    4. Existing foundations which interfered with the new scheme were then demolished, and new foundations poured. Existing wall framing and tower framing were removed to the bottom of the existing main floor; new concrete tower walls and stair walls were then poured. Next, new concrete slabs were poured and all foundation bolts to receive the new steel frames were installed. The existing wood floor diaphragm was then connected to the new concrete floor. The top portion of the existing wooden tower was cut off and removed as a single piece by crane.

    5. A steel frame was erected up to the level of the octagonal turret and a concrete floor at this upper platform level was poured. The large bell was then transferred along the needle beams into its final position. Temporary columns and bracing were removed and the load from the needle beams was transferred directly to the new upper slab.

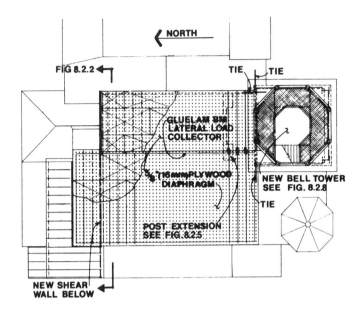

Figure 8.2.4--Partial roof framing plan.

Figure 8.2.5--Column extension.

Figure 8.2.6--Temporary supports for bell tower.

6. The wood framing and cladding with porcelain enamel for the octagonal portion of the new tower were completed on the ground and lifted into place as a single piece. After field welding of the octagon legs to the upper ring of the tower, the bells were lifted into their final vertical position and permanently secured.

The procedure described above maintained the structural integrity of the existing building, and permitted final integration of new and old construction, without major disruption of the building function. The church was used for services without any interruption during the construction period.

The concrete base from foundation to main floor level allowed direct lateral connection of the existing diagonally-sheathed main floor diaphragm to the base for transfer of total lateral forces in both directions to the base. Above the main level, the steel tower frame was designed to resist, at roof level, full lateral load in a north-south direction (16.97 kips) (75,490 N), and half lateral load in an east-west direction (6.57 kips) (29,220 N), in addition to supporting the weight of the bells and the enclosure (fig. 8.2.8).

374   WOOD STRUCTURES

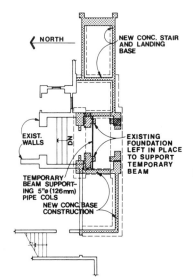

Figure 8.2.7--Partial foundation plan of new tower.

In order not to obstruct public circulation on the main floor, the tower frame was designed as a portal frame between the main floor and choir gallery level and as a braced frame above.

A glue-laminated beam supporting the new choir gallery roof (see discussion of Area "B") was developed as a major lateral load collector and was connected to the frame (fig. 8.2.9).

8.2.4 Damage to Wood Framing From Organisms

In all areas of work, deterioration of wood members was discovered. Decay alone, or in association with termite infestation, was the major problem. Not surprisingly, in all perimeter walls where concrete curbs were low and wood construction was in contact with the grade, wood members were damaged. Intrusion of rain water into the walls, due to faulty flashing and drainage, was a contributing factor to damage from decay.

In all cases, the affected portions of studs and posts were removed, concrete curbs built up, new pressure-treated sills fitted against cut ends of studs, and gaps between new sills and new curbs dry-packed. In most cases, the exterior redwood siding above the new curbs was left intact and treated with wood preservative. Where existing studs were damaged above the new curb line, new ones were fitted to replace them.

RECYCLING OLD STRUCTURES

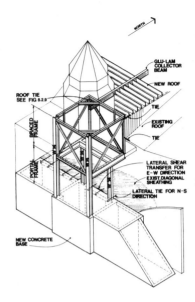

Figure 8.2.8--Schematic isometric view of new tower.

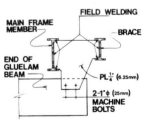

Figure 8.2.9--Roof tie construction at tower frame.

Main posts supporting the bells were affected by termites and decay so that only approximately 15 to 20 percent of the bearing area was effective. No live termites were discovered, although evidence of their presence in the past was extensive (figs. 8.2.10 and 8.2.11).

Figure 8.2.10--Termite and decay damage in bell tower.

Figure 8.2.11--Deteriorated studs, north wall.

Decay in the roof beam supporting the roof over the choir gallery was a surprising discovery. Apparently, the cathedral was extensively affected in the past by roof leakage, which caused the condition discovered. This beam was replaced by a glue-laminated beam positioned several feet above it.

8.2.5 Conclusion

The reconstruction and upgrading which have been described demonstrate conclusively the adaptability of wood buildings to the upgrading process. Two methods are illustrated:

> 1. Application of wood diaphragms--vertical or horizontal-- where unobstructed wall or roof/floor areas are available (north wall and roof diaphragms in the example above).
>
> 2. Construction of concrete or steel elements to accept concentration of high loads where limitations of space restrict solutions using low-stress material (concrete base and steel tower in the example above).

Work performed without disrupting the main church area and functions produced an upgrading of the total structure by positioning new elements for maximum effectiveness. Tracing the paths of acting forces and providing load transfer connections are important considerations if the effectiveness of any new work is to be maximized.

These methods are valid for upgrading and improving the lateral stability of any wood structure of reasonably sound initial construction.

8.3   Hi-Rise in Wood (Butler Square)*

8.3.1 Introduction

Modern code requirements have eliminated the construction of multistory wood structures almost entirely except for two- and three-story frame buildings. It was, therefore, with considerable interest and some trepidation that we approached the analysis required in remodeling an eight-story building with a full basement with eight floors and a roof all of heavy timber construction.

The building had been constructed for use as a warehouse, and floor-load capacities were more than adequate to consider remodeling into office and retail space. Except for one story of cast iron, the columns were all solid timber. The size of the columns varied at 2-inch intervals at each floor, starting with 24 inches square nominal at the lowest level and decreasing to 8 inches square at the top story. The columns were placed on a grid of 14 by 16 feet. Two inches of subfloor and 1-1/2-inches of finish flooring formed the wood floor of about 3-1/2 inches total thickness. This floor was carried by solid timber joists 8 inches by 12 inches in cross section and spaced about 4 feet

---

*Author: Frank Horner (1923-1980), Horner Associates, Consulting Engineers, Inc., St. Paul, Minn.

6 inches apart. The beams supporting the joists consisted of two solid timbers, each 6 by 18 inches in cross section, bolted together side by side with the 18-inch dimension vertical. The column transition at each floor was formed by a cast iron piece resting on the column below, supporting the two beams and the upper column, all held together by gravity alone. Exterior bearing walls of solid masonry, up to 34 inches thick at the lowest level, furnished the major elements for lateral stability of the building.

The fact that the building had been sparsely occupied for about 10 years and actually had been vacant for at least 2 years, indicated its obsolescence. Upon renovation, Butler Square, as it was subsequently named, became the first major recycled building in the Minneapolis, St. Paul area.

8.3.2 Structural Considerations

Structural inspection revealed that no substantial deterioration had occurred and that no major modifications were required to assure safety of the building. There were few interior partitions. Most of the brick walls and structural elements were unpainted and exposed. The cost of demolition and removal of unusable portions of the building was therefore only a small portion of the cost of renovation. No significant exterior modifications were allowed since the building was listed in the "National Register of Historical Buildings" (fig. 8.3.1).

Careful and detailed inspection was made of all the timber supporting members. The inspection had a two-fold purpose. The first was to determine whether or not there was any actual deterioration, such as rotting or splitting, or any other defects that would render the members unsafe. The second was to grade the lumber in place to determine if it had a live-load capacity sufficient for the intended use. Grading was done visually by evaluation of knots, splits, checks, etc., in accordance with "Grading and Dressing Rules for Douglas-Fir" published by the West Coast Lumberman's Association.

Existing steel banding of split columns was left in place with some additional banding done. Some of the defective timbers were replaced, while other members were reinforced by the use of steel reinforcing rods. A hole was drilled into the member perpendicular to the grain of the wood through the split or check to be repaired. A reinforcing bar was then inserted into this hole and embedded in epoxy adhesive. This type of repair was not done very extensively because of the cost.

The aesthetic appeal of the building's interior did not seem to be impaired by any of the repair techniques used since they all seemed to blend into the character of the structure. The visual strength of the exposed structure was such a dominant factor that column banding or other repairs were not visually detracting.

The architect determined that the massiveness of the building, 500,000 square feet, was undesirable for its intended use as premium office space, and consequently utilized the atrium concept to provide an openness to the interior space (fig. 8.3.2). This necessitated

RECYCLING OLD STRUCTURES 379

Figure 8.3.1--Exterior view of Butler Square. Photo courtesy of Phillip MacMillan James and Associates.

removing a substantial portion of each floor, thereby decreasing the structural stability and the diaphragm capacity of each floor. Restoring these constituted the major structural requirements (figs. 8.3.3 and 8.3.4).

Diaphragm capacity of the floor was restored by use of a new, raised floor. To form the floor, precast gypsum metal-edged planks were placed in steel supports carried by the wood floor below. The rigidity of the plank system aided in maintaining the stability of the building. In the open atrium, existing beams were left in place and strapped to the columns to maintain column stability. Beams removed from the atrium were placed at right angles to the existing beams to provide stability at 90° to the existing beams.

Structural elements are left exposed throughout the building (fig. 8.3.5). Where possible, beams removed from the atrium area were reused as required in construction. The space between the original and the raised floors was utilized for mechanical and electrical services, leaving the underside of the existing floor structure uncluttered for

Figure 8.3.2--Interior view of Butler Square showing one of the two atria. Photo courtesy of Phillip MacMillan James and Associates.

use as a finished ceiling. The wood structure and the interior surfaces of the brick walls were sandblasted and then left exposed. More elevator shafts existed than were necessary for the new occupancy, so some were utilized as mechanical shafts with existing elevator penthouses serving as mechanical rooms.

8.3.3 Provision of Fire Resistance

Since the building could not be brought into full conformance with current building codes, especially considering the addition of the atrium, it was decided to obtain equivalent fire resistive requirements by alternate methods. The new floor, consisting of gypsum plank, provided a substantial fire barrier. The entire building, including the space between the old and new floors, was sprinklered. Along with the inherent fire resistive capacities of the heavy timber construction, these additions made the building acceptable to the local building authorities, by fire safety standards.

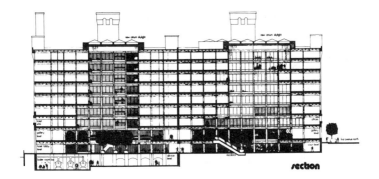

Figure 8.3.3--Sectional view of Butler Square showing the two atria. Photo courtesy of Phillip MacMillan James and Associates.

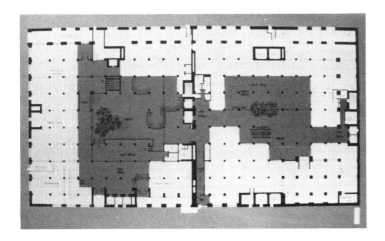

Figure 8.3.4--Main floor plan of Butler Square. Photo courtesy of Phillip MacMillan James and Associates.

382   WOOD STRUCTURES

Figure 8.3.5--Interior view of Butler Square showing exposed structural elements. Photo courtesy of Phillip MacMillan James and Associates.

8.3.4   Summary

Although only half of the building has been renovated at this time, the Minnesota Society of Architects has already given it an Honor Award.

Bids on construction were received in mid-1973. Total cost of reconstruction was $17.54 per square foot.

8.4   Dome Strengthening at Quincy Market (Boston, Massachusetts)*

8.4.1   Introduction

The Rotunda Dome for the Quincy Market Building at Fanueil Hall, Boston, was built in 1826 to the design and specifications of architect Alexander Parris. There is an exterior load-carrying dome from which an inner dome is suspended (figs. 8.4.1 and 8.4.2). From the architect's original "Specification of Particulars" we read:

---

*Author: William H. Connolly, Zaldastani Associates, Boston, Massachusetts.

Figure 8.4.1.--Quincy Market, suspended inner dome after restoration.

The seventy-seven ribs of the exterior dome and the forty-eight ribs of the interior dome are constructed of 2- by 12-inch boards with 1- by 12-inch boards nailed to either side. The tops of the ribs are bolted into a compression ring at the base of the lantern, while the bases of the ribs are seated on and spiked into a plate (4- by 10-inch tension ring). The interior dome is constructed in like manner excepting that the ribs are intersected by 6-inch coffers of ornamental plaster panels.

Curvature of the ribs was obtained by using short lengths of 2- by 12-inch and 1- by 12-inch boards and lap splicing this material with staggered splices (fig. 8.4.3). This type of framing appears to date back to the dome construction for the Halle de Bled in Paris in 1781.

The weak link in this technique, as far as Quincy Market is concerned, was the exterior dome tension ring construction. This consisted of sections of 4- by 10-inch material spliced by shiplapping with a single 1-inch diameter bolt through the lap (fig. 8.4.4). Expectedly, these splices have yielded over the years, inducing slight distortions into the exterior walls.

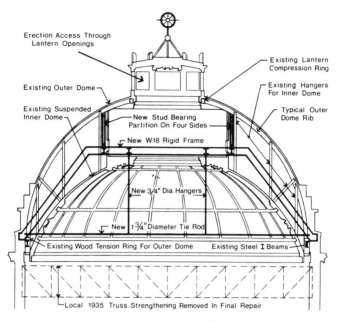

Figure 8.4.2.--Quincy Market dome construction.

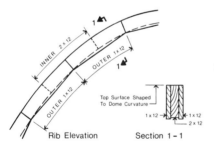

Figure 8.4.3.--Typical exterior rib construction.

In the 1930's, concern over exterior wall distortion plus local damage from a fire led to a decision to introduce local strengthening below the dome. Two steel trusses were erected across the narrow dimension of the building in 1935 and posts installed to support the dome construction. A suspended ceiling was attached to the bottom of the trusses and Parris' remarkable construction was removed from public view.

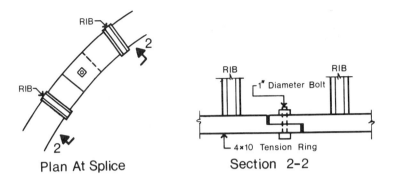

Figure 8.4.4.--Exterior dome tension ring splice.

8.4.2   Restoration

Forty years later, the Faneuil Hall Marketplace Restoration began, and with it a major effort to restore the original appearance of the Rotunda Dome. The final structural scheme involved two transverse rigid frames for strengthening with tie rods connecting the bottom legs (fig. 8.4.2). All of the rigid-frame construction, along with secondary beams and bracing, are concealed within the space between the two domes. Only the transverse tie rods for the frames are visible from below. The upper portion of the weights of the inner and outer domes is transferred to the rigid frames through new wood stud bearing partitions built on top of the new steel framing (fig. 8.4.2 and 8.4.5).

The lower portion of the dome weight is carried by an existing system of steel I-beams just below the existing 4- by 10-inch outer dome tension ring (figs. 8.4.2 and 8.4.5).

Erection of the rigid frames involved the preliminary use of full size templates to verify clearances. This led to the observation that the inclined rigid frame legs could not meet their original anticipated bearing locations at the exterior walls; accordingly, an offset bearing detail was devised as shown in figure 8.4.6. Access for installation of steel pieces was through the lantern windows at the top of the exterior dome. It would be difficult to imagine more difficult erection conditions than working with block and tackle and heavy steel pieces in the restricted space between the exterior and interior domes on this project.

Strain gage readings taken on the tie rods indicate that one rod had apparently stabilized at a load approximately 80 percent of calculated and the other at 40 percent. After some two years of use, the original glued-on strain gages were accidentally removed. For future monitoring, for example after a severe wind or snow storm, extensometer gage points have been punched on the rods near the exterior walls.

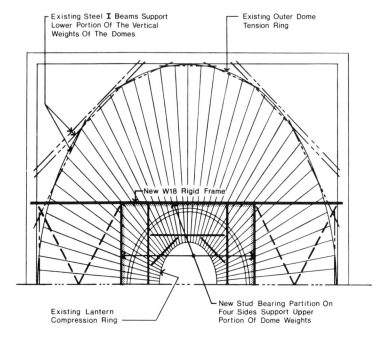

Figure 8.4.5.--New strengthening for outer dome.

Figure 8.4.7 gives a perspective view of the overall Quincy Market Complex.

8.5   Historic Restoration and Adaptive Use of
       Old Burke County Courthouse, Morganton, N.C.*

8.5.1 Introduction

The Old Burke County Courthouse, located on the public square in Morganton, N.C., has, from the time of its original construction in 1838, served as the focal point of the development of Morganton and especially the development of the center city. While several changes and modifications have taken place since the time of its original construction, the pivotal event in the physical history of the Burke County Courthouse occurred in 1903. It was at that time that the structure assumed its present appearance. Jeffersonian Classicism had become quaint and old fashioned and, consequently, the Board of County Commissioners hired one of the prominent architects of that era, Frank P. Milburn, to bring the courthouse up to date. Fortunately, Burke County did not choose to follow the lead of other communities in

---

*Author: Jack P. Hill, A.I.A., Jack P. Hill, Architect, Inc., Charlotte, N.C.

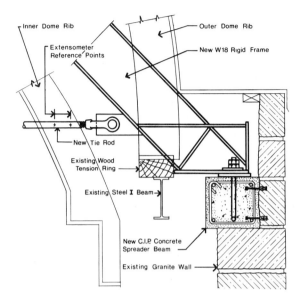

Figure 8.4.6.--Offset bearing at rigid frame base.

the state in demolishing the existing courthouse and erecting a new structure. Rather, they chose to modify the existing structure and to allow it to remain in active use, a pattern which has continued to the present time. In 1970 the Old Burke County Courthouse was placed on the National Register of Historic Places. In 1977 Burke County commissioned Jack P. Hill, Architect, Inc., Charlotte, N.C. to provide professional services for the historic restoration and adaptive use of the Old Burke County Courthouse. This work is now in progress.

The building itself is a two-story, stone, bearing-wall structure. The lower level, considered an "at-grade basement," contained four offices while the upper level served as the courtroom. The structural floor system for the upper level is entirely of wood construction bearing on the interior and exterior stone walls below. The structural roof system and the cupola are also of wood construction and are carried on two heavy timber trusses which span the width of the building. Trusses are supported by the exterior stone bearing walls. It should be noted that, before any of the current work was started, the building was thoroughly investigated in order to retain as much of its historical significance as possible. The complete structure of the building was investigated and reported on by Sutton-Kennerly and Associates, Consulting Engineers, Greensboro, N.C., in their report entitled Structural Report of the Old Burke County Courthouse, dated June 9, 1976. The historical significance of the building and its surroundings was investigated by Dr. Dan L. Morrill and

Figure 8.4.7.--Quincy Market complex. Dome at center, Faneuil Hall at bottom.

Mrs. Mary Lynn Morrill, Preservation Consultants, Charlotte, N.C., and the results presented in a report entitled A Physical History of the Old Burke County Courthouse, dated July 22, 1977. In addition to these reports, the architect for the current work made exhaustive on-site inspections and measurements of the building in order to prepare as-built drawings to document the building in its existing condition and to insure the authenticity of the restoration work to be undertaken.

8.5.2 History of the Building

The Burke County Court of Pleas and Quarter Sessions initiated a special tax in 1830 to raise funds for the construction of a new courthouse. The contract for erecting the building was let on March 8, 1833; construction was completed and the building occupied by 1838.

The evidence suggests that the initial renovation of the courthouse occurred in 1888. Stucco was placed on the interior and exterior stone walls of the courthouse. The North-South corridor in the full-height basement was sealed off, thereby creating two additional rooms. The inclined portion of the courtroom floor was installed.

The most substantial renovation of the Burke County Courthouse took place in 1903. An elaborate cupola was placed atop the structure at the center of the main roofline (fig. 8.5.1). Pedimented gables were

Figure 8.5.1--Ornate cupola added to Burke County Courthouse in 1903. Stucco was applied to exterior stone bearing walls in 1888.

constructed above the center of the northern and southern facades of the building. The porticos were raised approximately four feet to conform to the main roofline (fig. 8.5.2). The trim work on the exterior of the courthouse was altered by adding greater detail to the pediments, cornices, windows, and doorways. The rooms at the center of both halves of the full-height basement acquired arched brick ceilings and metal vault doors. Finally, plumbing (not including toilets), and electrical wiring were installed in the building at that time.

Figure 8.5.2--Portion of Burke County Courthouse with 1903 vintage trim at cornice, windows, doors, and balcony railing. In 1903, the roof of the portico was raised approximately 4 feet to line up with the roofline of the main building. This explains the square piers which support the original round columns.

Almost all of the alterations to the courthouse since 1903 have resulted from the additions of new plumbing, heating, and cooling systems. Steam heat was introduced into the courthouse in 1924. One can infer that this development caused the exposed portions of the fireplace chimneys on the roof to be removed and the fireplace openings in the full-height basement to be sealed off. The seats in the gallery of the courtroom were put in place in the same year. Several wooden partitions were erected in the full-height basement during the final years of its service as offices for the Burke County Court.

The historic significance of the Burke County Courthouse rests on four factors. First, it has been intimately associated with important developments and events in the evolution of Burke County and Western North Carolina. Between 1847 and 1862 the North Carolina Supreme Court held its summer session each year in the structure, this being the only location outside of Raleigh, N.C., ever used by the Justices. On April 17, 1865, Major General Alvin C. Gillem, commander of elements of the Union forces, directed by Major General George Stoneman, occupied the structure and destroyed most of the records located in the courthouse. Second, the Burke Courthouse and contiguous grounds have played an important part in the social history of Burke County and Morganton, N.C. For example, the public square was the customary location of ceremonies and celebrations. Political rallies, patriotic gatherings, religious convocations, recreational and mercantile assemblies, theatrical performances, all occurred in and around the building. Third, the Burke Courthouse possesses architectural significance. The initial style of the structure was an austere form of Jeffersonian Classicism. Although the building was renovated in the 1880's, it was in 1903 that the exterior and portions of the interior of the courthouse acquired the embellishments of Beaux Arts Classicism, the architectural style which continues to dominate the structure. Fourth, The Burke Courthouse is the only surviving element of the Antebellum Public Square of Burke Country and, therefore, constitutes the most important item in the historic townscape of Morganton, N.C.

8.5.3 Analysis of Existing Structure

In 1976 the Burke County Commissioners authorized a structural analysis of the existing courthouse building to determine the status of the structure and the work which might be required to restore the building. The report resulting from that analysis indicated that the purpose of the inspection was to examine and evaluate the condition of the existing structure. The structural system includes the foundation, the bearing walls, and the structural members which support the floor and the roof. For the purpose of this article, the major structural elements requiring attention at this time are the wood structure for the courtroom floor and the wood structure for the roof.

The inspection technique employed to produce the Sutton-Kennerly report was a visual "hands-on" procedure made possible by cutting holes in floors and ceilings. Access to the attic space is readily available so that inspection of the courtroom ceiling, the roof structure, and the cupola was accomplished in the same way.

This technique allowed field examination of a number of structural elements in various locations throughout the building. Newly cut slivers from the wood members were yellowish in color and uneven-grained with very distinct growth rings. In light of this evidence, and with previous experience with similar projects in this area, it was determined that the structural wood was southern pine. A review of the characteristics of the material, when compared to the "Standard Grading Rules for Southern Pine Lumber," 1970 edition, found a reasonable match in the description for Structural Joists and Planks, number 2 grade, medium grain, dry, material which allows for an extreme fiber stress in bending ("F_b") of 1,250 pounds per square inch.

The floor of the upper level (the courtroom floor), constructed in 1835 to 1838, was originally level and was framed with 4 by 12 nominal size timber joists spaced approximately 18 inches on center. It should be noted that these members are, in most cases, full size, and in some cases, oversize. The existing timber joists span from bearing wall to bearing wall with a maximum span of 14 feet -0 inches. These joists are in sound condition; however, they have twisted slightly and will require some corrective work such as shimming and bridging at the bearing points. Also, some additional lines of wood bridging will have to be placed between joists. When this work is complete, it has been calculated that the existing joist system will support a superimposed load of 150 pounds per square foot.

As part of the work of the 1888 renovation, an inclined floor was constructed over the top of the level floor and covered about half of the total floor area of the courtroom. The inclined floor begins at approximately the mid point of the building in a north-south direction and rises to a high point along the south wall. The inclined floor is framed with nominal 2 by 12 wood joist spaced approximately 18 inches on center. Total length of joists is approximately 25 feet and 0 inches with the low end tied to the level floor joists, bearing on a nominal 4 by 12 timber at the mid-point, and bearing on a wood ledger beam along the south wall. This system, too, is in sound condition; however, additional bridging, realignment, and anchorage will be required. Of prime concern is the fact that the 4 by 12 timber beam at the mid-point of the inclined joist span is capable of supporting a superimposed load of only 60 pounds per square foot on the inclined floor. Additional support for the inclined floor joists and/or stiffening for the existing 4 by 12 timber beam will have to be provided in order to increase the load-carrying capability of the system and meet the 100 pounds per square foot required by the current design. It is interesting to note that this situation was brought about by a consideration in the original design rather than a deterioration of the materials selected for construction.

The features of the roof framing system are the two heavy timber trusses which clear-span the building in an east-west direction and bear on the exterior stone walls. The sizes of the wood members and approximate dimensions of the truss are shown in figure 8.5.3. The top and bottom chord members and compression web members appear to be in excellent condition. There is some slight checking in the members, but this is minor and will not significantly reduce the capacity of the members. One of the trusses has a mid-span deflection of approximately 1-1/2 inches. This deflection could have been built into the truss or could have been caused by shrinkage of the wood truss members or slippage of the joints where members were inadequately fitted together. Calculations indicate that these trusses will safely support the in-place dead loads and the live loads required by the North Carolina State Building Code for the structure.

The joists for the courtroom ceiling are nominal 2 by 10 spaced at approximately 14 inches on center. Joists span from exterior wall to bottom chord of truss and from truss to truss. Maximum span is 22 feet and 0 inches. These joists appear to be sound and will safely support a superimposed load of 15 pounds per square foot. This load is limited by a deflection of L/360 or 0.75 inches.

Roof rafters are nominal 2 by 8 members spaced approximately 18 inches on center. Roof deck is nominal 1 by 6 material nailed directly to rafters. An interesting feature of the roof is the built-in gutter

RECYCLING OLD STRUCTURES 393

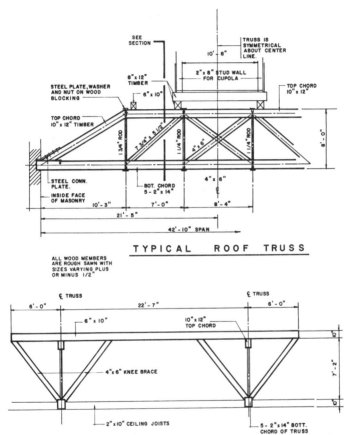

Figure 8.5.3--Typical roof truss, Burke County Courthouse.

which runs around the perimeter of the building. Wood rafters were notched out to receive a sheet metal gutter section which could not be seen from the exterior ground level and would not interfere with the ornate detailing at the cornice. Prolonged leakage and deterioration of certain sections of the gutter caused rotting of the overhanging portions of the wood rafters and deck and required replacement as a part of the restoration work completed in 1979.

The cupola for the building remains as the large and ornate structure which was added in 1903. Constructed of nominal 2 by 8 wood studs with 1 by 6 sheathing and sheet metal cladding, the entire cupola is carried on two nominal 8 by 12 timber beams which transfer the load to the two heavy timber roof trusses. These transfer beams remain in sound condition.

8.5.4 Restoration and Modification for Adaptive Use

It has been pointed out earlier that Burke County recognizes the historic significance of the courthouse building and is most anxious to restore the building to retain it as the focal point of the center city and to reinforce its rich heritage for the citizens of Burke County. After careful review of all of the facts concerning the building, it was decided that the historic restoration would be patterned after the 1903 remodeling which produced the most significant changes in the total appearance of the structure.

In addition to restoration work, the owners have decided that certain adaptive use of the interior spaces of the building would be advantageous in returning the building to an active role in the life of the local community. Rather than use the building as a museum-like show piece, it has been decided that the entire structure would be made available for various community functions. In the at-grade basement area it is anticipated that the original office spaces will remain. However, these offices will be restored and made available to various civic organizations within the county. In addition, complete new toilet facilities will be installed and a small passenger elevator from the at-grade basement up to the courtroom level will be provided in order to accommodate the handicapped.

The old courtroom will be modified in such a way as to provide facilities for little theater groups, musical groups, and civic gatherings requiring a seating capacity of up to 200 people. The existing assembly area will be restored while the court facilities such as judge's bench, jury rooms, witness stand, etc., will be removed and space made available for a portable stage, storage facilities, and toilet rooms. Further, if the situation arose, this room could always be temporarily modified to serve once again as a court facility.

The work of restoration and adaptive use is proposed to occur in five phases: (1) Restoration of the cupola, (2) Restoration of the exterior of the building, (3) Restoration and modification of the courtroom, (4) Restoration and modification of the full-height basement, (5) Restoration and modification of the public square. Of these, Phases 1 and 2 have been completed, with Phases 3 and 4 scheduled to

take place in 1980 and 1981. Funding for the project is primarily
through Burke County with assistance from the city of Morganton.
Grants-in-aid from the Department of the Interior, Heritage
Conservation and Recreation Service, under provisions of the National
Historic Preservation Act of 1966 have also contributed to the ongoing
of the project.

8.5.6   Conclusion

The historic restoration and modification for adaptive use of the Old
Burke County Courthouse in Morganton, N.C., has been an exciting chal-
lenge for this office. Our success with the project to date can be
attributed to the fact that we are dealing with a knowledgeable client
who is charged with the responsibility of producing an accurately
restored and modified project while keeping a careful check on the
expenditure of public funds. Also contributing to the success of the
project is the preliminary information that was gathered before any
work was actually done. These data, including a structural analysis,
physical history of the project, the physical measurements of the as-
built building, and the attention to detail existing on the building
all have contributed to a logical sequence of design and construction
with few surprises. While the building itself has fallen into dis-
repair over the past years, the basic elements of the structure remain
in sound condition, making for a feasible project.

8.6   Restoring Covered Bridges*

8.6.1   Introduction

Covered bridges have long been favorite symbols of rural America for
largely picturesque reasons. To the structural engineer, however, they
also represent the remarkable achievements of men who had little or no
exposure to even the rudimentary hand analysis methods available when
they built the bridges. There are nearly 850 covered bridges left of
the more than 20,000 that had spanned this nation's streams and rivers.
The older surviving examples have withstood 150 years of wind, snow,
and dead loads, in addition to heavier traffic than their builders ever
imagined and occasional mistreatment by floods and ice floes. Deteriorated bridges of all types and materials is an increasingly recognized
national problem. Strengthening our covered timber bridges to carry
contemporary traffic is one facet of a solution to this problem, which
at the same time preserves some of our structural engineering heritage.

The majority of America's covered bridges have either Burr or Town
trusses as primary structural elements. The Town truss is named after
its architect-inventor, Ithiel Town, but is also known as a lattice
truss because of its appearance. The Burr truss (patented by
Theodore Burr) is simply a Howe truss, with its diagonals in compression, that has superimposed arches. Surviving examples of both truss
types have clear spans of up to 200 feet. While the Town truss will
generally use more board feet of timber than the Burr to carry equal

---

*Author: R. L. Brungraber, Instructor, Department of Civil
  Engineering, Bucknell University, Lewisburg, Pennsylvania.

loads over the same span, it became more popular because of its simple and highly repetitive joinery details (3). The Town lattice is also just as strong in "negative bending" as it is under standard loading. This is very convenient while rolling it into place on falsework, or for continuous, multiple-span bridges. The evaluation and reinforcement procedures and examples covered in this paper are referenced to these two truss types, but are applicable to any of the other truss varieties favored by the original builders. Specific bridges are referenced with the numbering system employed in "The World Guide to Covered Bridges (2).

8.6.2   Evaluating A Bridge's Condition

The two basic investigations required for a thorough structural evaluation of a covered bridge are establishing its overall alignment and determining the condition of each member. These two will be related in that certain displacements can focus attention on specific elements, while seemingly small structural defects can explain large misalignments.

The bridge's position can be accurately determined with two simple surveying procedures. The elevation of each truss, at regular intervals (posts, in Burr trusses), is measured relative to a mark established with a surveying level. The camber (or sag) of each truss along the bridge's length, and any tilt across the bridge, are readily calculated with these data. The longitudinal alignment is measured relative to a theoretical centerline established at the piers. This line is transferred to each roof beam and the deck immediately below, using a transit set up on one of the piers. The gross displacement of the bridge's centerline from the theoretical one at deck level represents the bridge's sideway. The net lateral displacement of the roof beam level, relative to the deck below, represents any rack in the bridge at that cross section.

The other required investigation is an evaluation of the condition of every member and connection. A close inspection and careful recording of each connection's condition yields valuable data which are assessed in various ways discussed later. In particular, the amount of slip in each tension chord splice should be measured. Decay or insect damage of members is generally detected by visual inspection of the surface and by probing with a sharp instrument to detect internal flaws. For critical, but hidden regions, such as concrete-encased arch ends, samples may be obtained with a small auger.

Interpreting the data from these two investigations will give insight into the relative severity of a bridge's problems and the most effective means of remedying them. The following cause and effect relationships are the product of many field investigations, original literature searches, and the design and analysis of covered bridges, new and old. This is not intended to be a complete explanation of all the forces that may be at work in a covered bridge, but rather as the starting point for evaluating a bridge's condition as a function of its position.

A covered bridge's misalignment can be described in terms of sag, sway, tilt, rack, and settlement. The most common cause of longitudinal sag is failing tension chord splices. Sag across the deck is a sign of failed or inadequate floor beams. Sidesway indicates inadequate lateral bracing of the trusses and arches, at the roof and/or deck levels. A tilted deck or racked cross section are symptoms of unevenly sagged trusses and inadequate knee braces. A common cause of settling is rotted truss members where they contact the piers or abutments. If the bridge has arches, they will, as a result, often buckle severely where they first connect to the trusses because all the load on the bridge is now bearing on the arches at this point, instead of being shared with the truss ends.

Any displacements are almost always interrelated. A very common movement sequence is: tension splices stretching on one side more than on the other, allowing the bridge to tilt, which induces a sidesway which, in turn, leads to racked cross sections.

8.6.3 Repairing or Strengthening the Bridge

The distinction between repairing and strengthening is one based on desired load capacity. If the intent is only to save the bridge from imminent collapse or to carry light pedestrian traffic, it may be enough to replace deteriorated parts, correct any faulty details, and secure the structure within alignment tolerances. This last step has been expeditiously done with props in the streambed and cables to nearby trees. If, however, vehicular traffic is desired, additional strengthening may be required. Even these repaired and reinforced structures should be analyzed by a professional and the safe weight limit posted.

8.6.4 Structural Analysis of Covered Bridges

Covered bridge trusses are usually very indeterminate, with their analysis further complicated by large uncertainties and irregularities in material and connection stiffnesses. While a computer analysis can be invaluable in determining load distributions, the assumptions required to model a timber truss for program input are of an order sufficient to justify simpler, hand methods of analysis (4).

The degree of load sharing between an arch and the truss is the critical issue. The relative stiffness of the arch and truss can be used to apportion applied loads. Such an analysis usually resulted in 50 to 75 percent of the imposed loads being absorbed by the arch. Another method, and one used by early analysts, is simply to design the truss for the live loads and the arch for the dead load--which can be as high as 1,500 pounds per foot for longer-span bridges (1).

Whatever the analysis method and bridge condition, some elements will be revealed as weak links. Repair and reinforcement procedures will be presented for each major element. Whenever possible, specific implementations of the recommended methods are cited.

8.6.5 Compression Members

Truss diagonals and top chords, and any superimposed arches, are the principal compression members in the bridge. Decayed top chords are rare, but not unknown. The end-panel diagonals will occasionally decay at their lower ends, where water is retained in the dirt and debris that collects in the diagonal/vertical connection. The arches, on the other hand, are particularly prone to decay at the piers.

The arch ends are so vulnerable to decay because they are exposed to the wet-dry cycling caused by high water, pier runoff, and wind-driven rain. Casting rotted arch ends in concrete is a very common reconstruction mistake that only aggravates the situation by preventing drainage and inspection for the almost inevitable further decay. Most original builders, without this option of concrete encasement, butted the arches directly against large stones or timbers laid in the piers at the correct springing angle. Modern steel saddles with weep holes and through bolts will also provide the required drainage and were used on the recently built Hunsecker bridge in Lancaster County, Pa., (38-36-06 No. 2). A repair method is to cut the rotted ends off square and cast a concrete bearing pad out from the abutment to meet the shortened arch, without encasing it. The only conceivable benefit of encasing the arch ends is for holding down the bridge in flood waters. This function is better performed by rods or cables anchored in the piers.

Temporary support for the bridge will have to be provided with props, jacks, and cribbing while any rotted arch sections are removed and replaced. Seasoned timber of the same species as the rest of the arch should be used, if possible. The splice between the new and old pieces is typically lapped, with through bolts and possibly dowels lying in the shear plane (fig. 8.6.1). Epoxy consolidation is a relatively new

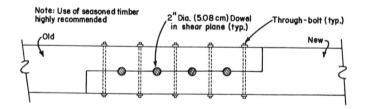

Figure 8.6.1--Lapped connection between new and old portions of a repaired member.

technique that might be used to strengthen decayed members in place, without removing the rotted sections. Any buckled arch sections should be straightened and refastened, once the bridge's load has been jacked onto temporary cribbing.

The arches were added in the first place by Burr himself, and by some subsequent rebuilders, in order to take load from the trusses. If analysis shows that desired traffic loads would overload a truss, arches may still be added or strengthened. The simplest of these supplemental arches is a segmental one of straight, solid-sawn timbers. It can be attached to the trusses directly above an existing arch, sometimes without interfering with the knee braces (38-22-04). Others have been bolted directly to the original arches that were laboriously hand hewn to the desired radius (38-06-06). A major consideration with any segmental arch is restraining it against the net outward radial force where the straight members butt at an angle in order to approximate a curve. If this connection is not carefully detailed (generally the case), the arch segments are subjected to high bending stresses at their midpoints (fig. 8.6.2).

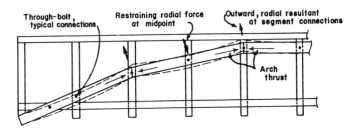

Figure 8.6.2--Radial forces at butted ends of sections of a segmental arch.

Smoothly curved arches are more effective, elegant, and difficult to fabricate. The 168-foot Hunsecker bridge (built in 1974) has a solid arch with a 210-foot radius. This was achieved by "bending or bandsawing, at the option of the contractor." A simpler, and therefore more common, method of fabricating a heavy curved arch is laminating one in place, with dimension lumber. The arch Milton Graton and Sons added to the Town Lattice in Newfield, N.Y., (32-55-01) is a fine example of this type. It was heavily through-bolted to prevent the interlayer slip which has permitted the drastic in-plane buckling suffered by the lightly bolted arches added to the Bartlett, N.H., (29-02-02) covered bridge.

Glulam arches would not be as prone to this buckling mode, but have not, as yet, been added to an existing bridge. The author did use them in a complete covered bridge design using modern timber technology, but limited the three-pinned arch to a 100-foot span so that the members would be readily transportable--a major consideration with any large members fabricated off-site.

8.6.6 Tension Members

The truss verticals and lower chords are the principal tension members to consider. Verticals will frequently decay at the lower-chord connection and may be replaced entirely or be spliced. Consistently high water in the spring can wear away the protruding bottoms of the posts, as in many Lancaster County, Pa., bridges. Saddle straps that wrap around the lower chord and up along the sides of the vertical can effectively replace the lost piece, if the bolts through the strap and vertical are sized properly. Verticals will also shear off where the diagonals frame into them. This is a function of eccentricity of that joint (5) and is usually worst in panels near the piers, where the diagonals are most heavily loaded. The later builders often compensated for this by installing props on the other side of the post that transfer load to the chords. When a second top chord was put on the Meem's Bottom bridge in Virginia (46-86-01), it was installed alongside the truss, level with the vertical/diagonal connection. This avoided eccentric loading on the vertical. During the same major restoration, required by its being burned by vandals, braces were also added at the lower ends of the vertical. These transferred the diagonal load more directly into the bottom chord.

The truss diagonals in the midspan panels (normally stressed in compression) are actually stressed in tension under certain asymmetric loadings. If they are not restrained with spikes against such stress reversal, they may bounce out of their intended position and become worthless in compression.

The tension chord is a notorious weak point of all heavy timber trusses, and covered bridges have been no exception. The lower chord is prone to decay at piers and connections with verticals, where water can accumulate and be absorbed. The tension chord has an even more basic flaw, however, in its splices. (Note: Some northwestern covered bridge builders were able to avoid the problem with full length chords--up to 98 ft). The splices tend to slip, no matter how carefully detailed, which results in a sagging truss. The original builders included arches, camber, and adjustable splice hardware to compensate for this sag, and modern rebuilders have employed many schemes to remove it.

Milton Graton and his sons have rebuilt (and built) more covered bridges than anyone still in the business. Before repairing the tension chord splices they jack the sag out of the trusses; using falsework in the streambed, when possible. On long spans over deep water, such as the Bedell bridge across the Connecticut River, they had to replace the falsework with temporary, but heavy, suspension cables. Anchored to "dead men" buried beyond the piers, the cables ran up over railroad tie towers at the piers, then down to support cribbing and jacks beneath the bridge's span. This bridge unfortunately blew down in Hurricane Frederic (September 15, 1979) soon after reopening. Once smooth camber is jacked into the bridge, they reinforce the tension chord splices with timber fish plates, shear blocks or dowels, and through bolts.

RECYCLING OLD STRUCTURES 401

Another method of strengthening tension chords involves adding steel rods that supplement the tensile strength of the timber chord, without changing the forces in the rest of the bridge's members. Basically, steel rods are placed along both sides of both tension chords and tightened to remove the load from the splices. The rods are divided into panel-length pieces and attached at each vertical/bottom chord connection. The rods are differentially post-tensioned to match the chord force as it varies along the span by torqueing nuts on the threaded rod ends at the hardware bolted to each panel point. Panel-length rods were installed by the author on the Sonestown, Pa., bridge (38-57-03) during its 1973 restoration (fig. 8.6.3). The Sonestown bridge was recently subjected to the traumatic removal of a vertical by an automobile driven by a patron of a newly-opened tavern near the bridge. While the truss did sag, the bridge did not collapse. The

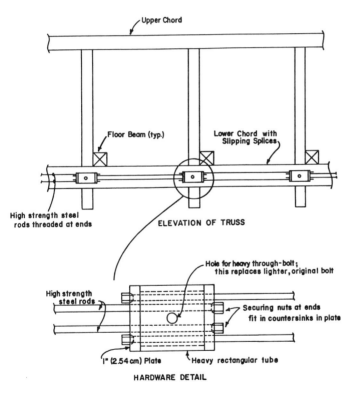

Figure 8.6.3--Method of reinforcing tension chords.

vertical was replaced, the floor beams shimmed to level the deck, and the bridge was reopened to traffic. The rods, and the similarly-acting welded strap used on a Forksville, Pa., bridge (38-57-01), are hidden by the cover, making this a very effective but inconspicuous method of strengthening this weak point of timber trusses.

8.6.7   Lateral Bracing

There are four major structural elements that keep the bridge in longitudinal alignment: roof beam bracing, floor beam bracing, knee braces, and the deck itself. Roof beam bracing restrains the top chord against lateral buckling as well as bracing the bridge against lateral wind loads. If the braces are wood, they should be snugged with pairs of wedges in the roof beam mortices, and the roof beams themselves firmly connected to the top chords for this bracing to function properly. An effective method, used by several original builders, of attaching the roof beams to the top chord can be seen on the Beaver Springs, Pa., bridge (38-55-02) and is illustrated in figure 8.6.4.

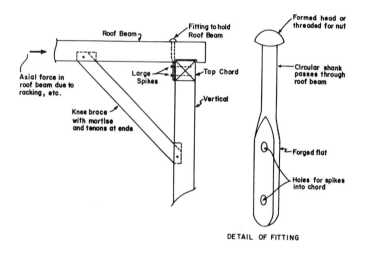

Figure 8.6.4--Method of tying roof beam to truss chord.

The bridge's principal lateral stiffness is provided at deck level by steel rods or timber x-braces between the floor beams. Additional, and quite substantial, strength is available in the floor deck which can easily be detailed to act as a diaphragm or deep lateral beam. Heavy timber decking, laid at 45° to the bridge axis in two perpendicular layers constitutes a double-diagonal-sheathed diaphragm--in effect a lateral Town Lattice truss. This deck system has been recommended to reinforce the 200-foot Red Mill bridge near Reading, Pa. (38-06-06). To avoid decay between layers, preservative treated timber is required. Glulam decking, if it were solidly attached to the floor beams, could act as a deep, lateral beam spanning the full length of the bridge.

Knee braces prevent the bridge from racking but are only as effective as their connections are rigid. At the ends of the knee braces the common, original mortice and tenon connection is generally fine, as are the split-ring connectors used on the Hunsecker bridge. However, if the roof beam is not solidly held down against thrust in the knee brace, it can be lifted off the top chord when the knee brace is heavily loaded in compression. The connection detail shown in figure 8.6.4 is an inconspicuous and original but very effective way to make this connection. The knee braces themselves are far more subject to impact damage by trucks than to decay. If traffic clearance is a problem, "ships' braces"--solid pieces cut from the crotches of trees-- can be bolted into the corners as substitutes, as Milton Graton did on the Waitsfield, Vt., bridge (45-12-02). Plywood gusset plates have been used over existing knee braces in order to stiffen or hold a racked bridge.

8.6.8   Floor Systems

Under heavy modern loading, original floor beams have become increasingly subject to failure. Covered bridges in the mid-Atlantic region were often built with heavy floor beams only at the panel points. The New England bridges, however, tended to have lighter floor beams distributed all along the bottom chord. This resulted in a stronger and more redundant structure, in which one weak floor beam was not as critical. One problem with the New England system is the resulting bending stress induced in the lower chords. If the steel rods discussed earlier are installed along these chords, however, the chords are not loaded in tension and therefore have increased bending capacity. The closely spaced floor beams do make x-bracing between them impractically complicated, though lateral strength can alternatively be provided by one of the strong decks also discussed earlier.

There are other methods of distributing concentrated traffic loads among several floor beams and therefore along the length of the trusses. So-called "distribution stringers" are relatively common in New England covered bridges. They are heavy longitudinal timbers firmly strapped to the underside of each floor beam, either along the center of the deck or separated to run under the wheel loads. Longitudinal glulam deck panels are also strong enough to distribute point loads more evenly to the bridge.

Steel has been used in floor systems in many ways. The Forksville, Pa., bridges now have transverse steel floor beams at truss panel points, which support steel longitudinal stringers and transverse laminated-on-edge dimension lumber as a deck. The floor beams are hung on steel channels bolted to the midpoint of each timber vertical post. Most of those Vermont covered bridges which, because of their location, must carry high loads, have had heavy steel girders installed under them to carry all the imposed loads. The timber trusses only hold up the now strictly ornamental cover. There are cases where even with steel girders, a center pier was added to halve the original clear span. This radical step is not only preferable to, but probably cheaper than, total replacement; and is indicative of the effort that has been expended recently to preserve these bridges.

8.6.9 Siding and Roofing

The siding and roof are what give the covered bridge its name. They were (and are) intended only to protect the timber from the elements. Repairing or replacing these coverings is really the only regular maintenance the bridge should need. Neglecting this relatively small job, however, will always lead to much more expensive rehabilitation later, or eventual collapse. The importance of keeping the roof and siding in good, water tight condition can not be overemphasized.

Many siding types were originally used including: clapboard, vertical board and batten, shingles, etc. The same is true of roofs although most of them imposed smaller dead loads than the slate roof on the Bangor, Maine, bridge (19-10-01). Any openings in the cover, such as windows or eave vents, provide needed light and ventilation, as well as beneficially decreasing the surface area exposed to wind pressure. Some early short-span bridges even used heavy siding to strengthen the bridge. This concept might be applied to strengthen existing bridges by using plywood in the siding which, if properly detailed and attached to the trusses, would act as deep, built-up timber girders to increase load capacity.

Historically, fire has been a major enemy of even well-maintained covered bridges. Heavy timber is actually very slow-burning, but the lighter siding and roof structure is highly flammable. The bridges can be recovered if the timbers have not been too weakened, as was recently done in Woodstock, Vt., (45-14-15) and Virginia (46-86-01). Since it is the lighter lumber that provides the fuel, consideration should be given to specifying fire-retardant treated siding, roofing, and even decking, if they are being replaced anyway.

8.6.10 Conclusion

America's surviving covered bridges represent some of the finer examples of this type of construction due to a process of natural selection through the passing of time and the imposition of ever-heavier loads. Beyond their obvious historic value and charm, therefore, they should be able to continue to perform their original and vital role, if properly restored and maintained. Through sensible and sensitive application of modern technologies, covered bridges can move into the future without any great loss of historic integrity.

8.6.11   References for Section 8.6

8.6.11.1 References

1. American Society of Civil Engineers.
    1976. American wooden bridges. Am. Soc. Civ. Eng.,
    New York, N.Y.

2. Donovan, R. T.
    1980. World guide to covered bridges. The Natl. Soc. for the
    Preservation of Covered Bridges, Holliston, Maine.

3. Graton, M. S.
    1978. The last of the covered bridge builders. Clifford-Nicol,
    Inc., Plymouth, N.H.

4. Kemp, Emory.
    n.d. Case study of Burr truss covered bridge. J. of Struc.
    Div., No. 11461, Am. Soc. Civ. Eng.

5. Smith, K. M., and J. E. Andrews.
    n.d. Restoration of the Meems bottom covered bridge. Va. Dep.
    Highways and Transportation, 1401 East Broad Street,
    Richmond, Va. 23219.

8.6.11.2 Additional Sources of Information

1. Adams, K. A.
    1963. Covered bridges of the West. Howell-North Books,
    Berkeley, Calif.

2. Allen, R. S.
    n.d. Covered bridges of the Middle Atlantic. Bonanza Books,
    New York, N.Y.

3. Allen, R. S.
    n.d. Covered bridges of the Middle West. Bonanza Books,
    New York, N.Y.

4. Allen, R. S.
    n.d. Covered bridges of the Northeast. Bonanza Books,
    New York, N.Y.

5. Allen, R. S.
    n.d. Covered bridges of the South. Bonanza Books,
    New York, N.Y.

6. Avent, R. R., P. H. Sanders, and L. Z. Emkin.
    n.d. Structural repair of heavy timber with epoxy. For.
    Prod. J. 29(3):15-18.

7. Barth, D. O.
    1980. America's covered bridges. Civ. Eng., p. 51-54.
    February.

8. Brungraber, R. L.
   1978. Design of a glulam covered bridge. Design presented to Colorado State University at Fort Collins, Colo., in partial fulfillment of the requirements for the degree of Master of Science.

9. Brungraber, R. L.
   1980. Revival of the covered bridge--case studies. Presented at the April 14-18, 1980, ASCE Convention and Exposition held at Portland, Oreg.

10. Congdon, H. W.
    1941. The covered bridge. Vermont Books.

11. Covered Bridge Topics.
    n.d. Official magazine for the Natl. Soc. for the Preservation of Covered Bridges, Inc., nonprofit organization, Joseph S. Cohen, ed., Holliston, Maine.

12. Graton, M. S.
    1980. The building of Zehnder's holz-brücke, Clifford-Nicol, Inc., Plymouth, N.H.

13. Holtman, D. F.
    1929. Wood construction. McGraw-Hill, New York, N.Y.

14. Hool and Kinne.
    1942. Steel and timber structures. McGraw-Hill, New York, N.Y.

15. Johnson, S. M.
    1965. Deterioration, maintenance, and repair of structures. McGraw-Hill, New York, N.Y.

16. Park, S. H.
    1980. Bridge inspection and structural analysis (handbook of bridge inspection). S. H. Park, P. O. Box 7474, Trenton, N.J. 08628.

17. Pavlak, E.
    1976. The bridge that could. Commonwealth of Pennsylvania, Dep. of Transportation, Harrisburg, Pa.

18. Phillips, M. W., and J. E. Selwyn.
    1978. Epoxies for wood repairs in historic buildings. U.S. Dep. of Interior, Washington, D.C., Stock No. 024-016-00095-1.

19. Prudon, T.H.M.
    1975. Wooden structural members: Some recent European preservation methods. Assoc. for Preservation Technol. III(1):5-11.

20. Structural Calculations and Working Drawings for a Covered Timber Bridge Over Conestoga Creek in Lancaster County, Pennsylvania.
    1976. Rummel, Klepper, and Kahl, Mechanicsburg, Pa.

21. Stumes, P.
    n.d. Testing the efficiency of wood epoxy reinforcement systems. Assoc. for Preservation Technol. VII(3):2-35.

22. Timber Construction Manual.
    1974. Amer. Inst. of Timber Construction, Englewood, Colo., 2d ed.

23. Wood Structures, A Design Guide and Commentary.
    1975. Am. Soc. of Civ. Eng., New York, N.Y.

8.7 Repairs to Wooden Trusses in Elastomers Finishing Building--Exxon Chemical Company*

8.7.1 Introduction

The purpose of this project was to assess the structural integrity of explosion-damaged wood trusses and to make recommendations for a repair method if repair appeared feasible.

The method of repair selected had to meet two major requirements of Exxon:

  a. Production lines below the roof trusses had to be maintained at all times with minimum interference.

  b. Production lines could not be exposed to the outside elements at any time.

These requirements precluded the removal of any major structural elements and roofing for replacement. The repair method selected had to allow the trusses to be repaired in place.

8.7.2 History of Project

The southern pine trusses shown in figure 8.7.1 were constructed in 1943 and apparently performed satisfactorily until January 1971, when an explosion occurred and truss T-2E failed, as shown in figures 8.7.2 and 8.7.3. The explosion occurred just east of the building, and the resultant blast caused the failure of this truss by splitting the wood members adjacent to the bolted connections. The truss did not collapse but sagged considerably and was shored up immediately. Exxon engineers checked the deflections of all 14 trusses in August 1971.

---

*Author: Robert L. Huck, Robert L. Huck Associates, Inc., Consulting Engineers, Baton Rouge, Louisiana.

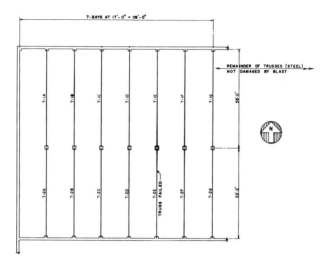

Figure 8.7.1--Roof framing plan.

Figure 8.7.2--Shoring under failed truss 2E.

Robert L. Huck and Associates, Inc. was employed by Exxon in January 1972 to make an engineering report on the trusses. The structural survey and report included evaluation of the trusses in their existing condition, recommended method of repair, and cost estimate of the repair work. Microfilms of drawings of the wood truss were furnished for preliminary office work prior to a field survey.

Figure 8.7.3--Fracture at bottom chord joint, truss, 2E.

8.7.3 Scope of Work

The investigation consisted of the following phases:

1. Complete field investigation of the 14 wood trusses.

2. Computer analysis on each type of truss.

3. Comparison of the original joint and member designs with the results of the computer analysis.

4. Preparation of record pictures noting the representative condition of the wood and joints at critical areas (fig. 8.7.4, 8.7.5, and 8.7.6).

5. Analysis of field data on existing condition of trusses.

6. Check of existing deflections of trusses.

7. Evaluation of general structural integrity.

8. Recommendation of method of repair.

9. Preparation of engineering drawings covering the repair.

8.7.4 Method of Field Inspection

Members were visually inspected at close range from a movable scaffold. All major defects were noted on individual truss tally sheets with descriptions and measurements. The field inspection included a check of the following:

1. Major splits or excessive checking in members.

2. Rusting of bolts in joints.

410 WOOD STRUCTURES

Figure 8.7.4--Split upper chord splice and split diagonal, truss T-2F.

Figure 8.7.5--Damaged lower chord splice, truss T-2D.

Figure 8.7.6--Split vertical member, truss T-2F.

3. Loose bolts.

4. Decay in any members.

5. Large knots in wood causing distress.

6. Moisture content of truss members.

7. Bowing of members.

8. Slippage in joints.

9. Separation between adjacent members at joints.

10. Major splice points.

    a. Separation between butt joint ends.

    b. Crushing of wood fibers in ends of top chord.

    c. Warpage of wood at joints.

11. End bearing condition.

12. Deflection of trusses.

8.7.5 Computer Analysis

The truss members and joints were marked for field inspection purposes and computer programming. All piping, heaters, sprinklers, and other external loads were located in the field. Applicable design dead and live loads were proportioned to each type of truss T-1 and T-2. Truss T-2 was similar to truss T-1 (fig. 8.7.7) except that it had 10 panels.

Each truss was programmed for the ICES STRUDL computer program. The maximum stresses in each member fell within the allowable 1943 NDS stresses as shown on the original drawings.

8.7.6 Deflection of Trusses

A timber member normally undergoes some shrinkage after assembly, due to continued seasoning. The shrinkage causes bolts to loosen; creates splits and checks; and as a result, the connections are loosened causing deflection, distortion, and weakening of assemblies. It was our opinion after the study that part of the deflection of the trusses was due to creep and part was due to the blast. Those trusses closest to the blast (origin of blast on east side) exhibited more deflection than did the others.

The original truss drawings indicated that a positive centerline camber of 1-1/2 inches was to be placed in the lower chord of truss type T-1 and 1 inch was to be placed in truss type T-2.

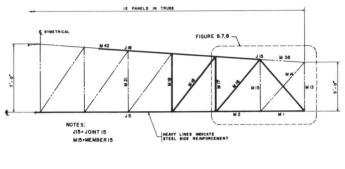

Figure 8.7.7--Schematic of truss T-1.

As a part of this investigation, Exxon surveyors provided elevations at each splice point in the bottom chord, at the centerline, and at each bearing end of the trusses. The mean deflections below the horizontal at the splice points and at the centerline of the trusses were then determined from these data. It was noted that the minimum centerline deflection on truss type T-1 (57-ft 0-in. clear span) was 2.34 inches and the maximum was 3.06 inches with an average of 2.73 inches. It was also noted that the minimum centerline deflection on truss type T-2 (52-ft 2-in. clear span) was 1.85 inches and the maximum was 4.31 inches with an average of 2.58 inches.

To get the actual maximum deflections that had taken place (assuming that the trusses were built with camber) reported initial cambers were added to calculated centerline deflections below the horizontal as follows:

   Truss T-1G (57-ft clear span) = 3.06 in. + 1.5 in. = 4.56 in.

   Truss T-2F (52-ft 2-in. clear span) = 4.31 in. + 1.0 in. = 5.31 in.

Refer to table 8.7.1 for the results of the deflection data taken on the project.

Since the recommended deflections of 2.85" and 2.61" mentioned above were much smaller than some of the measured deflections, some of the trusses were probably suffering from the effects of the blast more than others.

Table 8.7.1.--Summary of results of deflection measurements

Deflection characteristics	Truss type and clear span	
	Truss T-1 span = 57 ft 0 in.	Truss T-2 span = 52 ft 2 in.
1. Recommended deflection limitation of roof = L/240	2.85 in.	2.61 in.
2. No. of trusses with deflections exceeding L/240	3	2
3. No. of trusses with deflections of L/240	2	0
4. No. of trusses with deflections less than L/240	2	5

8.7.7   Investigation Analysis

All of the field inspection data was recorded on "Member and Joint Re-cap Sheets for Truss T-1 and T-2". These were summarized on another form "Results of Member Survey" which was a pictorial presentation, with members identified on each individual truss that needed some degree of repair, reinforcement, or replacement.

Some trusses had excessive deflections. The major cause for these deflections was loose splice joints in the upper and lower chords. Referring to figures 8.7.4 and 8.7.5 it is noted that large horizontal splits in the wood had occurred at these joints. In addition, a truss with loose joints is undesirable especially when subjected to impact load. This condition (loose joints) was probably present when the blast occurred. Obviously there will be more damage when an impact load is applied to a loosened joint than when it is applied to a tight joint. Accordingly, a bolted timber structure must be periodically tightened by retorquing the connection hardware.

Wood core samples (1/8 inch in diameter) were extracted from the wood trusses using a Djoa 400 increment borer. The samples were visually examined and most were found to be relatively free from decay. Truss T-2B showed evidence of decay in that portion of the truss that was partially embedded in the brick wall; a roof leak was evident in this area. The brick adjacent to the truss was removed, paint stripped from the truss, and this portion of the truss was treated with a 5% solution of pentachlorophenol to a 0.15 pcf retention. Treating holes 1/4 inch in diameter, spaced at 6" o.c. each way, were drilled from one side and plugged on the other side. After pressure treatment, the holes were plugged with wood dowels and the surface of the wood was flushed with the solution three times.

WOOD STRUCTURES

The wood core samples and visual inspection indicated that the wood trusses from an overall point of view were fairly sound and had a lot of useful life left in them. One truss, T-2E, had to be repaired to a greater extent since that truss had failed.

The major problems with the trusses were generally concentrated in two areas, the bottom chord and the two panels at one or both ends. The two end panels had bad splits in the members and the bottom chord had numerous splits and loose bolts. There were a few other members in some trusses that required either stitch bolts or side-scab wood members. Any repair method had also to remedy the bad splice joints previously mentioned.

8.7.8 Recommended Method of Repair

A typical method of repair for all trusses was designed which would generally strengthen the whole truss and prevent any recurrence of excessive deflections. One truss at a time was strengthened. The general method of repair was as follows:

1. Shore the upper chord of the truss to carry the roof loads, thereby removing the roof loads from the truss during the jacking procedure. Jack the members back to their original position (or to a lesser deflection) so that the bottom chord was horizontal. The elimination of deflection was to facilitate the placement of bolts and alignment of bottom-chord steel side plates. It was impossible to jack all the trusses back to their original positions due to permanent set.

2. Loosen all connections.

3. Reinforce the bottom chord by placing steel plates (11 inches wide x 1/4 inch thick) on both sides of the member. Templates were made for drilling holes in these plates to match the existing bolt holes, thereby utilizing the existing connection. See figures 8.7.7 and 8.7.8.

4. The method of installation of the bottom-chord steel side plates was as follows:

    a. Remove the nuts from the existing bolts and place the 11 x 1/4 plate on this side first.
    b. Replace the bolts one at a time and reverse the direction of each bolt.
    c. In like manner, place the 11 x 1/4 plate on the other side of the member.
    d. Tighten all bolts in the bottom chord member.
    e. Ten new 3/4" machine bolts were added to the bottom chord in addition to the replacement bolts.
    f. Referring to figure 8.7.7, two 4" shear plates were added to joint J5.

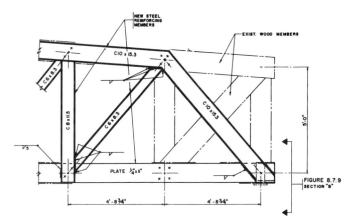

Figure 8.7.8--End panel repair.

5. Add new double diagonal steel members (C10x15.3) in the direction opposite that of the existing diagonal wood member in the first end panel of each truss. Referring to figures 8.7.7 and 8.7.8, the addition of this diagonal reduced the stress to practically zero in members 1, 2, 13, 14, 15 and 38. This generally was the most critical area of the truss.

6. In the top chord area (figs. 8.7.7 & 8.7.8) place two reinforcing channels (C10x15.3) on either side of the top chord wood member from joint J15 to just past top chord joint J18. The top chord channel and the diagonal end channel were welded together to form a single assembly. Here again, the existing bolts were replaced (similar to step 4) and a total of twelve (12) 4" shear plates with six (6) 3/4" machine bolts were added at the end of the reinforcing channel in member 42.

7. Weld the top chord and diagonal assembly (C10x15.3) to bottom side plates as shown in figure 8.7.8 and relocate existing clip angles on both sides, figure 8.7.9.

8. Referring to figure 8.7.7, reinforce the vertical members M17 and M19 with C8x11.5 side members and thereafter install reinforcement C6x8.2 on diagonal members M16 & M18. Field weld vertical and diagonal members to top and bottom chords as noted in figure 8.7.8.

9. In diagonal and vertical members where necessary, stitch bolts were placed where members were split on the ends.

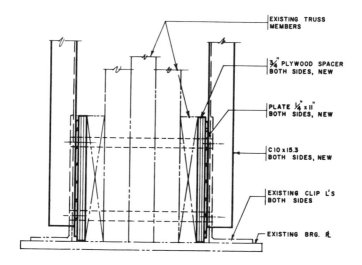

Figure 8.7.9--Section B (see fig. 8.7.8).

10. Referring to figure 8.7.7, reinforce vertical member M21 (3x6) with two 2x6, one each side of the existing 3x6. Attach with four 1/2" machine bolts and 2 1/2" diameter plate washers.

11. On the north end of truss T-15, side steel space plates were added to both sides of the truss to insure an adequate connection to the W12x27 vertical member (M13 in Figure 8.7.7). In this truss only, the W12x27 had been substituted for a wood member at some previous date.

12. All truss connections were re-tightened.

13. The wood lateral bracing had become ineffective as a result of sagging, splits in the wood, and loose bolt connections. A new system of bottom-chord bracing was designed between trusses T-1F & T-1G and T-2F & T-2G to carry any lateral loads into the side masonry walls. (The original roof truss system had very little resistance to lateral forces, i.e., blast).

14. Remove all existing shoring columns after the repairs had been completed.

RECYCLING OLD STRUCTURES 417

8.7.9   Cost of Construction

The total construction cost of repairing these 14 trusses and adding all new diagonal bracing was $180,000. The alternative to repair was replacement of the trusses at an estimated $750,000. However, the real cost to the owner in the replacement system would be the loss in production time and product.

This project was completed while the production line was maintained at all times.

8.8   Rehabilitation of Fire-Damaged Timber--The Filene Center*

8.8.1   Introduction

The authors of Chapter 2 have prepared an excellent dissertation on the effects of fire and high temperatures as they relate to the degradation of structural timber elements.

A good example of the application of these concepts to fire-damaged timbers was the rehabilitation of the structural glued laminated timber framing utilized in the Filene Center for Performing Arts at Wolftrap Farm, Virginia. This facility, which is operated by the National Park Service in cooperation with the Wolftrap Foundation, is the first national park for the performing arts. Construction of this showcase for the performing arts, which seats 3,500 persons and rises an equivalent of 10 stories above the ground at the stage area, was completed in 1971.

The horizontal framing for this project includes four unique glulam and steel queen post trusses spanning approximately 135 feet from the proscenium arch to the rear of the balcony. At the proscenium arch, these four queen post trusses frame into a 6-foot-deep steel plate girder which spans approximately 125 feet across the stage opening. At the rear balcony, these trusses are supported on steel pipe columns. Additional glulam and steel king post trusses frame intermediate spans of 60 to 100 feet. Sub-framing consists of glulam purlins with 2- by 6-inch heavy timber decking covering the entire roof area. Figure 8.8.1 illustrates this most unusual framing system.

Figure 8.8.2 is a construction photo illustrating the erection of one of the main queen post trusses. The vertical support elements at the perimeter of this structure shown in the background of figure 8.8.2 are steel H columns which were subsequently sheathed with 1-inch-thick cedar boards. These support members also serve as acoustical fins and are part of the overall sound design of this facility.

8.8.2   The Problem and the Solution

In March 1971, while this facility was under construction, a fire occurred. It was determined that the fire started at the base of

---

*Author: Thomas G. Williamson, Vice President, Laminated Fabricators, Inc. Indianapolis, Indiana.

Figure 8.8.1--King post and queen post trusses and other laminated members in the Filene Center.

several of the wood-sheathed steel H columns and spread quickly up these columns, completely consuming the cedar cover boards. The fire then progressed across the 2- by 6-inch roof deck and also spread across the entire face of the stage house which was sheathed in cedar boards. The fire had started to spread down the cedar-sheathed steel columns on the opposite side from where it started when firefighters were able to extinguish the fire.

Figure 8.8.3 is a graphic illustration of the fire damage showing the buckling of the steel H columns which resulted from the intense heat of the fire. These columns were classified beyond repair by inspectors, necessitating their replacement. While some of the wood deck and smaller glulam purlins were also damaged to the extent that complete replacement was required, it was determined that the most effective repairs to this facility would be to salvage the four large glulam queen post trusses, if at all possible.

The glulam compression chords of the queen post trusses were 12-1/2 by 78 inches in cross section prior to the fire damage. To replace these glulam and steel queen post trusses would have resulted in significant delays and additional costs, and would have made it impossible to meet previously established opening date requirements for this national performing arts facility. A thorough investigation of these glulam members indicated that they had suffered a maximum char depth of approximately one-half inch at the area of the most extensive damage--i.e., near their bearing at the proscenium arch. Figure 8.8.4 is a

Figure 8.8.2--Erection of one of the queen post trusses. Photo by Jack Rottier, National Park Service, U.S. Department of the Interior.

photograph showing that most of the damage was confined to this stage area thus making rehabilitation of the main glulam trusses feasible.

Engineers involved in the design of this facility re-analyzed the structural load capacity of the queen post trusses based on the section remaining after removal of the charred material by sandblasting. Allowable design stresses were established in a manner similar to those presented in Section 4.5.4. Reduced section properties were calculated, and based on these values it was determined that the members, even with reduced cross section, had sufficient residual structural capacity to safely resist the imposed design loads.

An additional area of concern confronted by the designers in evaluating the glulam members was the possible effects of the fire on the large steel plate connections used in these trusses. These heavy steel plate connections can be seen in figure 8.8.2. While these connections could not be completely removed to evaluate the fire damage behind them, field determinations indicated that the char on the wood behind these steel plates was negligible. There was also concern that heat might

Figure 8.8.3--Fire damage to steel H-columns. Photo by Jack Rottier, National Park Service, U.S. Department of the Interior.

have been transferred through the numerous bolts used in these steel connections to the interior of the wood members, further damaging the glulam. Selected bolts were removed to permit evaluation of the wood and it was again observed that negligible char had occurred on the wood around the bolts due to heat transfer. The glulam chords of the queen post trusses were therefore sandblasted in place and re-used. Smaller glulam members that had been damaged by the fire, and the damaged decking, were replaced.

To confirm that the fire had not affected the adhesive bonds in these glulam members, test specimens were cut from the small purlins which had been determined to be unsalvageable. Block shear tests performed on the gluelines indicated that no significant loss in glueline strength had occurred. By being able to salvage the main glulam-steel queen post trusses, the contractors on this project were able to assure that the opening night curtain was raised as scheduled, somewhat less than 3 months after this fire had occurred.

Figure 8.8.4--Fire damage at the stage area. Photo by Jack Rottier, National Park Service, U.S. Department of the Interior.

It is noted that in addition to the effect of the fire on the structural capacity of the queen post trusses, an additional problem confronted the designers. The heat from the fire had buckled the webs of the large steel plate girder spanning the stage area. After a thorough investigation of the effect of the heat on this member, it was determined that a steel truss should be installed just below the existing steel girder spanning the stage opening. This steel truss thus became a supplemental load-carrying element to relieve the loads on the girder. Figure 8.8.5 shows the installation of this steel truss. The laminated queen post trusses also shown in figure 8.8.5 are as they appeared following the sandblasting operation. The smaller purlins and the decking shown in this figure were replaced as necessary.

Numerous other examples of fire damage to structural glued laminated timbers which have been salvaged by removal of char through sandblasting and refinishing could have been cited. However, this project was used as it illustrates an unusual glulam framing system.

Figure 8.8.5--Supplemental steel truss installed below fire-damaged plate girder.

8.8.3 Summary

If engineers are familiar with the way in which fire affects the structural strength of wood and understand the fundamentals of assigning design stresses to structural timber, they can readily assess the residual strength capabilities of heavy timber members which have been damaged by fire. These evaluations can result in considerable savings in rebuilding the structures if the timber members can be salvaged by char removal. Sandblasting has been a common way to remove such char. The principles presented in this publication provide the basic concepts with which the designer must be familiar in order to properly rehabilitate heavy timber members which have been subjected to fire damage.

8.9   Reconstruction of Two Wood Bridges*

8.9.1 Longitudinal Stiffening of Sydney Creek Bridge, Highway 65, Ontario

The Sydney Creek bridge, shown in figure 8.9.1, is a 22.9-m (75-ft)-span wood truss, the design of which is common to Northern Ontario. The details of the bolt-laminated wood trusses vary slightly depending upon the span, but the deck systems are identical for the majority of these bridges. The deck system, as shown in figure 8.9.2, is composed

---

*Author: Raymond J. Taylor, Research Engineer, Ministry of Transportation and Communications, Downsview, Ontario, Canada.

RECYCLING OLD STRUCTURES

Figure 8.9.1--Overall view of Sydney Creek bridge, Highway 65.

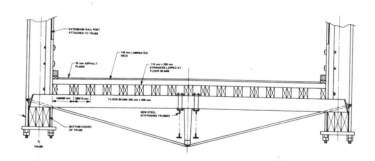

Figure 8.9.2--Cross section of truss span at Sydney Creek bridge.

of a transverse-laminated wood deck on longitudinal wood stringers, which in turn rest upon bolt-laminated wood floor beams. These floor beams, which are supported by the lower-chord truss members, are trussed by two steel bars and a central kingpost which provides camber to the deck above.

The problem with these structures originally arose when they began to exhibit continuous breakup of the asphalt wearing surface. In addition, visual observation indicated excessive deflections of the deck under heavy wheel loads. The Sydney Creek bridge was load-tested in 1976 using the Ontario Ministry of Transportation's testing vehicles. The results indicated that the trusses had adequate load-carrying capacity but that the deck system did not provide adequate longitudinal load distribution. This lack of distribution capability not only caused excessive relative deflections between adjacent floor beams, but caused over-stressing of the steel tension bars attached to them. The measured stresses on the bars reached near-yield conditions under 70 percent of the normal proof load of two 42,700 kg (94,000 lb) dual-axle tandems placed side by side.

To introduce more longitudinal stiffness and to increase load distribution, a pair of small steel stiffening trusses were designed and installed under the deck. These were placed on either side of the king posts and attached to each floor beam as shown in figure 8.9.2, and ran the full length of the bridge. The required stiffness of these steel trusses was determined by evaluating the deck system in conjunction with the load test results. After assuming a preliminary configuration, the final design of the steel trusses was formulated by means of a frame analysis, the floor beams to which they were attached being considered as spring supports. The flexibility of the floor beams, including the tension bars and king posts, was determined using an independent frame analysis.

The steel trusses were fabricated using only bolted connections and were constructed in situ as their position was within the structural system of the bridge deck.

The bridge was load tested again after reconstruction, as displayed in figure 8.9.3, and it exhibited a marked increase in longitudinal load distribution. The wearing surface has since displayed better resistance to breakup, and the tension bars attached to the floor beams registered only 128 MPa (18.6 ksi) under the two 42,700 kg (94,000 lb) dual-axle tandems.

Plans have now been made to reconstruct another bridge in Sioux Narrows, Ontario, which is believed to be the longest single-span wooden bridge in North America. This 64-m (210-ft)-span bridge has an identical deck system and displays the same breakup of the asphalt plank wearing surface.

8.9.2   Transverse Post-Tensioning of Hebert Creek Bridge, Highway 539, Ontario

The Hebert Creek Bridge, shown in figure 8.9.4, displayed the characteristic floor delamination problems caused by nail deterioration in longitudinally-laminated timber bridges. The structure was posted for a 15-ton (13,600-kg) load restriction and slated for replacement in 1976.

Figure 8.9.3--Load testing Sydney Creek bridge for a gross live load of 181.2 tonnes (200 tons).

HEBERT CREEK ELEVATION

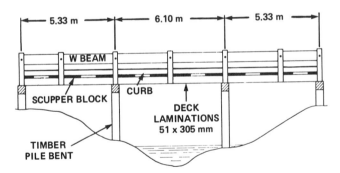

Figure 8.9.4--Elevation of Hebert Creek bridge, Highway 539.

To reconstruct the bridge, a method of transverse post-tensioning was devised, whereby interface friction between the laminations was to replace the deteriorated nailing. Prior to reconstruction, some tests were performed to establish relaxation curves for pressures applied perpendicular to the grain. In addition, various computer analyses were used to evaluate the transverse pressure required to ensure adequate load transfer between laminations. A post-tensioning system as detailed in figures 8.9.5 and 8.9.6, was developed which displayed adequate flexibility to maintain an acceptable transverse pressure, after relaxation and without restressing. Protection was provided by enclosing the bars within grease-filled P.V.C. piping as shown in figure 8.9.5.

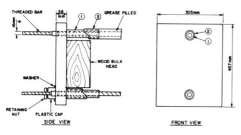

Figure 8.9.5--Details of anchorage system.

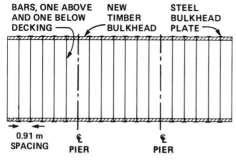

Figure 8.9.6--Plan layout of post-tensioning system.

Reconstruction was performed by cutting transverse grooves in the asphalt surface for the top bars, and attaching the bottom bars to the underside of the deck with standard galvanized brackets. The installation of the system, including removal and replacement of the curbs and guardrail posts, but not the stressing operation, took approximately 70 man-hours. The stressing operation itself took about 90 man-hours, but could have been done much more rapidly if more jacks or a stiffer anchorage bulkhead had been used.

The bridge was load-tested before and after reconstruction using the Ministry's test vehicles. Before post-tensioning, a maximum weight of 37,900 kg (83,500 lb) was applied by a dual-axle tandem. The test was terminated before the maximum load was applied due to inadequate load distribution capability between the timber laminations. This inability to transfer load to adjacent members resulted in the failure of a lamination. The maximum flexural strain was 1,470 µε and the maximum vertical deflection was over 25 mm (1 in.). However, after post-tensioning, the fully loaded test vehicle was applied, consisting of 42,600 kg (93,800 lb) tandems and the bridge displayed no signs of structural distress. This time the maximum flexural strain was 940 µε with a maximum vertical displacement of 16 mm (0.62 in.). Under a loading (37,900 kg, 83,500 lb) equal to that in the original test, the maximum flexural strain was 750 µε indicating an increase in flexural capacity of nearly 100 percent. Figure 8.9.7 displays the effects of

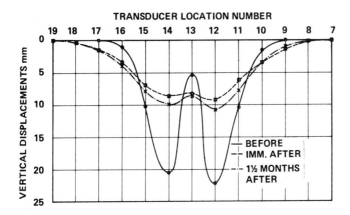

Figure 8.9.7--Vertical deflections at center of north end span under 33,000 kg tandem load.

post-tensioning on the vertical deflections taken across the center of the north end span under a tandem weight of 33,300 kg (73,300 lb). Figure 8.9.8 shows the post-tensioning transverse stress on the wood deck as recorded since 1976; note that the minimum acceptable level is about 275 kPa (40 psi) as determined by the original computer analysis.

The post-tensioning at Hebert Creek indicates that this method of reconstruction is more than adequate in reviving the load carrying capacity of these structures. In addition it is economical, eliminates the breakup of the asphalt surface and resists the penetration of moisture through the deck. Since the rehabilitation of Hebert Creek in 1976 two other bridges have been post-tensioned, one in

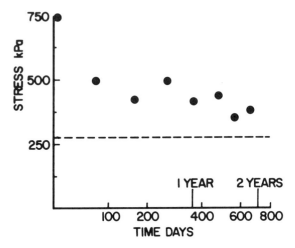

Figure 8.9.8--Applied transverse stress in Hebert Creek bridge.

Prince Rupert, B.C., and the other in Nanticoke, Ontario. Presently the Ministry is developing post-tensioning designs for new construction that will eliminate the need for nailing these decks, and has plans to build a prototype in 1980. This will take advantage of the additional strength available using this design with a span of up to 10 m (35 ft).